Advances in Spatial Science

Editorial Board

Manfred M. Fischer
Geoffrey J.D. Hewings
Anna Nagurney
Peter Nijkamp
Folke Snickars (Coordinating Editor)

AF800901

For further volumes:
http://www.springer.com/series/3302

Esteban Fernández Vázquez •
Fernando Rubiera Morollón
Editors

Defining the Spatial Scale in Modern Regional Analysis

New Challenges from Data at Local Level

Springer

Editors
Esteban Fernández Vázquez
Fernando Rubiera Morollón
Applied Economics (RegioLab)
University of Oviedo
Oviedo
Spain

ISSN 1430-9602
ISBN 978-3-642-44748-8 ISBN 978-3-642-31994-5 (eBook)
DOI 10.1007/978-3-642-31994-5
Springer Heidelberg New York Dordrecht London

© Springer-Verlag Berlin Heidelberg 2012
Softcover re-print of the Hardcover 1st edition 2012
This work is subject to copyright. All rights are reserved by the Publisher, whether the whole or part of the material is concerned, specifically the rights of translation, reprinting, reuse of illustrations, recitation, broadcasting, reproduction on microfilms or in any other physical way, and transmission or information storage and retrieval, electronic adaptation, computer software, or by similar or dissimilar methodology now known or hereafter developed. Exempted from this legal reservation are brief excerpts in connection with reviews or scholarly analysis or material supplied specifically for the purpose of being entered and executed on a computer system, for exclusive use by the purchaser of the work. Duplication of this publication or parts thereof is permitted only under the provisions of the Copyright Law of the Publisher's location, in its current version, and permission for use must always be obtained from Springer. Permissions for use may be obtained through RightsLink at the Copyright Clearance Center. Violations are liable to prosecution under the respective Copyright Law.
The use of general descriptive names, registered names, trademarks, service marks, etc. in this publication does not imply, even in the absence of a specific statement, that such names are exempt from the relevant protective laws and regulations and therefore free for general use.
While the advice and information in this book are believed to be true and accurate at the date of publication, neither the authors nor the editors nor the publisher can accept any legal responsibility for any errors or omissions that may be made. The publisher makes no warranty, express or implied, with respect to the material contained herein.

Springer is part of Springer Science+Business Media (www.springer.com)

Preface

Defining what the concept of "Region" exactly means represents a central theme in the field of spatial science. Since the beginnings of regional analysis, discussions about the proper way of delimitating the concept of region itself have been present. This is probably due to the fact that we lack a universal delimitation of the appropriate spatial scale in theoretical terms. This definition fundamentally depends on what we are searching for, or what our specific objective is.

Traditionally, regional scientists have been limited in our empirical analysis to working with data collected for administrative regions. These administrative divisions were normally designed on the basis of historical and/or political factors. We have had no option but to take the information available at that scale as the input for regional modeling. Modern regional analysis, however, requires thinking at a different spatial scale. For example, endogenous growth theory focuses attention on the local level rather than national or larger regional levels. New Economic Geography, moreover, claims that most of the economic dynamics take place in local environments such as cities in which spillovers, agglomeration economies, or connections via networks are present and play a fundamental role.

The new requirements of information demanded by more recent approaches in regional science stimulated the generation of statistical information at a smaller scale. Data availability has increased enormously over recent years, together with the development of techniques to estimate new information from previous data resources. Thus, one of the most important challenges for modern regional analysis is to take advantage of this more disaggregated information in order to more precisely define the spatial scale and delimitation of our studies. The main objective of this book is the exploration of ways of defining the appropriate spatial scale and how to use or generate local data to build regional models at the local level.

We organize the book into three parts that deal with three different but related general issues, which are listed below:

1. The first one is focused on suggesting appropriate definitions of the spatial scale to use. It comprises four contributions that reflect on (1) the proper way of defining economic areas from the information collected at the local level and (2)

which spatial scale is the most suitable depending on the specific objective of the analysis. Different concepts such as local labor markets, metropolitan areas, functional areas, or analytical regions are considered and discussed in these four chapters, giving a complete idea of the different approaches and perspectives regarding the identification of local and regional dynamics.

2. The second issue deals with estimation of alternatives when the required data are not directly observable. The second section of the book presents some options to recover data at a small scale. Several approaches to estimate information at the local level or to integrate different spatial levels of data are revised and illustrated with applications. Techniques based on Entropy Econometrics, ISQ procedures, and Disequilibrium Adjustment framework, among others, are proposed. The last two chapters explore how to maximize the information contained in local data by means of dynamic panel models and spatial econometrics.
3. The third general issue is related to economic analysis and modeling from data collected at different spatial scales, with a special focus on data at the local level. The book finishes with a section that studies this issue by presenting several empirical applications with real-world examples, on which several common problems in regional and urban economics are approached from models built on spatially disaggregated data.

Building Economic Areas from Local Data

The classic contributions of Von Thünen, Lösch, and Christaller among others gave a definition of an economic region consistent with the Economic theory of land, markets, and distance costs. Since these first approaches were proposed, many new ideas have emerged in regional and urban economics. The New Economic Geography (NEG) pointed out the importance of agglomeration economies and the distinction between central and peripheral areas, thereby introducing new challenges to the way of defining regions with an economic meaning. Many researchers have developed quantitative techniques with the aim of identifying coherent local areas. Different names were given to these areas, such as *Functional Economic Area*, *Labor Market Area*, etc., but they all mean a territory that internalizes the home-to-work daily journeys of their residents.

In accordance with this approach, the first chapter of the book, written by Sforzi, starts by defining and calculating regions that satisfy the idea of internalizing the home-to-work daily journeys of their residents. To begin with, from the methodological point of view, Sforzi expands traditional approaches to the identification of metropolitan areas to introduce a new category he labels the "network approach," which explicitly recognizes that metropolitan areas are cliques of networks of cities, and can be monocentric or polycentric. He goes on to provide two sets of metropolitan areas for the case of Italy, which are identified by means of rigorous and replicable standards that can be used by other researchers in subsequent investigations.

Rubiera and Viñuela start from a definition of local units and propose a subsequent aggregation of these local units into aggregated regions. This aggregation covers all the space of a country and is organized in accordance with NEG postulates. The classification of the territory carried out by these authors is consistent with the classical point of view and also with the relevance of agglomeration and scale economies pointed out in recent contributions. This chapter applies the concept of incremental distance in order to analyze the relevance of the position of each territory with respect to the metropolis. As an empirical illustration, Rubiera and Viñuela apply this re-aggregation of basic spatial units into analytical regions to the Spanish economy and compare them with other administrative or historical regions that usually have no economic meaning and, consequently, are not appropriate for analyzing the economy from a regional/spatial perspective.

The third chapter, by Boix, Veneri, and Almenar, explores ways of aggregating units that form a metropolitan area, which are used as the reference point for the previous approach of Rubiera and Viñuela. They propose a procedure in line with Sforzi but which take into account the specific connections that emerge in a big metropolis and its influential area. The metropolitan units identified have three basic purposes: to provide a general view of the characteristics of each country's metropolitan reality, to compare the metropolitan processes of both countries, and to identify metropolitan units that could be used in subsequent analysis. This has been done by the authors comparing two functional approaches to the concept of metropolitan area: a general methodology applicable to most European Union countries based on the concept of the *Functional Urban Area*, and their proposed *Dynamic Metropolitan Areas*, which are specifically designed to deal with the particular characteristics of networking and polycentricism.

Finally, from the base of the previous chapter, Brezzi, Piacentini, and Sanchez-Serra present the results and conclusions of some recent work carried out at the OECD to develop an international methodology for measuring the socioeconomic and environmental performance of urban and metropolitan areas. In their chapter, the authors propose an international methodology for the definition of urban areas that is applied to 27 OECD countries. The methodology identifies urban areas as functional economic units, characterized by densely inhabited "urban cores" and "hinterlands" whose labor market is highly integrated with the "cores" by commuting flows. The development of a harmonized functional economic definition overcomes previous limitations related to administrative definitions by increasing cross-country comparison. The definition of urban areas that they use takes into account the possibility of polycentric development, since more cores physically separated can be included in the same urban area. They integrate information from geographical sources with population data to get a better understanding of urban forms and urbanization processes. Finally, the proposed methodology identifies for each country all urban systems with a population of at least 50,000, enabling analysis of medium-sized urban areas and not only of large metropolitan areas.

Estimation of Spatial Disaggregated Data

The problem of having the information required to define different regional units is always present. This justifies the second part of the book, which focuses on different techniques applied to estimate disaggregated data from observable aggregates.

This section starts with a chapter by Fernández-Vazquez and Garduño, where they present an application of an estimation procedure based on entropy econometrics in order to infer data on wages for Mexican municipalities using aggregated information. Mexico is a specially useful case because official data are available with a high level of disaggregation. Specifically, their objective is to estimate wages paid by industry and municipality for a group of four Mexican states. The information required to put into practice the technique proposed are data that reflect a priori beliefs about the possible wage distribution across municipalities and industries, as well as observable municipal and industry aggregates. From these two pieces of information, they apply a Cross Entropy adjustment and compare their estimates with the actual values observed in the Mexican Economic Census for 2009. Given that assuming perfectly observable information on the municipal aggregates was quite unrealistic, the standard adjustment is extended in order to consider the possibility of errors in the margins.

In the following chapter, Lemelin and Mainguy explain the so-called ISQ method and use it to estimate the GDP of the subareas that form the province of Quebec, Canada. The ISQ technique is presented as a top-down method, which consists of allocating total labor income and net income of unincorporated business (NIUB) by industry among regions. For this purpose, they use allocators constructed from fiscal data on wages and salaries and NIUB obtained from the Quebec Ministry of Revenue. For each industry, other components of value added are then distributed in proportion to the sum of total labor income and NIUB. The key ingredients in the method are a compilation of fiscal data on incomes and reliable home-to-work commuting tables by industry. As the next natural step following their estimation exercise, the authors analyze the estimates to examine the recent evolution of the geographical pattern of economic activity in Quebec.

Kim and Hewings show in their chapter an application of the Disequilibrium Adjustment Framework to small region forecasting for the metropolitan area of Chicago. This chapter highlights the dynamic nature of adjustment models as one of its key advantages and explores the possibility of further applications beyond its existing uses for empirical assessments. Specifically, the authors apply a spatial econometric version of the regional disequilibrium adjustment model for small area socioeconomic forecasting and impact analysis. Rather than using the standard adjustment model only for forecasting purposes, the chapter introduces the idea of combining it hierarchically with a regional econometric input–output model (REIM) that provides a long-term trajectory of economic growth that is not reflected by the adjustment model alone. Further, they present an application of the proposed combined framework for a small area population and employment forecasting under various scenarios.

In the same fashion as the first chapter in this second section, Bernardini-Papalia suggests the use of entropy-based estimation applied to recover geographically disaggregated data. More specifically, she proposes a generalized cross-entropy estimator that allows for modeling heterogeneity in subgroup indicators by addressing the spatial dependency problem. The approach proposed in her chapter offers a tractable framework for modeling the underlying variation in small area indicators, in particular when the data set contains outliers. The basic idea is the proposal of an estimator based on an entropy measure of information which provides an effective and flexible procedure for reconciling micro and macro data. A maximum entropy (ME) procedure gives the possibility of including out-of-sample information which can be introduced as additional constraints in the optimization program or specifying particular priors for parameters and errors. The proposed method of estimation is presented as capable of yielding disaggregate data consistent with prior information from different data sources in the absence of high-quality and detailed data. Problems of collinearity and endogeneity are also tackled without imposing strong distributional assumptions. Within this framework, the author shows how partial information at the disaggregated level can be combined with aggregated data to provide estimates of latent variables or indicators which are of interest at the small area level.

The chapter by Mayor and Patuelli estimates a dynamic panel model using a spatial filter in order to account for spatial heterogeneity and/or spatial autocorrelation in unemployment. They compare different methods for obtaining short-run unemployment forecasts in small geographical units and observe their performance between different countries. Their empirical application analyzes regional unemployment rates at the NUTS-3 level for two countries: Spain and Switzerland. Taking advantage of the strong heterogeneity in terms of population or area size across NUTS-3 regions between these two countries, they investigate the variation in the performance of different spatial econometric methods. They further evaluate the forecasting performance of two competing econometric methods: a spatial vector autoregressive (SVAR) model and a dynamic heterogeneous-coefficients panel data model based on an eigenvector-decomposition spatial filtering (SF) procedure.

Finally, the last chapter in this section, by Paez and Mur, presents a methodological contribution to the literature on estimation of spatial models from data at small scale. The authors focus on Geographically Weighted Regression (GWR) techniques and propose several modifications to the conventional specification. The standard GWR analysis is based on a global perspective, but the authors argue that spatial location matters. In their chapter, they suggest different strategies for constructing local weighting matrices that reflect the local surrounding of each observation. Specifically, they study three different but related questions: the development of a GWR test to compare local versus global estimates; the definition of the bandwidth, which they solve in a fully adaptative framework; and the nonuniqueness of the GWR estimates, which follows from the uncertainty in relation to the selection of the kernel.

Applications of Spatial Analysis with Small Area Observations

In the final section of the book, five applied studies are presented where economic analysis is carried out for functional regions and small-area data are used as inputs for the empirical analysis.

Paredes, Lufin, and Aroca contribute with the first chapter in this section. The authors present a study where they apply the concept of Functional Urban Area for quantifying the size and scope of agglomeration economies in Chile. Their hypothesis is that thicker labor markets generate a wage premium in comparison with thin labor markets. To carry out their testing strategy, a redefinition of the geographical space to identify functional thick labor markets is required, and they use spatial and social network analysis to identify functional areas. Once their functional regions are identified, three empirical issues are discussed. The first is the issue raised when the aggregated wage at regional level is used as a dependent variable. The authors add an independent variable indicating the size of the region and its coefficient identifies the wage premium. A second issue is how the literature has defined the geographical scale of a thick labor market. They approach this problem by proposing a stepwise tool for spatial delineation of thick labor markets following a combination of methods based on commuting flows among counties. The third discussion emerges when the wage differential is considered as a causal effect of thick labor market instead of other control variables such as human capital. The authors propose finding identical workers between a thick and thin labor markets using Coarsened Exact Matching (CEM) as a previous step to estimating the wage differential.

Duque, Royuela, and Noreña present a case study of Medellin (the second largest city in Colombia) using a stepwise procedure that efficiently delimits intra-urban slums. The exact delimitation of the borders of these areas has not been a matter of study for local authorities. However, a proper identification and delineation of these areas is a central issue, as it can help prevent poverty traps and crime nests. The authors argue that using administrative areas implies a simple identification but that this procedure can be severely inefficient since in many cases statistical inference based on normative regions may be strongly affected by aggregation problems. They claim that analytical regions, on the contrary, prevent several statistical problems and ensure homogeneous impacts of specific public policy actions.

López, Angulo, and Artal present an empirical application that studies the patterns of co-localization of the economic activities in the Mediterranean Axis in Spain. They base their analysis on statistics built on the concept of spatial dependence that usually characterizes spatial processes. They present an analysis different from traditional studies on cluster analysis, which usually employs official data on employment levels, number of enterprises, or business volume for geographical units such as the county, municipality, region, or country. The authors use data taken at the greatest level of spatial integration (post code level) and they identify clusters of companies dedicated to the same economic activity. Then, they propose applying to these highly disaggregated data a methodology borrowed from epidemiology studies (Kulldorff Scantest) to improve the detection and identification of clusters.

Chasco and Le Gallo provide an analysis with a multilevel perspective using techniques capable of dealing with data at various spatial scales and supports. The authors explain that the combination of data at different levels creates two categories of problems: (1) statistical problems due to using aggregate or disaggregate data from different levels that could imply a loss of representativeness and (2) interpretation problems, which may arise if we use a wrong method of aggregation, generating an "ecological fallacy" when the level of aggregation is higher than normal, or an "atomistic fallacy" when the level of disaggregation is lower that normal. In their chapter, they propose using explanatory variables at the aggregate level that serve as "moderators" of the relationships present at the individual level. They illustrate how this procedure works with an example: a hedonic price model that measures the impact of air and noise pollution on housing prices. In this example, multilevel models could be very useful because effects on prices operate at different levels and they must be taken into consideration jointly.

Pablo-Martí and Arauzo-Carod contribute with a chapter that closes this section and the book itself. The objective of their contribution is to identify possible patterns of concentration of manufacturing and services industries in Spain. The authors divide Spain into homogeneous cells and check whether each industry follows a concentrated or dispersed pattern and whether collocation exists for pairs of industries so that clusters made by different industries can also be identified. The methodology they apply allows the overcoming of previous drawbacks in terms of data constraints, computing limitations, and border's definitions.

Some Final words

This is an applied research book about spatial analysis characterized by its dual nature, both in regard to the authors and to the contents. The list of contributors is a mixture of well-known names in regional science together with younger researchers in the field of regional and urban economics and spatial econometrics. Regarding the contents of their contributions, most of the chapters present technical or methodological contributions, but always illustrated with useful empirical applications. Armed with this background, we face two fundamental challenges for modern spatial analysis: spatial scale definition and the use of data at small scale.

Contents

Part I
Building Economic Areas from Local Data

Chapter 1
From Administrative Spatial Units to Local Labour Market Areas

Some Remarks on the Unit of Investigation of Regional Economics with Particular Reference to the Applied Research in Italy

Fabio Sforzi

1.1 Introduction

The region as the unit of investigation is what distinguishes Regional Economics from other branches of Economics. Edgar M. Hoover, the founder of the Regional Economics, devoted a chapter of his book *An Introduction to Regional Economics* (Hoover 1971) to the region and its definition.

The region in Regional Economics is different from the region in Economic Geography (which corresponds to a productive agglomeration) and from the region in Political Science (which corresponds to a political-administrative entity). Moreover, while for Economic Geography and Political Science the region itself is the object of study, for Regional Economics it is the instrument used to investigate economic phenomena and take policy decisions.

Regional economists are aware of this difference, but they have shown a regrettable tendency to ignore it, and to use administrative units as a proxy of the region in their empirical analyses. Sometimes, they claim that this choice is due to the lack of alternatives, or the impossibility of having a region consistent with the theoretical assumptions of Regional Economics. But they should not underestimate the fact that if the unit of investigation they adopt is not consistent with the theoretical framework of the discipline, empirical results will be meaningless.

A number of researchers – initially in the United States, in the 1960s (Fox and Kumar 1965), then in Europe, since the 1970s (Smart 1974) – have devised quantitative techniques for the identification of regions consistent with the theoretical framework of Regional Economics. Different names were given to these regions, such as Functional Economic Area and Labour Market Area, but they all signified a region that internalizes the home-to-work daily journeys of its residents.

F. Sforzi (✉)
University of Parma, J.F. 8, Parma, Italy
e-mail: fabio.sforzi@unipr.it

E. Fernández Vázquez and F. Rubiera Morollón (eds.), *Defining the Spatial Scale in Modern Regional Analysis*, Advances in Spatial Science,
DOI 10.1007/978-3-642-31994-5_1, © Springer-Verlag Berlin Heidelberg 2012

In Italy, the debate on the unit of investigation started when regional economic policy appeared in the 1970s.[1]

The problem was to define the appropriate territorial unit for economic research to support policy decisions (Sforzi-IRPET 1977).

This debate in the political field was influenced by the parallel debate taking place in academia about the unit of investigation of Industrial Economics (Becattini 1979). It was the culmination of theoretical thinking on the concept of industry begun in the 1960s (Becattini 1962) and continued in the 1970s with the industrial district-based interpretation of the economic development of Tuscany (Becattini-IRPET 1969, 1975), connected with a new reading of Alfred Marshall's writings and thought(Becattini 1962, 1975).

Research on the economic development of Tuscany (entrusted in 1972 to IRPET by the Tuscany Region[2]) was to be the prerequisite for the formulation of policies for place-based economic development by the regional government (Sforzi 2007).

The implementation of these policies required the definition of appropriate territorial units. That is, a place where people share a common interest.

IRPET had a leading role in defining these territorial units. It established criteria and methodology for carving up the 'political region' (Tuscany) into a meaningful set of 'functional regions' (places) (Sforzi-IRPET 1977; Regione Toscana/ Dipartimento Programmazione-IRPET 1978; Regione Toscana 1979; Sforzi 1980, 1982).

The functional regions were composed of groups of municipalities (the basic units of analysis) merged together by daily journeys from home to work. Data were gathered via the 1971 Population Census.

A few years later, thanks to a research agreement with the National Institute of Statistics (ISTAT), the IRPET approach to functional regionalization was applied to Italy as a whole (ISTAT-IRPET 1986). The 1981-based functional regions were called *Sistemi Locali del Lavoro* (in English, Local Labour Market Areas: LLMAs). After that, LLMAs were updated on the basis of data collected by the Censuses of 1991 and 2001 (ISTAT 1997, 2005).

This updating showed that the boundaries and number of LLMAs were modifiable. LLMAs thus proved that stable administrative units are not useful for the understanding of the multiple paths of industrialization. This became apparent both in the old industrialized areas of Northern Italy and in those newly industrialized of the South. LLMAs substantially changed the knowledge of how the Italian economy was spatially organized. And to some extent, the attention of

[1] The regions (NUTS level 2 in the Eurostat classification) were established by the Italian Constitution of 1948, but they were instituted in 1971. Their main task was the design and implementation of policies for the development of the regional economy.

[2] IRPET was, and still is, the Regional Institute for Economic Planning of Tuscany. It was founded in 1968 with the aim of carrying out studies on the region's economy and society. From 1968 to 1973 Irpet was directed by Giacomo Becattini. In 1974 became a public body under Regional law 48/1974 enacted by the Tuscany Region.

academics and policymakers has shifted from a sector approach to a place-based approach.

Over time, LLMAs have become an instrument of analysis: (a) for collecting official statistical data; (b) for studying patterns and changes in the Italian economy.

After this Introduction, the remainder of the chapter is organised as follows. Section 1.2 summarises the background of the Italian experience of regionalization. Section 1.3 introduces the philosophy of regionalization. Section 1.4 describes the data and the methodology used for the identification of LLMAs, and the results. Section 1.5 is devoted to a short description of the main applications of LLMAs for research and policy. The concluding section contains some final remarks.

1.2 Background

The Italian experience of functional regionalization is rooted in Tuscany's regional economic policy and preparatory studies conducted by IRPET between the late 1960s and the mid-1970s (Becattini-IRPET 1969, 1975). Aimed at interpreting the industrialization of the region, these studies overturned the approach of mainstream economic research. The mainstream approach began with the analysis of industrial sectors on a regional scale, and proceeded to their territorial distribution in search of industrial agglomerations and the related location factors. The IRPET approach, by contrast, started from the places to explain the underlying causes of their change and the effects of such changes on the economic and social structure of the region.

The change in each place – it was argued – was determined not only by internal forces (economic, social, and institutional), but also by external circumstances, including the concurrent behaviour of other places (near and remote) with which the places of the region had economic and social exchanges (i.e., flows of goods and people).

The places represented the unit of investigation of economic research, and were conceptualized as 'economic regions' within the 'political region'. They did not coincide with the boundaries of territorial administrative units (municipalities and provinces: LAU level 2 and NUTS level 3 in the Eurostat classification) located within the political region.

This conceptualization of the place had equally important: (a) the field of economic research, and (b) the field of political action.

1.2.1 Political Action

In the 1970s IRPET was a 'hybrid' research centre as it was located at the intersection between academia and politics (Sforzi 2008). The two worlds cooperated rather than conflicted. Research did not underestimate the operational need of politics, while politics did not underestimate the need for rigor in research.

As a matter of fact, it called for rigorous answers to its demand for knowledge before taking the final policy decisions. These were under the sole responsibility of politics. So, the two worlds influenced each other.

The design of regional economic policy was based on the principle of inter-institutional cooperation between the region, as the institution responsible for regional economic policy, and municipalities, as the institutions representing the common interest of local population. Municipalities had to contribute to the determination of objectives of the regional economic policy (Regione Toscana/Dipartimento Programmazione IRPET 1978). However, this institutional role could not be carried out effectively because their political boundaries no longer corresponded to the economic ones.

The changes in the urban-rural and organization of production, the changes in spatial relationships of supply and demand for labour and, thus, for services extended the space of everyday life of the local population. This was highlighted in the industrial change of Tuscany by the IRPET study (Becattini-IRPET 1975).

So the common interest of the local population needed to be identified at a supra-municipality level. This raised the question of redrawing the boundaries of municipalities to enable them to effectively play the role of local partners of the region in economic policy making. In other words, it raised the question of the functional regionalization of Tuscany for economic policy purposes. The guiding principle was obvious: the appropriate territorial unit for the regional economic policy needed to represent – through its own institutions – the common interest of its people (Sforzi-IRPET 1977).

This conceptual framework led to a new way of thinking about region and regionalization:

- The region should not be conceived as a territorial unit to be built by means of a regionalization procedure, but as a place (or local community), historically and geographically determined. It needs to be a reality for local community. The crucial question is how it emerges;
- Regionalization should not be conceived as a procedure to be executed efficiently by means of given criteria, but as the research for appropriate criteria and related strategy to better approximate the identification of something that in social reality already exists (Regione Toscana/Dipartimento Programmazione IRPET 1978).

The criteria used to regionalise Tuscany or divide it into 'places' are briefly summarized as follows:

- The municipality is the basic unit of analysis;
- A place must be identified on the basis of daily journeys from home to work, because labour has a basic role in people's life and guides their territorial behaviour with regard to the municipality where they live and work. This may be the same or different;
- A place must correspond to the area where the local population develops most of its economic and social relationships;

- A place must allow for the common interest of the local population to be identified as a whole.

Although statistically a place is a proximate and interconnected group of municipalities, it mirrors a local community of people that work and live within it. This is why places became the reference for the Tuscany region in the process of institutionalization of 'inter-municipal associations' (Regione Toscana/Consiglio Regionale 1978). A place in this sense was the appropriate territorial unit for designing and implementing regional economic policy (Regione Toscana 1979).

1.2.2 Economic Research

After his work at IRPET and coordinating the two studies on the industrialization of Tuscany (Becattini-IRPET 1969, 1975), Giacomo Becattini illustrated the theoretical background of these studies in his famous essay *Dal 'settore' industriale al 'distretto' industriale. Alcune sull'unità d'indagine dell'economia industriale – From the industrial 'sector' to industrial 'district': Some remarks on the unit of investigation of industrial economics* (Becattini 1979). The theoretical framework focused on the subject of local-external economies underpinning the Tuscan model of industrialization, because the small manufacturing enterprises embedded in specialized local communities played a leading role in the model. The essay paved the way for the discovery of the importance of small business for the whole Italian economy, and for the later conceptualization of the industrial district as a 'model of production' (Becattini 1989) and a 'way of industrialization' (Bellandi and Sforzi 2001).

It is also especially important as Becattini addressed the problem of the appropriate unit of investigation for economic research and policy decisions. To do this he looked back of Marshall's writings on economic nations.

Marshall in fact introduced the concept of economic nation in *Industry and Trade* (Marshall 1920b: p. 20), as follows:

> If the local spirit of any place ran high: if those born in it would much rather stay there than migrate to another place: if most of the capital employed in the industries of the place were accumulated from those industries, and nearly all the income enjoyed in it were derived from its own resources: — if all these conditions were satisfied, then the people of such a place would be *a nation within a nation* [emphasis added] in a degree sufficient to render propositions, which relate to international trade, applicable to their case from an abstract point of view; though in the absence of any statistics of the imports and exports of the place, they would to some extent still lack reality.

What attracted the Becattini's interest in the concept was the 'sense of belonging' that holds together these places. The sense of belonging in which objective components of a common interest and subjective components of a historical-cultural nature are blended together (Becattini 1979: p. 52).

If is true that in Regional Economics "the normal attribute of a region is the general consciousness of a common regional interest" by its people (Hoover: p. 152), in Becattini's interpretation, places ('economic nations' within have the same characteristics as the region in Regional Economics.

The sense of belonging is one of the possible founding criteria for re-conceptualizing industry, going beyond the more traditional criterion of technology. It makes possible the conception of industry through the "awareness of economic agents (employers and employees) of belonging to a particular industry" (Becattini 1962), and that is a place-based concept.

The awareness forms inside the place where "a set of productions have the common characteristic of occurring under the same technical conditions, that is characterized by the same production process" (Marshall, as quoted in Becattini 1962: pp. 22–23). This sense of belonging embeds the technology in the place where the production occurs "since neither the production technique nor the relations of competition or alliance towards the counterpart from feelings of rivalry and solidarity that arise (e.g. between employers and employees) can be strictly separated from the culture and social relationships that go with" (Becattini 1962: p. 28).

Along this line of thought the industrial district was proposed as the unit of investigation of Industrial Economics (Becattini 1979), and more generally of the national economy within the analytical framework of Applied Economics (Becattini 1987).

1.3 Philosophy of Regionalization

In Italy, the philosophy of regionalization based on the common ground of regional economic policy and economic research, entered the nascent field of Regional Science through the key role played by the regional research institutes like IRPET and other similar institutions. Although it was not a shared widely followed philosophy, but many regional scientists looked on it as a cornerstone of Italian Regional Science (Bianchi and Magnani 1985).

Within the field of Italian Regional Science there was a heated discussion about where this new field of research lay within Italian social sciences. The opinion of one founder (Bianchi 2009: p. 136) was that

> Regional Science – under the impulse of Isard's spatial economics – was becoming a branch of mathematical economics, while our purpose was to join economists, urban planners, sociologists, historians, geographers, lawyers etc.; that is, everybody, both academics and practitioners, that meant to grapple with the territory.

It was claimed that Regional Science was neither a branch of Economics nor a new discipline, but a common house of the different disciplines that shared interest in the territorial organization of society. This criticism was concretely expressed in the identification of the daily urban system (a conceptualisation of functional region) as the unit of investigation for studies of Regional Science.

As is widely known, the notion of daily urban system was introduced by Torsten Hägerstrand in his presidential address to the Ninth Congress of the European Regional Science Association in Copenhagen in 1969. The address was entitled *What about people in Regional Science?*, and later published in the Papers of the Association (Hägerstrand 1970).

At the beginning of his address, Hägerstrand (1970: p. 7) pointed out "a difference in emphasis or tone between the European and North American meetings" of Regional Science:

> When looking over the proceedings of the sixties, one gets the impression that participants in this part of the world have preferred to remain closer to issues of application rather than to issues of pure theory. We in Europe seem to have been looking at Regional Science primarily as one of the possible instruments with which to guide policy and planning. I have chosen to proceed along this line by suggesting that regional scientists take a closer look at the problem which is coming more and more to the forefront in discussions among planners, politicians, and street demonstrators, namely, the fate of the individual human being in an increasingly complicated environment or, if one prefers, questions as to the quality of life. The problem is a practical one and, therefore, for the builder of theoretical models, a 'hard nut to crack'.
>
> Now, first of all, does the problem fall within the scope of Regional Science? I think it does. A forest economist remarked some time ago that, 'forestry is people, not trees'. How much more accurate it would be to say that Regional Science is about people and not just about locations. And this ought to be so, not only for reason of application. Regional Science defines itself as a social science, thus its assumptions about people are also of scientific relevance.

So according to Hägerstrand, Regional deals with people, not just with industrial location, and deals with them as local community (forestry) and not as individuals (trees). In this statement there is the premise of the argument that leads Hägerstrand to define the way in which a local community circumscribes itself, that is, through the behaviour of people moving from the locality where they live to the locality where they work, coming back home at the end of the workday. (This is developed later in Hägerstrand's address).

This behaviour bounds a place that is conceptualized as daily urban system. The boundaries of such a place are given by economic relations, but most of everyday social relationships also develop within it. The daily urban system is open, since it trades people, goods and knowledge with other places, near or remote. Besides being permeable, the boundaries also change over time, to the extent to which businesses change their productive organization and people change their orientation towards labour.

So there is a clear difference between frameworks of Hägerstrand and Isard. The American economist supported a spatial reorientation of Economics through a general theory relating to industrial location (Isard 1956), while the Swedish geographer argues for overturning this setting, by placing the local community of which industry is an attribute at the centre.

Let me give an example to clarify this point. Langhirano is a place well known to gourmets because it produces ‘Parma Ham’. Langhirano can be seen either as one of the places where the food industry is localized, or as a local community near Parma which procures what it cannot produce itself by specializing in what it makes best. In the first view, the unit of investigation is the food industry, and a study of its spatial location across Italy reveals Langhirano. In the second view, the unit of investigation is the local community of Langhirano and research on the production structure of the place brings into focus the food industry. In the first view, the socio-economic reality is seen as ‘an array of interrelated industries’ and in the second view it is ‘a system of places’ (or local communities).

The similarity between Hägerstrand’s conception of Regional Science and Marshall’s conception of Economics is remarkable. Marshall considered Economics more important as part of the study of man – not in the abstract, but in relation to a given place and time – than as a study of wealth (Marshall 1920a).

The similarity is even more compelling when one considers that both Hägerstrand and Marshall put at the centre of their thinking labour and place, place being bound by labour. According to Marshall, social, and therefore economic change, mostly occurs in place, through the formation and enhancement of human abilities.

In Marshall’s social philosophy, labour occupies a central position (Becattini 1962). It is “the essential purpose of life”; indeed – as Marshall says – the labour “is life itself”. In particular, “labour is a necessity for the formation of character and, therefore, for progress”, because it is labour that educates and exercises human abilities, and allows their development (Sforzi 2008).

Human abilities are made up of all skills needed to carry out an economic activity: ranging from technical skills to the entrepreneurship. They contribute to the productivity of people.

Human abilities “are a means of production important as any other kind of capital” (Marshall 1920a: p. 347). Their increase is crucial for economic development. This statement follows from the role that Marshall attaches to knowledge among the agents of production, and the relationship he establishes between knowledge and organization, when he claims that “organization aids knowledge” (Marshall 1920a: p. 238). This aid varies according to different forms of production that the organization takes in different places.

Through the development of human abilities, the individual changes the place where s\he lives and, at the same time, it changes itself (Raffaelli 2001), conferring advantages to the individual. Among these advantages, Marshall lists the development of specialized skills: “so great are the advantages which people following the same skilled trade get from near neighbourhood to one another” (Marshall 1920a: p. 395). Some of these advantages concern the circulation of knowledge so that “the mysteries of the trade become no mysteries, but are as it were in the air” (Marshall 1920a: p. 395), becoming a place-specific public good.

To sum up, it can be argued that the place conceptualised as daily urban system by Hägerstrand and as economic nation by Marshall share the same conceptual framework.

1.4 The Unit of Investigation: Local Labour Market Areas

The *sine qua non* for regionalizing a national economy in a meaningful set of functional regions as conceptualised above is the accessibility to suitable data.

In Italy, this opportunity arose in the mid-1980s when ISTAT processed the 1981 Census data on daily journeys to work between municipalities. Thanks to the experience matured by IRPET in the functional regionalization of Tuscany (Sforzi-IRPET 1977; Sforzi 1980, 1982) and the reputation acquired for its studies on regional economic development exploiting local census data (IRPET 1976, 1980, 1983), ISTAT accepted the proposal to sign a research agreement for regionalizing Italy.

The functional regionalization of the Italian economic system was performed by means of a multi-stage analytical strategy consistent with the philosophy of regionalization discussed above (ISTAT-IRPET 1989). The definition of this strategy of regionalization had benefited from a survey of various available strategies and quantitative methods performed by applying them to Tuscany (Sforzi 1980, 1982; Sforzi and Martinelli 1980; Sforzi et al. 1982; Sforzi and Openshaw 1983).

The places identified through the functional regionalization were officially named Sistemi Locali del Lavoro (in English, Local Labour Market Areas: LLMAs).

The 1981-based LLMAs were presented in 1986 in a joint ISTAT-IRPET seminar, and the full research was published 3 years later (ISTAT-IRPET 1989). The research agreement was renewed for the following Censuses, so that regionalization was updated in 1991 and in 2001 (ISTAT 1997, 2005). Figures 1.1, 1.2, and 1.3 show the functional regionalization of 1981, 1991 and 2001.

As mentioned above, the boundaries of places change over time. This change is confirmed by the modifications in boundaries and number that LLMAs show from 1981 to 2001. There were 955 LLMAs in 1981, 784 in 1991 and 686 in 2001 (ISTAT 1997, 2005). These modifications occurred to varying extents across Italy, and do not follow a general rule.

LLMAs can be classified by their main productive activity or demographic size. A LLMA can be urban or rural, industrial or agricultural, but also specialized in consumer or business services which offer support to the industrial sector across the country. These varied criteria mean that a LLMA is not characterised by predefined economic features. A LLMA is not dominated by either small or large businesses, so that it can subsequently be characterized as an industrial district (i.e., a manufacturing LLMA dominated by SMEs) or as a industrial pole (i.e., a manufacturing LLMA dominated by one or a few large firms). Finally, a LLMA can also be classified as metropolitan, based on criteria such as an established population size or demographic density, or the localization of national level services. In short, a LLMA can be classified on the basis of localization economies in the form of Jacobian diversity externalities.

To illustrate this, let me comment on a couple of figures. These show LLMAs identified as local economies specialized in manufacturing industry (Fig. 1.4) and in consumer services or the hospitality industry (Fig. 1.5).

Fig. 1.1 Local labour market areas, 1981 (Source: ISTAT-IRPET 1989)

Figure 1.4 shows the widespread presence across the country of local economies typified by SMEs; that is, industrial districts dominate the industrial landscape of Northern and Central Italy, although they are also located in the South (Sforzi 2009).

To some extent Fig. 1.5 is complementary to Fig. 1.4. The local economies typified by the hospitality industry are mainly located in poorly industrialized areas. Particularly apparent is the localization of these local economies in the extreme North-East (South Tyrol). An economist (or a tourist) can easily recognize well-known Italian resort areas like Capri (Campania), Taormina (Sicily) and the Costa Smeralda (Sardinia).

Fig. 1.2 Local labour market areas, 1991 (Source: ISTAT 1997)

1.5 Main Applications

LLMAs reflect the methodology of aiming for a single consistent pattern of units of investigation for economic research and policy decisions. They have been very widely used in academia and institutions and the number of studies and applications is so high that it is impossible to compile an exhaustive bibliography.

In Italy, LLMAs were used, and are still commonly used today, for studies on industrial economics, urban and regional development, local development, sociology of innovation, among other fields. They have made possible a substantial

Fig. 1.3 Local labour market areas, 2001 (Source: ISTAT 2005)

advancement of knowledge in methodological, theoretical and factual issues. Below are some of the main applications.

In industrial economics the identification of industrial districts is important.

Before the mapping of the Italian industrial districts was completed (Sforzi 1987; Goodman et al. 1989; Pyke et al. 1990), the district was a concept in search of empirical evidence in order to be recognised as an effective 'instrument of analysis' for the interpretation of Italian economic development based on small businesses. Until then, the importance of small businesses for economic development was based on anecdotal evidence. The empirical evidence, supported by a statistically reliable approach, showed that industrial districts accounted for a large

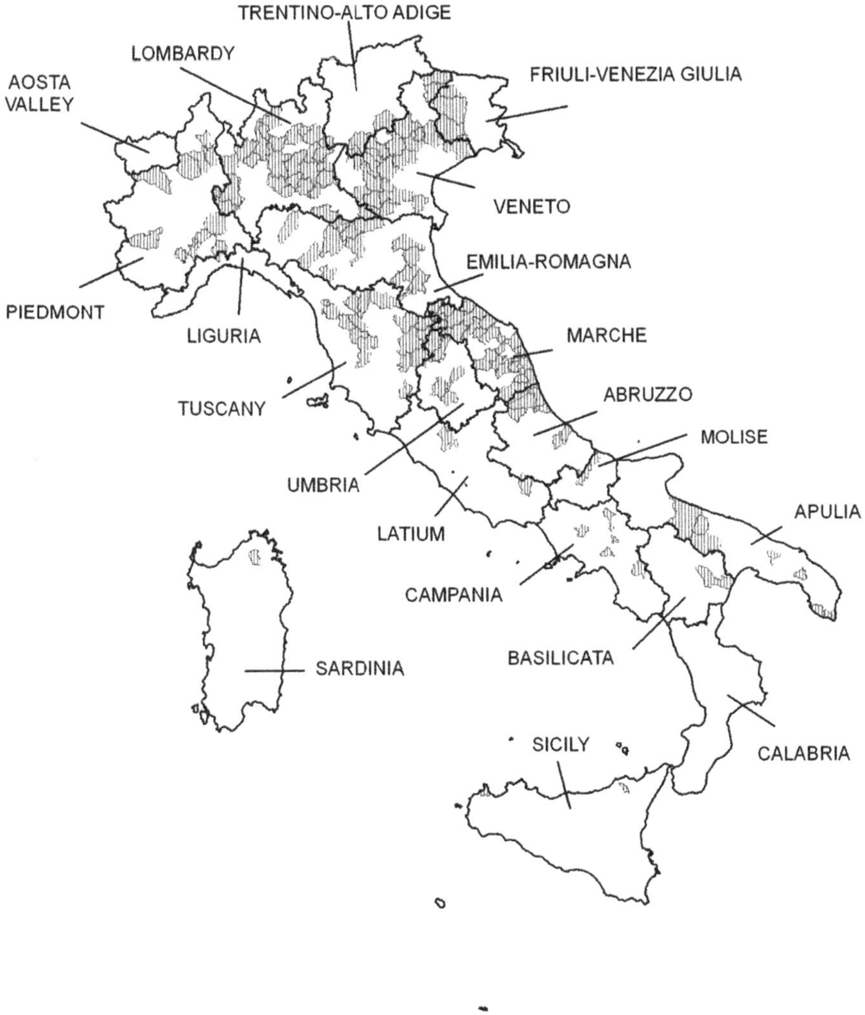

Fig. 1.4 Industrial districts, 2001 (Source: Sforzi 2009)

part of the Italian economy, and that small businesses, when they were organized according to the district model, could be competitive as a single large company.

The mapping proved that the district could also be a potential 'policy instrument' to support the competitiveness of SMEs. This recognition had important policy implications because in the early 1990s the Italian government enacted a law where the industrial district was constituted as a legal instrument to foster technological innovation of SMEs (Ministero dell'Industria 1991, 1993). The regions were to implement this innovation policy by identifying their own districts on the basis of the LLMAs.

Fig. 1.5 LLMAs specialised in consumer services, 2001 (Source: ISTAT 2005)

After this, studies on industrial districts were widespread and varied (Signorini 1994; ISTAT 1996; Brusco and Paba 1997; Fabiani et al. 1998; Casavola et al. 1999; Signorini 2000; Tessieri 2000; Ministero delle Attività Produttive 2002). Within the same strand of research, LLMAs studies led to the identification of other types of industrial areas, such as local production systems (Burroni and Trigilia 2001). The OECD territorial review of Italy used LLMAs to assess changes in the North–divide (OECD 2001) began to monitor industrial districts with the dissemination of data from the 2001 Census of industry and services (ISTAT 2006; Fazio and Pascucci 2006–2007).

Studies on urban and regional development have used the LLMAs for analysing the Italian urban system and European integration (Dematteis and Bonavero 1997) and the production system in the framework of the political geography of Italian regions (Conti and Sforzi 1997).

The analysis of Italian industrialization carried out through LLMAs highlighted a geographical pattern different from the model of the 'Three Italies' (Bagnasco 1977). Note that the unit of investigation used by Bagnasco was the region at NUTS 2 level. This is further evidence of the superiority of the place-based approach to territorial administrative units.

A more recent application of the LLMAs is the mapping of creative cities (Lazzeretti et al. 2008).

The availability of an analytical grid of LLMAs for the whole country benefited studies on local development. The possibility to conceptualize the term 'local', releasing it from the constraint of the territorial administrative units was important. Before the term local was conceptualized as a 'place', and place was empirically established as LLMA, local could mean a municipality, an urban district, or even a group of municipalities clustered to provide public services. Once the correspondence between local and place (or local community) was established, it was possible to give theoretical concreteness to the debate. For example, local development was no longer automatically associated with administrative devolution and people's participation. Local development was no longer seen as a reaction to globalisation, confusing it with political localism. It was no longer indentified with municipalism, spatial planning policies and investment in social overhead capital undertaken by municipalities. However, it was not possible to prevent that local development from being associated with LLMAs dominated by SMEs and, therefore, with industrial districts (Signorini 2000). As a consequence, it became necessary to clarify that districts were only one among the multiple paths of local development (Bellandi and Sforzi 2001).

In the field of sociology of innovation, LLMAs have been adopted to identify areas of high-tech activity (Biagiotti et al. 2011) and which are more conducive to innovation, as measured by the number of patents (Gherardini 2010).

A new frontier of the applied research using LLMAs are the studies on international migration linked to entrepreneurship. This field of research is of increasing interest in Italy because many local economies are experiencing a process of ethnicisation of their industrial structure (Barberis 2008). Migrant entrepreneurs are producing 'made in Italy' goods but with a lower integration with the local business community and a higher integration with their community of origin in the motherland. This phenomenon is summarized by the words 'made in Italy by …', where migrant entrepreneurs who 'make' are mostly Chinese (Lombardi et al. 2011).

The decision to apply the Italian LLMA approach to Spain (Boix and Galletto 2005) paved the way for international comparisons. This has affected both the industrial districts (Boix and Galletto 2008) and rural areas (Boix and Vaillant 2010).

LLMAs are now part of the ISTAT information system. They are used as territorial units to produce statistics on national economic performance, enterprises and their local units, and exports. These place-based statistics are disseminated online through the website and are freely available to researchers and practitioners, public and private, as well as to policymakers. The boundaries of LLMAs are a component of the digital cartography (ISTAT) and form an information layer of the Statistical Atlas of Municipalities, helping to meet the increasing demand for local data by people involved in analysis, design and evaluation of local development policies (ISTAT).

1.6 Final Remarks

The place-based approach to economic phenomena has brought academics, practitioners and policymakers a new way of interpreting economic reality and its change, and has proved a useful support to policy decisions.

Until ISTAT was able to provide national place-based statistics, the new approach lacked concreteness. However, it is equally true that without a single and coherent set of LLMAs that had proven its reliability for applied research the supply of place-based statistics would not have found any justification.

The development and dissemination of *ad hoc* statistics – such as place-based statistics – has a cost, since they are not required for institutional duties. If little or ineffective use is made of them the national statistics office may decide to 'close the tap', and stop producing them.

It can be argued that the relation between production and use of place-based statistics is a circular process: 'applied research/knowledge/policy decisions'. The way in which this circuit is activated and whether it starts from research or policy is less important than how feedback is incorporated.

The lesson from the Italian experience is that the demand of knowledge for policy purposes motivated economic research to look for the appropriate unit of investigation. Research had the strength to resist the political influence, and freely developed its theoretical thinking and empirical applications with the intention of generating new knowledge, aware of its policy implications. It should be recognized, however, that economic research was not caught unprepared: its critical thinking on the subject was already in progress.

References

Bagnasco A (1977) Tre Italie. La problematica territoriale dello sviluppo italiano. il Mulino, Bologna

Barberis E (2008) Imprenditori immigrati. Tra inserimento sociale e partecipazione allo sviluppo. Ediesse, Roma

Becattini G (1962) Il sistema marshalliano. In: Becattini G (ed) Il concetto di industria e la teoria del valore. Boringhieri, Torino, pp 88–125

Becattini G (1975) Introduzione. Invito a una rilettura di Marshall. In: Marshall A, Marshall P. (eds) Economia della produzione, a cura di G. Becattini. ISEDI, Milano, pp IX–CXIV

Becattini G (1979) Dal 'settore' industriale al 'distretto' industriale. Alcune considerazioni sull'unità d'indagine dell'economia industriale. Rivista di economia e politica industriale 5 (1):7–21

Becattini G (1987) Introduzione. Il distretto industriale cronaca di un ritrovamento. In: Becattini G (ed) Mercato e forze locali: il distretto industriale. il Mulino, Bologna, pp 7–34

Becattini G (1989) Riflessioni sul distretto industriale marshalliano come concetto socio-economico. Stato e Mercato 9(25):111–128

Becattini-IRPET (1969) Lo sviluppo economico della Toscana: un'ipotesi di lavoro. Il Ponte 25 (11–12):4–31

Becattini-IRPET (1975) Lo sviluppo economico della Toscana, con particolare riguardo all'industrializzazione leggera. Guaraldi, Firenze

Bellandi M, Sforzi F (2001) La molteplicità dei sentieri di sviluppo locale. In: Becattini G, Bellandi M, Dei Ottati G, Sforzi F, a cura di. Il caleidoscopio dello sviluppo locale. Trasformazioni economiche nell'Italia contemporanea. Rosenberg & Sellier, Torino, pp 41–63

Biagiotti A, Burroni L, Faraoni N, Gherardini A (2011) I sistemi locali dell'alta tecnologia in Italia. In Burroni L, Trigilia C, a cura di. Le città dell'innovazione. Dove e perché cresce l'alta tecnologia in Italia. il Mulino, Bologna, pp 41–59

Bianchi G (2009) I quarant'anni dell'Irpet. Quaranta ma non li dimostra. Il Ponte 65(1–2):135–140

Bianchi G, Magnani I (1985) Introduzione. Discutendo di sviluppo multiregionale. In: Bianchi G, Magnani I (eds) Sviluppo multiregionale: teorie, metodi, problemi. Collana Scienze Regionali. FrancoAngeli, Milano, pp 21–26

Boix R, Galletto V (2005) Identificación de Sistemas Locales de Trabajo y Distritos Industriales en España. MITYC-Secretaría General de Industria, Dirección General de Política para la Pequeña y Mediana Empresa, Madrid

Boix R, Galletto V (2008) Marshallian industrial districts in Spain. Scienze Regionali-Italian J Sci 7(3):29–52

Boix R, Vaillant Y (2010) Industrial districts in rural areas of Italy and Spain. Sviluppo Locale 14 (35):73–114

Brusco S, Paba G (1997) Per una storia dei distretti industriali italiani dal secondo dopoguerra agli anni novanta. In: Barca F (ed) Storia del capitalismo italiano dal dopoguerra a oggi. Donzelli, Roma, pp 265–333

Burroni L, Trigilia C (2001) Italy: local development through local economies. In: Crouch C, Le Galès P, Trigilia C, Voelzkow H (eds) Local production systems in Europe. Rise or demise? Oxford University Press, Oxford, pp 46–78

Casavola P, Pellegrini G, Romagnano E (1999) Imprese e mercato del lavoro nei distretti industriali italiani. Sviluppo Locale 6(10):41–59

Conti S, Sforzi F (1997) Il sistema produttivo italiano. In: Coppola P (ed) La geografia politica delle regioni italiane. Einaudi, Torino, pp 278–336

Dematteis G, Bonavero P (eds) (1997) Il sistema urbano italiano nello spazio unificato europeo. il Mulino, Bologna

Fabiani S, Pellegrini G, Romagnano E, Signorini LF (1998) L'efficienza delle imprese nei distretti industriali italiani. Sviluppo Locale 5(9):42–73

Fazio NR, Pascucci C (2006–2007) Le esportazioni dei sistemi locali nel 2005, con particolare riguardo ai distretti industriali. Sviluppo Locale 12(29–30):3–35

Fox KA, Kumar TK (1965) The functional economic area: delineation and implications for economic analysis and policy. Papers and proceedings of the Regional Science Association 15:57–85

Gherardini A (2010) La valutazione dei brevetti. In: Ramella F, Trigilia C (eds) Invenzioni e inventori in Italia. il Mulino, Bologna, pp 209–249

Goodman E, Bamford J, Saynor P (eds) (1989) Small firms and industrial districts. Routledge, London

Hägerstrand T (1970) What about people in regional science? Pap Reg Sci Assoc 24(1):7–21
Hoover EM (1971) An introduction to regional economics, 2nd edn. Knopf, New York
Regione Toscana/Dipartimento Programmazione-IRPET (1978) Criteri per l'individuazione delle unità intercomunali. In: Regione Toscana/Dipartimento Programmazione-IRPET, Proposta di istituzione e individuazione delle unità intercomunali. Regione Toscana/Giunta Regionale, Firenze
IRPET (1976) Caratteri e prospettive dello sviluppo toscano. IRPET, Firenze
IRPET (1980) Nuovi contributi allo studio dello sviluppo economico della Toscana. IRPET, Firenze
IRPET (1983) TFM 1981. Modelli per l'analisi regionale e sub-regionale: il caso della Toscana. Le Monnier, Firenze
Isard W (1956) Location and space-economy: a general theory relating to industrial location, market areas, land use, trade and urban structure. The MIT Press, Cambridge, MA
ISTAT (1996) Rapporto annuale. La situazione del Paese nel 1995. ISTAT, Roma
ISTAT (1997) I sistemi locali del lavoro. 1991, a cura di F. Sforzi, Collana Argomenti 10. ISTAT, Roma
ISTAT (2005) I sistemi locali del lavoro. 2001, research report prepared by A. Orasi and F. Sforzi, published online: http://dawinci.istat.it/daWinci/jsp/md/download/sll_comunicato.pdf. Accessed 21 July 2005
ISTAT (2006) Le esportazioni dai sistemi locali del lavoro. Dimensione locale e competitiva dell'Italia sui mercati internazionali, a cura di S. Menghinello, Collana Argomenti 22. ISTAT, Roma
ISTAT Confini amministrativi e dei sistemi locali del lavoro. www.istat.it/it/strumenti/. Accessed 12 Oct
ISTAT Atlante statistico dei comuni, Collana Informazioni. ISTAT, Roma
ISTAT-IRPET (1986) I mercati locali del lavoro in Italia. In: Seminario su Identificazione di sistemi territoriali, analisi della struttura sociale e produttiva in Italia, Roma, 3–4 December
ISTAT-IRPET (1989) I mercati locali del lavoro in Italia, a cura di F. Sforzi. FrancoAngeli, Milano
Lazzeretti L, Boix R, Capone F (2008) Do creative industries cluster? Mapping creative local production systems in Italy and Spain. Ind Innovat 15(5):549–567
Lombardi S, Lorenzini F, Sforzi F, Verrecchia F (2011). Chinese entrepreneurship in context: localization, specialization and their impact on Italian industrial districts. In: Paper presented at the 51st ERSA Congress, Barcelona, 30 August–3 September
Marshall A (1920a) Principles of economics, 8th edn. Macmillan, London
Marshall A (1920b) Industry and trade, 3rd edn. Macmillan, London
Ministero dell'Industria (1991) Legge 5 ottobre 1991, n. 317, Interventi per l'innovazione e lo sviluppo delle piccole imprese. Gazzetta Ufficiale della Repubblica Italiana, Serie Generale n. 237 – Supplemento Ordinario n. 60, Roma
Ministero dell'Industria (1993) Decreto Ministeriale 21 aprile 1993. Determinazione degli indirizzi e dei parametri di riferimento per l'individuazione, da parte delle regioni, dei distretti industriali. Gazzetta Ufficiale della Repubblica Italiana, Serie Generale n. 118 – Supplemento Ordinario n. 51, Roma
Ministero delle Attività Produttive (2002) L'esperienza italiana dei distretti industriali. IPI, Roma
OECD (2001) OECD territorial reviews: Italy. OECD, Paris
Pyke F, Becattini G, Sengenberger W (eds) (1990) Industrial districts and inter-firm co-operation in Italy. IILS, Geneva
Raffaelli T (2001) Il ruolo del concetto di distretto industriale nel sistema teorico marshalliano. Sviluppo locale 7(17):16–32
Regione Toscana (1979) Istituzione delle associazioni intercomunali, L.R. 17 agosto 1979, n. 37. Bollettino del Consiglio regionale, pp 2706–2711
Regione Toscana/Consiglio Regionale (1978) Istituzione delle associazioni intercomunali, Documenti di Toscana/Consiglio Regionale 4, Firenze

Sforzi F (1980) Identificazione di sistemi sub-regionali in Toscana: comparazioni di strategie analitiche. In: IASA-IASI/CNR-IRPET (ed) Analisi e pianificazione dei sistemi regionali e sub-regionali (mimeo). IRPET, Firenze

Sforzi F (1982) Identificazione degli ambiti sub-regionali di programmazione. In: Bielli M, La Bella A (eds) Problematiche dei livelli sub-regionali di programmazione. FrancoAngeli, Milano, pp 81–102

Sforzi F (1987) L'identificazione spaziale. In: Becattini G (ed) Mercato e forze locali: il distretto industriale. il Mulino, Bologna, pp 143–167

Sforzi F (2007) Presentazione. In: Becattini G, Scritti sulla Toscana. Vol I – La ricerca sul campo e la 'libera scuola' di Artimino, a cura di F. Sforzi. Le Monnier, Firenze, pp V–XX

Sforzi F (2008) Il distretto industriale: da Marshall a Becattini. Il pensiero economico italiano 16 (2):71–80

Sforzi F (2009) The empirical evidence of industrial districts in Italy. In: Becattini G, Bellandi M, De Propris L (eds) A Handbook of industrial districts. Edward Elgar, Cheltenham, pp 327–342

Sforzi F, Martinelli G (1980) La configurazione della struttura nodale in un sistema regionale: aspetti concettuali e metodi di analisi. In: IRPET, Nuovi contributi allo studio dello sviluppo economico della Toscana, cap. 8. IRPET, Firenze

Sforzi F, Openshaw S (1983) Sistemi sub-regionali in Toscana: dalla teoria alla pratica. In: IRPET, TFM 1981. Modelli per l'analisi regionale sub-regionale: il caso della Toscana. Le Monnier, Firenze, pp 199–251

Sforzi F, Openshaw S, Wymer C (1982) La delimitazione di sistemi spaziali sub-regionali: scopi, algoritmi, applicazioni. In: Paper presented at the 3rd, Venezia, 10–12 November

Sforzi-IRPET (1977) La questione comprensoriale e l'individuazione di aree funzionali in Toscana, Documenti di lavoro 1 (gennaio), Firenze

Signorini LF (1994) Una verifica quantitativa dell'effetto distretto. Sviluppo Locale 1(1):31–70

Signorini LF (2000) Lo sviluppo locale. Un'indagine della Banca d'Italia sui distretti industriali. Donzelli, Roma

Smart MW (1974) Labour market areas: uses and definition. In: Diamond DR, McLoughlin JB (eds) Progress in planning, vol 2. Pergamon Press, Oxford, pp 239–353

Tessieri N (2000) Multinazionali e distretti industriali in Italia. Sviluppo Locale 7(13):71–99

Chapter 2
From Local Units to Economic Regions in Spain

Where Agglomeration Economies are Meaningful

Fernando Rubiera-Morollón and Ana Viñuela

2.1 What is a *Region*?

Over the last 100 years, Urban and Regional Economics has grown spectacularly as an applied field of Economic Science, using the terminology proposed by Schumpeter in 1954. This discipline has been capable of providing theories and analysis that go far beyond its strict goals. Some of the most interesting economic theories of the past 20 years have been proposed from the perspective of Regional and Urban Economics and we have seen extraordinary growth in the developmenzt of statistical tools for empirical study. However, all this theoretical and technical development has not been accompanied by a clear definition of the fundamental concept of *region* in Regional Science (Behrens and Thisse 2007).

In this first section we review how the evolution of Regional Science itself has been accompanied by a continuing reassessment of the concept of *region*. Each of the new ways of understanding what a *region* is and how it may be defined provides important nuances and is underpinned by a different view of the aspects that are to be highlighted. At the same time as carrying out a synthetic review of the evolution that has taken place, we shall also attempt to deduce the fundamental elements of each approach to the concept of *region*.

F. Rubiera-Morollón (✉)
REGIOlab – University of Oviedo, Oviedo, Spain
e-mail: frubiera@uniovi.es

A. Viñuela
Facultad de Economía y Empresa, Dpto. de Economía Aplicada, Universidad de Oviedo,
Avda. de El Cristo, S/N, Oviedo 33006, Spain

E. Fernández Vázquez and F. Rubiera Morollón (eds.), *Defining the Spatial Scale in Modern Regional Analysis*, Advances in Spatial Science,
DOI 10.1007/978-3-642-31994-5_2, © Springer-Verlag Berlin Heidelberg 2012

2.1.1 The Weakness of the Concept of Region Since the Origins of Regional Science

The existence of a multidisciplinary field known as Regional Science can lead us to assume that the concept of *region* should be clearly defined. However, this is not so; the definition of *region* has been an ever-present weakness for all the disciplines that may be grouped together under the name of Regional Science.

Geography was defined by Kant, Humboldt and Hetter as *the Science of dividing the landscape into regions* (see Lancaster 1939). The development of Geography in the eighteenth and nineteenth centuries already raised the need to delimit the total surface area into spatial areas –i.e. regions– so as to facilitate their study. The *region* thus becomes a tool of the geographer to encompass and study the totality of space. Starting from known geographic references, new spatial areas that are homogeneous with respect to one another according to diverse criteria are progressively explored and described (see Claval 2007). These early geographers soon became aware of the difficulty of defining the concept of region insofar as the variables that could be used to define regional areas may be highly diverse. This might in turn give rise to highly diverse ideas and could even place the analysis outside the field of Geography itself when employing mainly geological, cultural or social criteria, among others (Claval 2003).

Whatever the case may be, the complexity of the definition grew with increased knowledge of physical space, as the emphasis of Geography and regional delimitation began to fall more on social or cultural aspects than on geological, topographical or physical criteria. This actually made it even more difficult to reach consensus regarding a single concept of region. Authors such as Jean-Louis Giraud-Soulavie in France, Alexander von Humboldt in Germany and William Marshall in Britain were aware that the concept of region must necessarily incorporate cultural, economic and social variables. The key element, from these first regional delimitations that fully incorporate multi-dimensionality in the discipline, is to maintain the essence of delimiting homogeneous spatial areas. These pioneering studies already revealed the limitation of the administrative division of territories, as they gave rise to heterogeneous spaces in terms of scientific criteria (Claval 2007).

2.1.2 The Incorporation of Economics in the Definition of the Concept of Region

At the beginning of the twentieth century and subsequent to the decisive contributions of Humboldt, Weber and Von Thünen, Economics was the discipline that began to show the most interest in approaching the definition of homogeneous regions using specific criteria from Economic Science.

The work of the three aforementioned authors clearly established the key concept of *centrality*, i.e. there exists a central place to which businesses and individuals come to perform their necessary market exchanges. By clearly accepting the existence of this central place to which all such businesses and individuals gravitate, we can build a view of economic space in connection with that place. For example, Von Thünen, in his work *The Isolated State* of 1826, links the value of land with its situation relative to the *centre*, where the market is located (Von Thünen 1966). The distance of each place with respect to the closest market implies a different value or rent to the land, *ceteris paribus* other features of the place, where the differences derive precisely from the savings made in transportation and opportunity costs. The land closer to the centre will be more highly valued. All activities will be willing to pay more for these locations, but those activities that can generate more value per square metre of surface area and which value proximity to the centre most will manage to become established there, because either their transportation costs or their opportunity costs are higher. In other words, activities that are more extensive but of lesser added value will be located farther from the centre, while intensive activities in terms of the use of space and high added value will be able to acquire the land closest to the centre. This behaviour is repeated throughout different territorial units around different centres that present similar dynamics in their surroundings.

Within this context, August Lösch made a major contribution in 1938 with his work *The Nature of Economic Regions*. The starting point of Lösch's line of reasoning is the acceptance that regions conceived exclusively in terms of geographical or cultural criteria are –from a purely economic point of view– artificial units with no interest for Economics, unless they possess the capacity to effect significantly different economic policies by constituting themselves as administrative regions with far-reaching or total (country-level) independence in their political actions. The concept of *region* proposed by Lösch rests on the distribution of land in terms of transport and opportunity costs. Different centres or markets are progressively generated over a homogeneous space. The inhabitants or economic activities of the region will come to the market, the *centre*, which, thanks to lower transport costs, will enable them to cover their needs and distribute their products. Obviously, the world is not a homogeneous plane that may be distributed in perfect hexagons but it can actually be very well explained by means of the regions suggested by Lösch.

Lösch's approach gains in value when combined with the proposal formulated almost concurrently by the German geographer Walter Christaller in 1935. Although Christaller's idea is less precise in defining how space is structured for economic reasons, it provides a complement that Lösch did not fully take into account. In addition to the existence of the allocation of space explained by Lösch, Christaller notes that there is a *hierarchy of central places*; i.e. not all central points or cities are equal, as there are *higher-order* centres, with a greater concentration of activities, and other *lower-order* places. Some basic needs that require very frequent journeys are spread over space in *lower-order central places*, while other, less frequent consumer or exchange activities may become concentrated in *higher-order central places*.

In 1949, Zipf identified an empirical regularity that relates size with position in the *hierarchy of central places*, providing this approach with more power to define the concept of *region* in terms of economic criteria.

2.1.3 Geographical Communities, Functional Regions, Local Labour Markets and Metropolitan Areas

While the view that emerges from combining the ideas of Humboldt, Weber, von Thünen, Lösch, Christaller and Zipf was to remain in force throughout the twentieth century and even today, Regional Science has continued its search for the definition of a *region* and its physical limits.

The combination of Geography and Sociology, among other disciplines, gave rise to the concept of *Geographical Communities*, spaces in which strong social and economic interactions are produced that eventually generate identities and commonalities (Poplin 1979). Based on the idea of the existence of *Geographical Communities*, different authors have sought ways to define sets of elements that delimit space which has led to the idea of *functional regions;* i.e. regions defined regardless of administrative borders which are constructed according to diverse economic, social, cultural or geographical factors.

Some studies have followed this line in Economic Science, especially within the field of Labour Economics, in addition to that of Urban and Regional Economics itself. Authors in these disciplines began to be concerned with defining geographical areas that share a *single local labour market*. As cities grow and expand, their geographical boundaries also move further afield. Consequently, *metropolitan areas* often cover several administrative divisions. This expansion of the *metropolitan area* leads to the ever-increasing phenomenon of *commuting*, i.e. daily displacements between the place of residence and the workplace, which are sometimes located in different administrative regions. Thus, the coordination of local policies and actions in terms of transportation or urban planning becomes a must and some procedures have to be designed to define and identify a space in which most of the population residing there also works there: what is known as the *local labour market* (in what follows, *LLM*).

The drawback of this approach lies in establishing the criteria and ways of defining to what extent a place belongs to a particular *LLM* or not. Numerous methodologies have been developed to define the geographical area that can be considered a *LLM*. Worth highlighting in this regard are the methodologies designed by Sforzi (1987, 1990), Serra et al. (2002), Rozenblat and Cicille (2003) and ISTAT (1997, 2006). The previous chapter of this book, by Sforzi, summarizes the different applications of all these methodologies in diverse countries.

2.1.4 External Economies and Industrial Districts

A somewhat related, though differentiable line of research on *functional regions* is the definition of *industrial districts* mentioned in the previous section. In 1820, Alfred Marshall published his seminal work *Principles of Economics* in which he presented the nowadays-essential concept of *external economies* (Marshall 1890). The concentration of highly specialized activities in a reduced geographical area leads to the unleashing of a set of cumulative processes and increasing returns. Such effects are not caused by the firm's scale of production, but by its interactions with other similar firms within its setting and the scale of the industry (sector) in the location as a whole. Thus, the relations of the companies that coexist in a place generate increasing returns that cannot be explained from within the firms themselves, but rather from the industry (sector) and the place where a variety of similar firms are located.

The concept of *external economies* is obviously fundamental within the field of Industrial Economics, but also in that of Regional and Urban Economics, which finds a different way to that of the classical approach of addressing the connections between space and economic activity. In 1979, Giacomo Becattini published an article in the *Rivista di Economia e Politica Industriale* which marked the beginning of a new branch of the literature that contains a new approach to the relationship between economy and space. This author put forward the concept of the *industrial district*, which is not a concentration of businesses or a network of firms, but rather the productive manifestation of local society. The concept of industrial district is defined as a "[. . .] socio-territorial entity which is characterised by the active presence of both a community of people and a population of firms in one naturally and historically bounded area" (Becattini 1991).

When identifying the existence and extent of *industrial districts* in practice, it becomes impossible to use administrative boundaries in view of the fact that an industrial district may go beyond such limits or be contained within them. Furthermore, industrial districts are a dynamic concept, while administrative limits do not usually change over time. An *economic region*, as defined by classical authors, structured around a single centre, can also contain diverse *industrial districts*. Unlike other approaches, it may be stated that the identification of industrial districts does not necessarily cover the entire territory. Once their presence and limits have been established, there are "empty" areas in which no industrial districts have been found to be present. In short, this is not an approach that seeks to cover the whole country, but only to detect those places where the spatial concentration of certain sector generates *Marshallian external economies*.

It is clear that the key to applying the concept of *industrial district* to an observable empirical reality subject to analysis lies in delimiting its outline. The need to start out from a territorial unit whose extension is not constrained by administrative boundaries and which may change over time is consistent with the concept of *local labour markets (*Sforzi and Lorenzini 2002). This is precisely the way in which Sforzi (1987, 1990) and ISTAT (1997 , 2006) put forward the methodology for defining industrial districts. First, *LLMs* are used as a territorial unit for identifying such districts. Second, each *industrial district* is identified from its socioeconomic features that distinguish them from other *LLMs*.

2.1.5 A Novel Approach Provided by the New Economic Geography: Economies and Diseconomies of Agglomeration

In the last third of the twentieth century, a vast body of diverse research began to take shape starting in 1991 with the pioneering publication *Increasing Returns and Economic Geography* presented by Krugman (1991) in the *Journal of Political Economy* (further extended by Krugman 1995) and which was fully developed with the work of Fujita et al. (1999), Baldwin et al. (2003) and Ottaviano and Thisse (2004), among others. We refer here to the New Economic Geography (NEG), whose development is fundamental in the field of Regional and Urban Economics as well as in that of International Economics.

The NEG contributes a number of novel aspects to the discussion of the concept of *region,* excellently summarized by Behrens and Thisse (2007). As these authors point out, this theoretical framework is the first to provide an explanation as to why strong economic disparities across space exist between regions and are maintained or even grow despite the fact there may be free movement of the factors of production, which, according to neoclassical economic theory, would result in their progressive disappearance.

In their simplest formulation, the NEG models propose the existence of two regions: *centre* and *periphery*. The idea of *centre* and *periphery* is defined here in a different way to the classical approach, although there are links between the two views. The NEG focuses on understanding *centripetal forces*, which tend to concentrate activities in the *central* region, and *centrifugal forces*, which tend to disperse them towards the *peripheral* region.

The gains derived from large-scale production and from the *positive externalities* associated with *size* lead to the concentration of economic activity in *central* locations from which the largest possible market is accessible. Transportation costs constrain this concentration behaviour, but the strength of this limitation depends on the consumption characteristics of the activity. Consistent with the classical approach, those activities that require intense personal interaction between consumers and producers (which includes many services) and/or are consumed daily or very frequently will display quasi-equal distributions over space. In contrast, activities that are tradable over broader distances, not requiring proximity to the point of consumption, and/or are demanded less frequently will concentrate their production in a limited number of *central* locations. As distance costs fall and trade increases, larger concentrations should normally grow in *size*. A shift in the national economy towards agglomeration-sensitive goods and services (and away from, say, agriculture) also favours the growth of larger concentrations (Parr 2002).

As large concentrations grow, diseconomies naturally appear, producing an expulsion effect for some activities. Wages and land prices are in part a function of city size. Wage-sensitive and space-extensive activities will be pushed out by what is sometimes called the *crowding-out effect* of rising wages and land prices in large metropolitan areas. This *crowding-out effect* will most notably be felt by medium-technology manufacturing –which has less need for the highly skilled labour in large cities (Henderson and Thisse 1997)– and also by wholesaling and distribution, which are extensive consumers of space, giving rise in turn to the growth of smaller cities.

On the other hand, the *agglomeration economies* associated with urban concentration lead to firms within the same industry benefitting through lower recruitment and training costs (shared labour-force), knowledge spillovers, lower industry-specific information costs and increased competition (Rosenthal and Strange 2001; Beardsell and Henderson 1999; Porter 1990). The increasing size of the metropolis makes certain infrastructures possible, such as international airports, post-graduate universities, research hospitals, etc. The recent literature stresses the positive link between productivity and the presence of a diversified, highly qualified and versatile labour pool (Duranton and Puga 2002); Glaeser 1998; Glaeser et al. 1995). As highlighted by Hall (1998), Eaton and Eckstein (1997) and Castells (1976), large metropolises stimulate the exchange of knowledge, while the link between urban agglomeration and economic growth has been explored by Polèse (2005). Activities that are characterized by the need for high creativity and innovation will in general choose to locate in or near to major metropolitan areas (Desmet and Fafchamps 2005).

It is reasonable to infer that the trade-off between the *centrifugal* and *centripetal* forces that push economic activities towards large cities or drive them out should give rise to an economic landscape characterized by regularities in industrial and employment location patterns based on the size of and distance from some other (larger) cities (Redding and Venables 2004).

The transition from a simple formulation of NEG models with two regions to a more complete formulation with *n* regions is by no means simple. The presence of more than two regions complicates the simple perception of the reality described by the basic models due to opening up the possibility of the existence of intermediate regions as well as more complex interactions. In fact, as pointed out by Behrens and Thisse (2007), one of the empirical and theoretical challenges of the broadening of the NEG framework requires the definition of models that encompass the complexity resulting from addressing the presence of multiple regional realities.

The perspective of the NEG brings several new dimensions to the discussion regarding the definition of the concept of *region*: (1) the connection between the concept of *region* and the explanation of existing regional differences; (2) the possibility of proposing regional classifications that do not present spatial contiguity, but rather common features that induce similar *centripetal* or *centrifugal* dynamics; and (3) the notion of the *region* as an economic unit that interacts economically with others and whose main feature is economic opening (no borders to the movement of goods and services or factors of production).

2.2 From the Theoretical to a Practical Concept of a Region

Clearly, the fundamental challenge is to find a regional classification that synthesizes the different conceptual approaches to the theoretical concept of *region* and which has a relatively easy empirical interpretation and construction. Ideally, we should have an empirically analysable *region* containing units consistent with the classical view of Lösch, Christaller and those who followed them which is at the same time the aggregation of functional units, though without creating scientifically irrelevant divisions. In short, a region that empowers the study of key issues in Regional Economics such as the effects of location- or urbanization-type *external economies of agglomeration*.

Thus, in this section we suggest an integrating proposal based on three major lines of empirical research developed over the past 25 years: (1) the literature on how to define *local labour markets* (*LLMs*) so as to define the base unit, the methodology designed and applied mainly by Sforzi in Italy and by Boix in the Spanish case; (2) the work by Coffey and Polèse in the late 1980s and early 1990s in which they propose a classification based on (population) size and the *distance to size* in order to capture the effects of *agglomeration economies*; and (3) the recent development of the concept of *incremental distances* proposed by Partridge, Rickman and others, which has links to the classical regional literature.

2.2.1 The Basic Unit of Analysis: Local Labour Markets

Insofar as we are attempting to define a concept of *region* based on economic criteria, it would be inconsistent to use administrative divisions, no matter how disaggregated they are. The starting-out point must be that of defining the basic spatial unit, delimited by boundaries that guarantee its consistency, comparability and meaning.

The regionalization method developed by Sforzi and Lorenzini (2002) and ISTAT (2006), applied to Spain by Boix and Galleto (2006), identifies *local labour markets* (*LLMs*) through a multi-stage process to then use them as the basic spatial unit to define industrial districts. Defining *local labour market areas* requires information on residence-work mobility so that they contain a geographic area within which most of the population lives and works. *LLMs* are created from the information on municipalities, a very small administrative unit, combining data on the resident employed population, total employed population and displacements from the place of residence to the workplace.

2.2.2 Re-aggregating the Basic Units According to Size and Distance to Size

Having defined the basic spatial unit, *LLM areas*, in order to include the importance of *agglomeration/urbanization* and *distance* to the major population concentrations, the next step is to classify these basic units in order to incorporate these concepts.

There are many ways to re-aggregate the areas, but we propose a regional classification wholly based on the key aspects of modern Urban and Regional Economics. Following Polèse (2009), we consider a number of key aspects:

(1) *location* matters, because industries (and therefore economic activity and employment) are always drawn to places best suited for commerce and interaction with markets; and (2) *size* matters, because dynamic industries, or the most advanced in each epoch, are naturally drawn to large cities and places within easy reach. A corollary can be deduced from (1) and (2), namely: (3) *proximity to size* also matters. Another basic idea of Regional Economics is: (4) *cost* matters, because without adequate size or a propitious location, places will grow if they have a clear labour cost advantage or, alternatively, an exceptional resource endowment.

Thus, Coffey and Polèse (1988), Polèse and Champagne (1999) and Polèse and Shearmur (2004) for Canada and the subsequent application of their ideas to the Spanish case by Rubiera (2006), Polèse et al. (2007) and Viñuela et al. (2010) propose a classification of space that takes into consideration the existence of agglomeration economies (*size*) and *distance* as key factors. However, the classifications of these authors always adopt administrative regions as the basic unit of analysis. Here we propose the classification or re-aggregation of more cohesive and meaningful basic units, *LLMs,* based on the criteria of population *size* and *distance*. To illustrate this approach, Fig. 2.1 shows a schematic representation for an idealized national space economy. Each cell is a municipality (administrative local unit), with various municipalities being aggregated into *LLMs* (blue line). The reader will undoubtedly note the resemblance to the classic idealized economic landscapes of Christaller, Lösch, and Von Thünen, all of which posit one metropolis or marketplace at the centre. Thus, Fig. 2.1 represents a large *LLM*, in terms of population, at the *centre* (the main metropolis, which includes different municipalities), but also four smaller *urban LLMs* of different population *sizes* around it, the rest being considered *rural* in terms of population size. Regardless of their size, *LLMs* can also be *central* (close to the main metropolis), or *peripheral* (located at some distance from the metropolis).

To summarize, this idealized space may be classified first by *size* into:

1. *Metropolitan areas*: *local labour markets* with more than a certain population size.
2. *Urban areas*: *LLMs* that are urban, though not large enough to be considered a metropolitan area.
3. *Rural areas*: *LLMs* that are clearly not urban.

A parallel distinction, based on proximity to the major metropolis, is applied to all non-metropolitan *LLMs*:

1. *Central areas*: *LLMs* "close" to the large metropolitan area.
2. *Peripheral areas*: *LLMs* located "far" from the metropolitan area.

The 1-h drive criterion takes into account several factors such as road conditions (e.g. motorway or not), the spatial limits of metropolitan areas and the distinctive characteristics of the area being classified. Thus, as can be seen in Fig. 2.4, *central* areas do not necessarily form perfect rings around metropolitan areas. The 1-h threshold, also used in other applications, has been found to be very robust and a

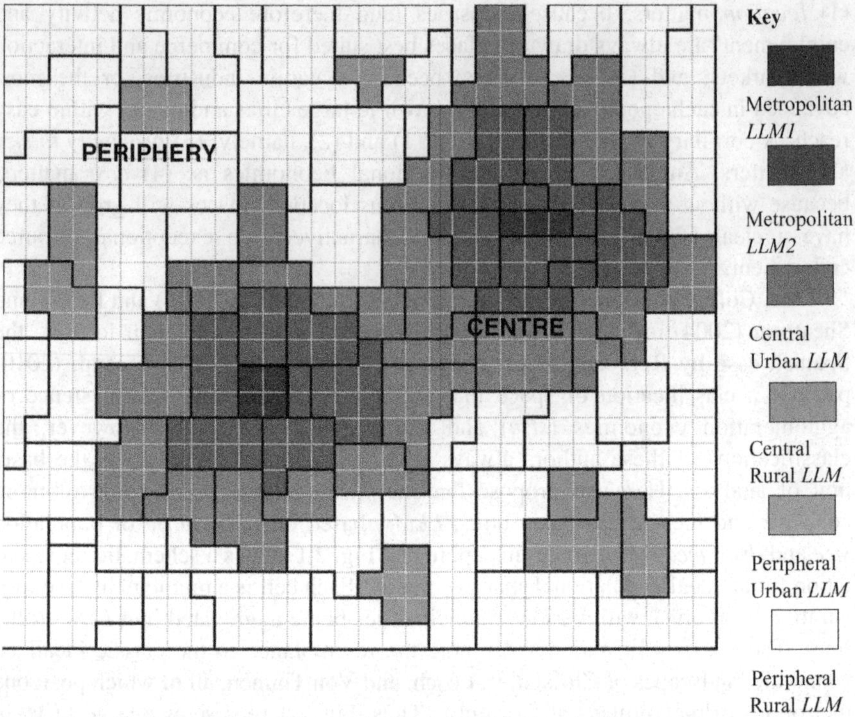

Fig. 2.1 Schematic representation of the classification of spatial units (Source: Own elaboration based on Viñuela (2011))

good indicator of the range within which spatial interaction with the metropolis remains fairly easy, especially for face-to-face relationships related to the consumption of higher-order services (see Porter (1990) and McCann (2007), among others).

2.2.3 The Measurement of Distance: The Incremental Distance

One of the major problems when classifying an area as *central* or *peripheral* is the subjectivity of the criteria used when choosing the distance metrics (linear distance, Euclidean distance, distance by road, time, etc.) and also the distance threshold.

However, taking into account Christaller's ideas (1935) on the hierarchies of places and the connection between urban size and the position in the hierarchy of places of each city from Zipf (1949), we know that only large cities are able to offer a full range of goods and services. If we only consider the distance to the central place (the metropolitan area), the higher place in Christaller's hierarchy, we somehow make a mistake by forgetting that certain goods and services *are* offered in smaller urban places. One way of solving this problem is to define a set of incremental distances to each tier (size) of urban area so that we first quantify the distance to the next tier, where *some* additional and higher-order goods and services are produced, and then quantify the incremental distance to the next higher urban tier, maybe a metropolitan area, where higher-order services and urban amenities are located. This idea of *incremental* distances, suggested by Partridge et al. (2008, 2009), brings together the effect of the *distance* to large agglomerations: individuals and businesses need access to the higher-order services, urban amenities, higher qualified jobs and lower cost products that are only present in large urban agglomerations due to the presence of strong agglomeration economies. Thus, we can measure the distance from a large agglomeration as a "penalty" to access the goods and services present in such an agglomeration.

The proposal set out in the previous section allows us to have a set of regions that do not necessarily have any spatial contiguity, but whose conceptual definition is strongly consistent. Although these regions can be incorporated into the analysis as dichotomous variables, they may be incorporated in empirical models in a different way via the use of *incremental distances*.

Note that the distance criteria proposed by Partridge et al. (2008, 2009) essentially consists in establishing a hierarchy of *LLMs* such that each *LLM* is associated not only with a *size* value (population both residing and working within the *LLM*), but also with a *distance* value that reflects the total cumulative disadvantage due to distance measured across all urban *LLM* tiers (Fig. 2.2).

2.3 *Economic Regions* Based on *LLMs*: A Proposal for the Spanish Territory

Spain is divided administratively into 8,105 municipalities which are aggregated into 50 provinces (NUTS III level), excluding Ceuta and Melilla, and 17 Autonomous Communities or NUTS II regions (Fig. 2.3 – Maps 1 and 2). The number of municipalities within each province ranges from 34 (Las Palmas) to 371 municipalities (Burgos). Besides, there are Autonomous Communities with several provinces, such as Andalusia with eight provinces, and others with only one, like Asturias. Furthermore, simply for the sake of comparison with certain other European member-states, the 17 Autonomous Communities can be aggregated into seven administrative regions or NUTS I regions (Fig. 2.3, Map 3), which have no real internal political or administrative meaning.

2.3.1 First Step: *Defining Local Labour Market Areas for Spain*

Applying an algorithm that consists of four main stages and a fifth stage of fine-tuning, Boix and Galleto (2006) aggregate the 8,106 Spanish municipalities into 806 *LLMs*. The algorithm starts out from the municipal administrative unit and

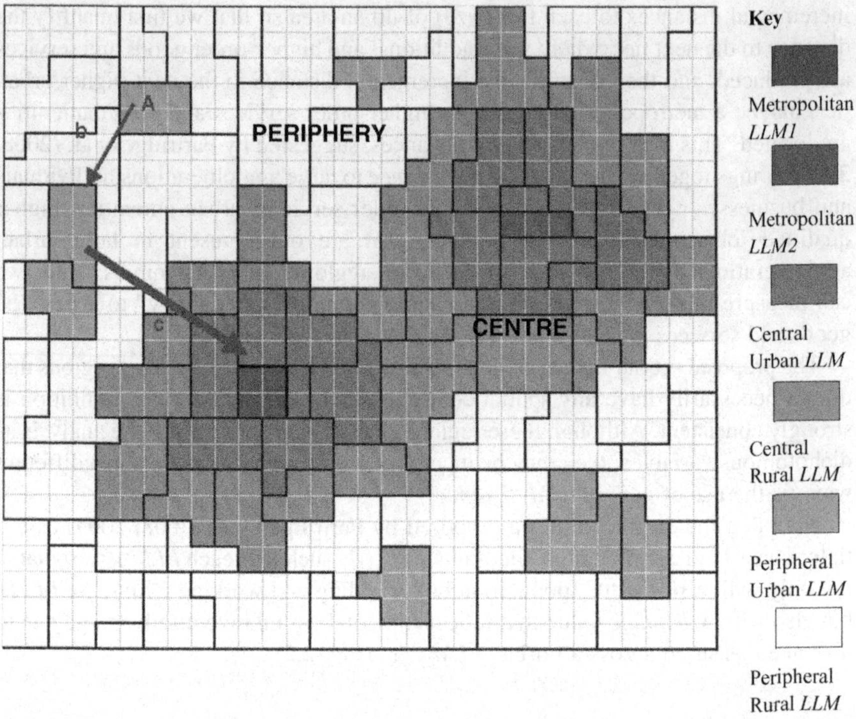

Fig. 2.2 Schematic representation of incremental distances. (From point *A*, the distance to the closest Urban *LLM* is *b* and only *c* to the Metropolitan *LLM* (the distance to the Metropolitan *LLM* is $b + c$ but $-b$, which is the distance already computed from *A* to the nearest urban area) (Source: Own elaboration based on Partridge et al. (2008, 2009))

Fig. 2.3 Spanish administrative division of the territory into Provinces (NUTS III), Autonomous Communities (NUTS II) and NUTS I

generates the *LLMs* using data from the 2001 Spanish Census on the resident employed population, total employed population and displacements from the place of residence to the workplace. A detailed explanation of the methodology is

Fig. 2.4 Spanish division of the territory into *Local Labour Markets* (Source: Own elaboration based on Boix and Galleto (2006))

available in Boix and Galleto's report, from 2006. Map 4 shows the 806 *LLMs* defined by these authors (Fig. 2.4).

2.3.2 Second Step: From Local Labour Markets to Economic Regions (Size and Distance to Size Criterions)

After defining the local employment systems, we can now classify these basic spatial units first according to *size* and then according to *distance to size*. Table 2.1 shows the distribution of local labour markets by population size in Spain, where five tiers or levels are defined.

The two first tiers, *LLM1* and *LLM2*, correspond to metropolitan areas or *centres* in Christaller's nomenclature. Given the major difference in size between the metropolitan areas of Madrid and Barcelona and the rest (with more than 500,000 but less than 2,500,000 inhabitants), we consider it appropriate to distinguish these two levels. The next lower urban tiers, *LLM3* and *LLM4*, basically include cities of more than 100,000 but less than 50,000 inhabitants and between 50,000 and 100,000 inhabitants, respectively. Finally, those *LLMs* with less than 50,000 inhabitants are considered *rural* areas (*LLM5*).

After classifying the *LLMs* in terms of population *size*, we propose to improve the classification by discriminating by *distance to size*, taking the metropolitan areas *LLM1* and *LLM2* as the highest tiers of the hierarchy.

Table 2.1 Distribution of *LLMs* by population *size* (2001)

	Number of *LLMs*	Number of municipalities	% of total population
LLM1 > 2,500,000 inhabitants	*Madrid*	*153*	*20.51 %*
	Barcelona	*51*	
2,500,000 inhabitants > *LLM2* > 500,000 inhabitants	*Valencia*	*52*	*16.49 %*
	Sevilla	*39*	
	Bilbao	*59*	
	Malaga	*20*	
	Zaragoza	*96*	
	Palmas de Gran Canaria	*15*	
	Sabadell	*17*	
	Santa Cruz de Tenerife	*17*	
500,000 inhabitants > *LLM3* > 100,000	*60 LLMs*	*2,102 municipalities*	*31.20 %*
100,000 inhabitants > *LLM4* > 50,000 inhabitants	*50 LLMs*	*666 municipalities*	*8.56 %*
LLM5 < 50,000 inhabitants	*686 LLMs*	*4,822 municipalities*	*23.23 %*
Total	***806 LLMs***	***8,106 municipalities***	***40,533,475 inhabitants***

Source: 2001 Spanish Census, INE

Two distance criteria are tested here when defining a *LLM* as *central* or *peripheral*: the one hour's drive criterion (Fig. 2.5 – Map 5) and the linear distance criterion (Fig. 2.5 – Map 6). A summary of the resulting classification can be seen in Table 2.2 Depending on the size of the population, an urban *LLM* can be classified as *LLM3* or *LLM4* and then, according to its location or distance (A or B criterion) from the metropolitan areas, it may be further classified as *central* (*LLM3C* or *LLM4C*) or *peripheral* (*LLM3P* or *LLM4P*).

The *Economic Regions* resulting from the classification of the *LLMS* according to size and distance can be seen for the Spanish case in Fig. 2.5 – Map 1 (1 h drive criterion) and Fig. 2.5 – Map 5 (linear distance criterion). The criterion chosen does not seem to significantly change the *central* or *peripheral* character of the *LLMs*.

2.4 Evaluation: *Administrative Regions* Versus *Economic Regions*

After defining the *Economic regions* based on *LLM* size –in terms of population– and location and adapting them to the Spanish case, it is necessary to evaluate the performance of such regions compared to administrative regions. According to Fischer, (1980), an optimal *region* should fulfil at least one of two principles: *internal homogeneity*, whereby individual regions should be as homogeneous in the attribute space as possible; and *external separation*, whereby different regions

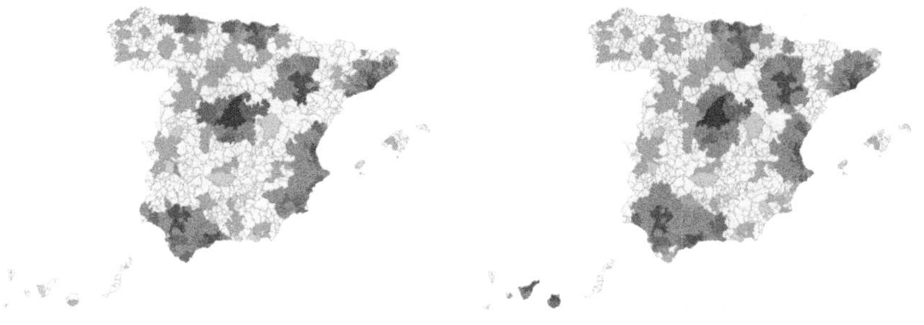

Map 5: Economic Regions based on *LLMs*, *size* and *distance to size*: one hour's drive criterion.

Map 6: Economic Regions based on *LLMs*, *size* and *distance to size*: linear distance criterion.

Fig. 2.5 Spanish division of the territory into *Local Labour Markets* (2001) (Source: Own elaboration with data from 2001 *Spanish Census*, published by INE (2007), and the Boix and Galleto (2006) methodology)

Table 2.2 *LLM* classification by *size* and *distance to size* adapted to the Spanish case

LLM1		
Local labour markets that constitute a metropolis of more than 2,500,000 inhabitants (Metropolitan Areas of Madrid and Barcelona)		
LLM2		
Local labour markets that constitute a metropolis of between 500,001 and 2,500,000 inhabitants		
	Central *LLMs*	**Peripheral *LLMs***
	(A) No more than one hour's drive from a *LLM1* or *2*	(A) More than one hour's drive from a *LLM1* or *2*
	(B) No more than 100 km, linear distance	(B) More than 100 km, linear distance
Urban *LLMs* Between 100,001 and 500,000 inhabitants	***LLM3C***	***LLM3P***
Urban *LLMs* Between 50,001 and 100,000 inhabitants	***LLM4C***	***LLM4P***
Rural *LLMs* Less than 50,000 inhabitants	***LLM5C***	***LLM5P***

Source: 2001 Spanish Census, INE (2007)

should be as far apart in the attribute space as possible. Pursuing these principles, in the next two sections we shall evaluate the robustness of the proposed *Economic Regions* versus the administrative regions commonly used (NUTs regions at different levels).

To evaluate the homogeneity of the regions, we shall use the well-known Theil inequality index (Theil 1967), frequently applied to the distribution of income and

Table 2.3 *Economic regions* versus *administrative regions:* the Theil index and the *within* component, both total and by gender, in Spain (2001)

Theil index		Theil WITHIN component *LLMs* (806 units)	Economic regions (8 units)	NUTS III provinces (50 units)	NUTS II Aut. Comm. (17 units)	NUTS I (7 units)
Total	*2.4340*	1.4279	1.8553	1.8435	1.9474	1.9995
Male	*2.3148*	1.3494	1.7819	1.7513	1.8522	1.9042
Female	*2.6401*	1.5601	1.9836	2.0008	2.1101	2.1628

Source: 2001 Spanish Census, INE (2007)

wealth. The index can be decomposed as the sum of the *between* and *within* components. The *within* component will be useful to quantify the intraregional homogeneity of the regions in relation to the spatial distribution of employment or economic activity. Given the characteristics of the Theil index, if the internal homogeneity of the regions increases (a decrease in the *within* component), this necessarily implies that the heterogeneity between regions increases (a rise in the *between* component).

Table 2.3 shows the *within* component of the Theil index for the administrative/political regions –NUTS I (7 regions), NUTS II (17 *Autonomous Communities)* or NUTS III (50 *Provinces)*– and also for the 806 Spanish *LLMs* and the proposed *Economic Regions* (8 regions) based on these *LLMs*. The table tests the homogeneity of the employment distribution, both total and by gender.

Despite the scale effect, i.e. *ceteris paribus*, intraregional inequality decreases with the number of regions, the *within* component for the eight *Economic Regions* is clearly lower than for any of the NUTS regions. In other words, the proposed classification shows a higher degree of internal homogeneity in the distribution of employment, i.e. the local labour markets grouped under size and distance criteria are more coherent (even by gender) than any other political-administrative division of the territory.

Understandably, the *Economic Regions* formed by grouping the 806 *LLMs* are less homogeneous in terms of employment distribution than the *LLMs* considered independently, such heterogeneity being mainly due to the lower tiers of *LLMs*, i.e. *LLM4* and *LLM5*, where industry-specific factors play a greater role in the distribution of employment. We have to bear in mind that *LLMs* are originally built to specify homogeneous local labour markets and therefore any other technical clustering, even those based on these *LLMs* as is the case here, will result in less homogeneous regions.

However, it must be stressed that the eight *Economic Regions* resulting from aggregating the *LLMs* , in contrast with any type of administrative division of the territory, have economic meaning and incorporate relevant approaches from Regional and Urban Economics. This regionalization scheme offers the researcher a concept of the Economic Region that has a better interpretation and analysis in terms of the Urban and Regional Economics literature.

Table 2.4 *Economic regions* versus *administrative regions:* the Theil index and the *within* component by industry, in Spain, 2001

		Theil WITHIN component				
Theil index		Economic regions (8 units)	*LLMs* (806 units)	NUTS III provinces (50 units)	NUTS II Aut. Comm. (17 units)	NUTS I (7 units)
Total	*1.0176*	0.9777	0.1826	0.4905	0.5489	0.6338
Agriculture, hunting and forestry activities	*3.7286*	3.5730	0.8932	1.7720	2.2061	2.4371
Fishing	*3.0814*	2.8681	0.9450	1.7421	2.1803	2.4068
Extractive industries	*2.3362*	1.8050	1.2637	1.6909	1.8861	1.9688
Manufacturing	*2.6969*	2.0901	1.5886	2.0602	2.1955	2.2682
Production and distribution of energy	*2.0300*	1.6178	1.1236	1.4548	1.5679	1.6292
Construction	*2.5604*	1.9500	1.4719	1.9139	2.0522	2.1082
Minor sellers; repairs	*2.5406*	2.0265	1.3782	1.7941	1.9193	1.9189
Hotels and restaurants	*2.9777*	2.0502	1.6686	2.1410	2.2983	2.3183
Transportation, storage and communications	*3.4599*	2.4150	2.0852	2.6019	2.7499	2.8005
Financial intermediation	*3.4331*	2.3208	1.9631	2.5118	2.6880	2.7438
Education	*2.7992*	2.1429	1.7305	2.1803	2.2864	2.3337
Health and veterinary activities	*2.9176*	2.2251	1.8073	2.3000	2.4296	2.4927
Other social activities and services for households	*3.0352*	2.2775	1.8804	2.4261	2.5642	2.6192
Household activities	*2.9100*	2.0937	1.6680	2.1373	2.2871	2.3427

Source: 2001 Spanish Census, INE (2007)

An additional analysis is carried out in Table 2.4 to evaluate the spatial patterns of distribution of employment for the same regions, though this time by industry. The 2001 Spanish Census offers employment figures for sixteen (16) types of industries.

The results for the Theil index once again show that *LLMs* are the best possible aggregation in terms of homogeneity, while *Economic Regions* are generally the second best. The *primary sector* and *extractive industries* tend to be highly concentrated in specific territories, and therefore neither the size nor location of the *LLM* determines its industrial specialization, but rather the location of natural resources. In other studies, the NEG fails to explain the distribution of economic activity in these cases. That is why *provinces* are found to be more homogeneous areas than the proposed *Economic Regions* for these industries. However, for the remaining sectors, i.e. those for which the location does not depends on the primary location of natural resources and where scale and agglomeration economics are especially relevant, *Economic Regions* are more homogeneous than *provinces* despite there being only 8 classes versus the 50 *provinces*.

It is widely known that *provinces* and *autonomous communities* are an administrative division of the territory that respond to political, sociological or historical will, but have no economic meaning, as they could be interpreted as the sum of different local or provincial governments. In other words, they are not homogeneous.

In the light of the analysis of homogeneity in terms of the distribution of employment, we conclude that *Economic Regions* are as good as *provinces* and clearly better than other administrative regions in pure terms of internal homogeneity and *between* class heterogeneity. However, it should also be recalled that *Economic Regions* have an added advantage: they have economic meaning and allow regional researchers to interpret the results obtained in line with the New Economic Geography and previous contributions of the regional and urban literature.

To conclude this evaluation, we would like to show some examples of spatial location of industries in Spain using the proposed *Economic Regions*. A Location Quotient (*LQ*) is calculated by *LLM* to identify the degree of specialization in the different types of industries. The *LQ* is the simplest way of measuring the specialization of a territory and can be calculated as:

$$LQ_{Xa} = \frac{\frac{\left[\sum_{i=1}^{n} e^{a}_{xi}\right]}{\left[\sum_{i=1}^{n} e^{a}_{i}\right]}}{\frac{E_x}{E}},$$

where:

LQ_{xa} = location quotient of sector x in synthetic region a,
n = number of spatial units in synthetic region a,
$e^{a}{}_{xi}$ = employment in sector x in spatial unit i of synthetic region a,
$e^{a}{}_{i}$ = total employment in spatial unit i of synthetic region a,
E_x = total employment in sector x in Spain, and
E = total employment in Spain.

The following maps show the *LLMs* specialized in Business Services and Real Estate Activities (Fig. 2.6), Financial Services (Fig. 2.7) and Manufacturing (Fig. 2.8), i.e. those areas where the *LQ* index is higher than one.

First of all, it should be stressed that in many cases the industries are highly concentrated in local areas that do not coincide with any of the administrative divisions commonly used, even at the highest level of disaggregation, i.e. municipalities. This calls for a higher level of internal coordination between municipal and even provincial and autonomous community institutions in the

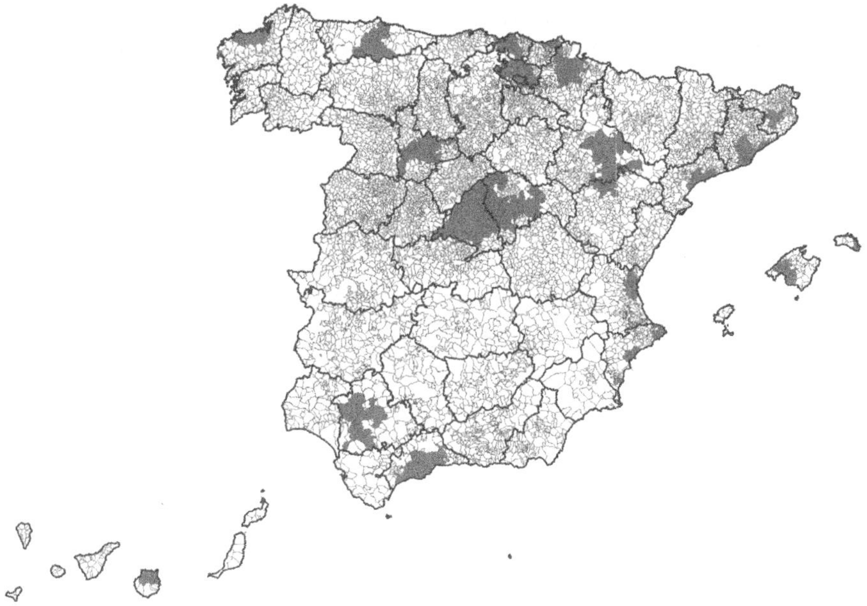

Fig. 2.6 *Local labour markets* specialized in Business Services and Real Estate Activities, Spain, 2001 (Source: Own elaboration with data from the 2001 Spanish Census, INE (2007))

Fig. 2.7 *Local labour markets* specialized in Financial Services, Spain, 2001 (Source: Own elaboration with data from the 2001 Spanish Census, INE (2007))

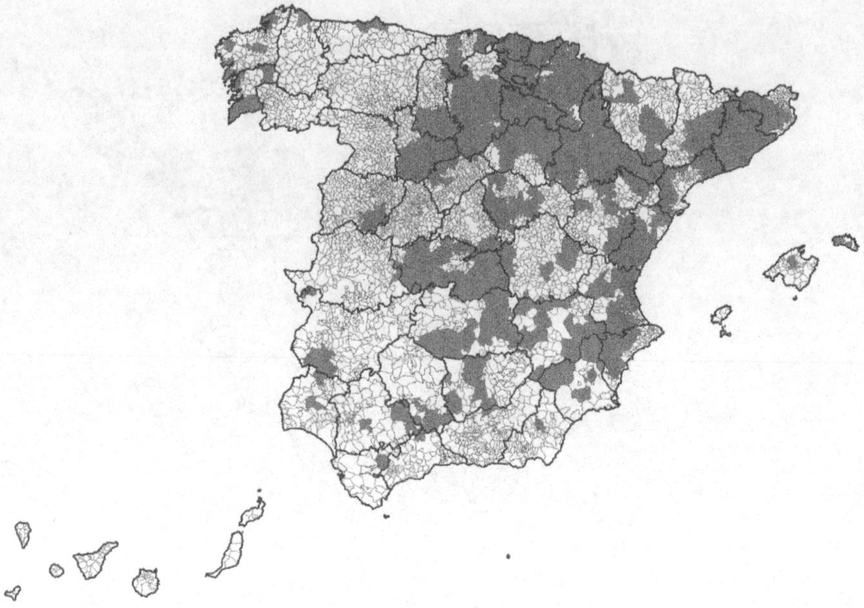

Fig. 2.8 *Local labour markets* specialized in Manufacturing, Spain 2001 (Source: Own elaboration with data from the 2001 Spanish Census, INE (2007))

design and implementation of, industrial, employment, infrastructure or transportation policies, among others.

Distance and *size* affect location patterns, and this can be easily proved by comparing the patterns of specialization maps with the map of the proposed Economic Regions, i.e. *LLMs* classified by *size* and *distance*. We can observe a strong concentration of higher-order services, such as Business Services and Real Estate Activities, in the larger metropolitan areas and a similar pattern of distribution for Financial and Insurance Activities, with a major preference for metropolitan areas and the next lower-tiered cities. On the other hand, as theory predicts, the higher wages and land rent in the bigger cities push manufacturing activities out of the major agglomerations, but these activities still tend to locate close to them. That is, distance to the market matters.

In summary, this is an example of how patterns of concentration can be analysed using the proposed *Economic Regions* and how the same type of analysis could be very limited if based on any other administrative division of the territory, whether *provinces*, *autonomous communities*, NUTS I or even municipalities.

References

Baldwin RE, Forslid R, Martin P, Ottaviano GIP, Robert-Nicoud F (2003) Economic geography and public policy. Princeton University Press, Princeton

Beardsell M, Henderson V (1999) Spatial evolution of the computer industry in the USA. Eur Econ Rev 43:431–56

Becattini G (1979) Dai Settore Industriale al Distretton Industriale. Rivista di Economia e Politica Industriale 1:1–8

Becattini G (1991) Il Distretto Industriale Marshaliano Come Comcetto Socio-Economico. In: Pyke E, Becattini GY, Sengenberger W (eds) Industrial districts and interfirm cooperation in Italy. Banca, Toscana

Behrens K, Thisse JF (2007) Regional economics: a new geography perspective. Reg Sci Urban Econ 37:457–465

Boix R, Galleto V (2006) Identificación de Sistemas Locales de Trabajo y Distritos Industriales en España. Dirección General de Política de la Pequeña y Mediana Empresa, Ministerio de Industria, Comercio y Turismo

Castells M (1976) La Cuestión Urbana. Siglo XXI Editores, México

Christaller W (1935) Die Zentralen Orte in Süddeutschland. Fischer, Jena

Claval P (2003) Geographie Régionale. De la region au territore. Armand Colin, Paris

Claval P (2007) Regional geographic: past and present (a review of ideas, approaches and goals). Geogr Pol 80(1):25–42

Coffey WJ, Polèse M (1988) La Transformation de L'espace Économique Canadien: ¿Assistons-nous à un Mouvement Centre-Péripherie?. Revue d'économie régionale et urbaine 1:1–32

Desmet K, Fafchamps M (2005) Changes in the spatial concentration of employment across US countries: a sectoral analysis (1972–2000). J Econ Geogr 5:261–284

Duranton G, Puga D (2002) Diversity specialization in cities: why, where and does it matter. In: McCann P (ed) Industrial localization economics. Edward Elgar, Cheltenham, pp 151–186

Eaton J, Eckstein Z (1997) Cities and growth: theory and evidence from France and Japan. Reg Sci Urban Econ 27(4–5):443–74

Fischer MM (1980) Regional taxonomy: a comparison of some hierarchic and non-hierarchic strategies. Reg Sci Urban Econ 10:503–537

Fujita M, Krugman PR, Venables AJ (1999) The spatial economy. Cities, regions and international trade. The MIT Press, Cambridge, MA

Glaeser EL (1998) Are cities dying? J Econ Perspect 12(2):139–160

Glaeser EL, Scheinkman J, Shleifer A (1995) Economic growth in a cross-section of cities. J Monetary Econ 36:117–143

Hall P (1998) Cities and civilization. Phoenix Giant, London

Henderson JV, Thisse JF (1997) On the strategic community development. CEPR discussion papers, 1550

ISTAT (1997) I Sistemi Locali del Lavoro 1991. Instituto Poligrafico e Zacca dello Stato, Rome

ISTAT (2006) Distretti Industriali e Sistemi Locali del Lavoro 2001. Collana Censimenti, Rome

Krugman PR (1991) Increasing returns and economic geography. J Polit Econ 99:483–499

Krugman PR (1995) Development, geography, and economic theory. The MIT press, Cambridge, MA

Lancaster PA (1939) The nature of geography: a critical survey of current thought in light of the past. Association of American Geographers, Lancaster

Lösch A (1938) The nature of economic regions. Southern Econ J 5(1):71–78

Marshall A (1890) Principles of economics. McMillan, London

McCann P (2007) Globalization and economic geography: the world is curved, not flat. Camb J Reg Econ Soc Camb Polit Econ Soc 1(3):351–370

Ottaviano GIP, Thisse JF (2004) Agglomeration and economic geography. In: Henderson JV, Thisse JF (eds) Handbook of regional and urban economics. North-Holland, New York

Parr J (2002) Agglomeration economies: ambiguities and confusions. Environ Plann A 34:717–731

Partridge MD, Rickman DS, Ali K, Olfert MR (2008) Lost in space: population growth in the American hinterlands and small cities. J Econ Geogr 8:727–757

Partridge MD, Rickman DS, Ali K, Olfert MR (2009) Agglomeration spillovers and wage and housing cost gradients across the urban hierarchy. J Int Econ 78:126–140

Polèse M (2005) Cities and national economic growth: a reappraisal. Urban Stud 42(8):1429–1451

Polèse M (2009) The wealth and the poverty of regions: why cities matters. University of Chicago Press, Chicago

Polèse M, Champagne E (1999) Location matters: comparing the distribution of economic activity in the Mexican and Canadian urban systems. Int J Sci Rev 22(1):102–132

Polèse M, Shearmur R (2004) Is distance really dead? Comparing the industrial location patterns over time in Canada. Int Reg Sci Rev 27(4):1–27

Polèse M, Shearmur R, Rubiera F (2007) Observing regularities in location patters. An analysis of the spatial distribution of economic activity in Spain. Eur Urban Reg Stud 14(2):157–180

Poplin DE (1979) Communities. Macmillan, New York

Porter M (1990) The competitive advantage of nations. Free Press, New York

Redding S, Venables A (2004) Economic geography and international inequality. J Int Econ 62 (1):53–82

Rosenthal SS, Strange CW (2001) The determinants of agglomeration. J Urban Econ 50 (2):191–229

Rozenblant C, Cicille P (2003) Les villes éuropeénnes: analyse comparative. Datar

Rubiera F (2006) Ciudades, crecimiento y espacialización territorial. Dinámicas espaciales de concentración del empleo y la producción en España. Colección de Estudios CES

Schumpeter JA (1954) History of economic analysis. Oxford University Press, New York

Serra J, Otero M, Ruiz R (2002) Grans aglomeracions metropolitanes Europees. IRMB

Sforzi F (1987) L'indentificacione spaziale. In: Becattini G (ed) Mercado e Forze Locali: il Distretto Industriale. Il Mulino, Bologna

Sforzi F (1990) The quantitative importance of Marshallian industrial districts in the Italian economy. In: Pyke E, Becattini GY, Sengenberger W (eds) Industrial districts and interfirm cooperation in Italy. Banca, Toscana

Sforzi F, Lorenzini F (2002) I distretti industriali, L´esperienza Italiana dei Distretti Industriali. Instituto per la Promozione Indistriale (IPI)

Theil H (1967) Economics and information theory. North-Holland, Amsterdam

Viñuela A (2011) Surpassing the administrative division limits on regional analysis: three essays on urban and regional analysis. Ph.D. thesis, University of Oviedo (European mention)

Viñuela A, Rubiera F, Cueto B (2010) An analysis of urban size and territorial location effects on employment probabilities: the Spanish case. Growth Change 41(4):495–519

Von Thünen JH (1826) Der Isolierte Staat in Beziehung auf Landwirtchaft und Nationalökonomie. Frederich Perthes, Hambourg

Von Thünen JH (1966) The isolated state. English translation of the original work from 1826. Pergamon Press, New York

Zipf GK (1949) Human behaviour and the principle of least effort: an introduction to human ecology. Addison Wesley Press, Cambridge, MA

Chapter 3
Polycentric Metropolitan Areas in Europe: Towards a Unified Proposal of Delimitation

Rafael Boix, Paolo Veneri, and Vicent Almenar

3.1 Introduction

Metropolitan areas concentrate the main share of population, production, and consumption in developed countries. They are likely to be the most important units for economic, social, and environmental analysis as well as for the development of policy strategies.

The metropolitan area does not fit well with administrative boundaries. Metropolitan areas change over space and time, reflecting the evolution of economy and society. Its assimilation with the administrative city, region or province usually introduces severe drawbacks when the metropolitan area is only part of this territorial unit, or when it considerably exceeds administrative boundaries.[1]

Unfortunately, discussion about the boundaries of a metropolitan area does not restrict the accuracy of the indicators. It does affect the welfare of residents when

[1] An example of the first problem is the assimilation of the metropolitan area of Barcelona to the province. Province data averages the results of the indicators and dissolves some of the potentialities and problems of the real metropolitan area. On the other hand, Milan and Madrid constitutes an example of the second case, where the limitations of data force the use of the province, too small to capture the real extension of both areas. In this case, the areas have expanded out of the administrative boundaries, and we could erroneously conclude that there is a reduced presence of some activities or maybe even their disappearance altogether if they moved out of the administrative limits.

R. Boix (✉)
Departament d'Estructura Econòmica, Facultat d'Economia, Universitat de València.
Avda dels Tarongers, s/n, Valencia 46022, Spain
e-mail: rafael.boix@uv.es

P. Veneri
Regional Development Policy, OECD. 2, rue André Pascal, Paris Cedex 16 75775, France

V. Almenar
Departament d'Estructura Econòmica, Facultat d'Economia, Universitat de València.
Avda dels Tarongers, s/n, Valencia 46022, Spain.

E. Fernández Vázquez and F. Rubiera Morollón (eds.), *Defining the Spatial Scale in Modern Regional Analysis*, Advances in Spatial Science,
DOI 10.1007/978-3-642-31994-5_3, © Springer-Verlag Berlin Heidelberg 2012

the definition of "metropolitan area" is transformed in policies affecting the basic pillars of competitiveness, social cohesion, environment, quality of life, and governance.

A second issue arises from the fact that the comparison between metropolitan units identified in different countries is difficult as countries use different methodologies. In several countries no definition of metropolitan units has been carried out at all. International institutions, more than others, have tried to find general methodologies to map metropolitan areas (OECD 2006; ÖIR 2006), though this represents a difficult aim due to the existence of different territorial structures across countries.

This significant problem appeared when we tried to perform comparative research between Spain and Italy using metropolitan areas as units of analysis. No official definition of "metropolitan area" was available and the few available approximations made by researches or institutions, when conceptually feasible, where not comparable. Considering these problems, this chapter aims to identify metropolitan areas in Spain and Italy using similar methodologies. The identified metropolitan units have three basic purposes. The first is to provide a general view of the characteristics of each country's metropolitan reality. The second is the comparison of the metropolitan processes of both countries, and the third is the identification of metropolitan units that can be used in subsequent analysis. This has been done focusing on two approaches to the concept of metropolitan area. First, a general methodology applicable to most European Union (EU) countries is used, in this case, the Functional Urban Area (FUA) methodology as proposed by GEMACA (1996). Second, a native methodology Dynamic Metropolitan Area (DMA) specifically designed to deal with the particular characteristics of networking and polycentricity has been used. We assess the results of their application to Spain and Italy, two very similar countries in terms of social, economic, and territorial structures, and expand the discussion to their use in other countries.

The research proposes two contributions. Firstly, from the methodological point of view, we expand traditional approaches to the identification of metropolitan areas to introduce a new category we named "network approach", which explicitly recognises that metropolitan areas are cliques of networks of cities, and can be monocentric or polycentric. This fact is recognised in the proposal of a methodology of identification named DMA. Secondly, there is a lack of detailed empirical comparative studies on the identification of metropolitan areas in different countries using similar methodologies. The lack of official definitions as well as the scarcity of studies to identify metropolitan areas in Spain and Italy is perceived as a severe drawback that dissuades from the use of metropolitan areas as units of analysis in both countries This chapter provides two sets of metropolitan areas, identified using rigorous approaches and replicable standards that can be used by other researchers in subsequent investigations.

The chapter is structured as follows. The second section discusses the approaches used to identify metropolitan areas in European countries. The third provides a review of previous studies in the identification of metropolitan areas in Spain and Italy. The fourth proposes two methodologies for the identification of metropolitan areas in the two countries. The fifth section presents the results of their application to the two countries. The work ends with concluding remarks and a short discussion about an agenda towards unified identification of metropolitan areas in Europe.

3.2 General Approaches to the Definition of Metropolitan Areas

Identification of metropolitan areas can be carried out using four basic approaches. Three have been identified in the Espon 1.4.1 Report (ÖIR 2006):

1. The "administrative" approach identifies metropolitan areas on the basis of the status of previously definite legal or administrative units. It is conceptualised as an instrument for purposes of governance and control. The identification departs from local or provincial boundaries and applies criteria to distinguish between metropolitan and non-metropolitan units (population thresholds, governmental decisions, historical reasons, etc.). Examples of administrative criteria can be found in OECD reports (OECD 2006) and in empirical applications of the ESPON FUAs (Table 3.1).
2. The "morphological" approach identifies metropolitan areas as those continuous urban settlements that reach certain thresholds of density, dimension or degree of urbanisation. The metropolitan area is conceptualised as a physical object, without referring to any relational consideration. Serra et al. (2002) provide an example of the application of this criterion (Table 3.1), and another example can be found in a study by Rozenblat and Cicille (2003).
3. The "functional" approach defines metropolitan areas as economic and social entities, not as mere geographical areas (ÖIR, 2006 – p. 17). Administrative boundaries are no longer a priority criterion and the focus is shifted to the functional relationships between the units that form the metropolitan area. Using this approach, a metropolitan area is defined as an area of interaction between a "core" (which may be defined using morphological criteria as population or employment thresholds) and its hinterland of neighbouring municipalities, which show a significant relationship with the core (usually approximated with travel-to-work commuting flows). Examples of this criterion can be found in the Functional Urban Regions (FUR) identified by the GEMACA group (1996) and the USA metropolitan areas (Table 3.1).[2]

We propose a fourth "network" approach, which recognises that the interactions characterising a metropolitan area takes place in multiple directions and levels, so that the the area is defined by a complex and multidirectional network of interactions between actors placed in several interconnected layers. The basic

[2] Freeman (2005) provides a comparison between the characteristics and results of the US metropolitana areas, GEMACA and Urban Audit.

Table 3.1 Empirical application of methodologies for the identification of metropolitan areas in the UE and USA

Identification method	Description of the method	Source and kind of data	Advantages	Disadvantages	On the whole
OCDE's Metropolitan regions (2006)	NUTS3/LAU3	Administrative definition	Simplicity Availability of many typologies of data Units of analysis steady in time	Socio-economic dynamic doesn't coincide with administrative definition Static unit of analysis steady in time and space	On the whole, the province seems to identify territory which is too wide for metropolitan areas, with the exception of some of the biggest areas
FUAs ESPON 2006	A FUA is composed of a core and a neighbouring area that is economically integrated with the core Due to the difficulties associated with their identification, FUAs tend to be approximated with NUTS3, to have more than 20,000 inhabitants	Various sources, usually census data on population, employment, and commuting at municipal and NUTS3 level. When a NUTS3 is adopted, a simple administrative definition is used. FUAs are usually determined on a functional approach	Little information needed	Spatially static units of analysis There is no identified method which can be applied to every country, so the method used could be administrative, morphological, or functional. As a matter of fact, a few times the identified area corresponds with the area of expansion of economic flows	On the whole, the method used is not clear or univocal. Project 1.1.1 proposed a methodology that cannot be applied to many countries, including Italy and Spain, due to lack of available data. The biggest unit of ESPON (MEGAs) are often similar to provinces and take with them all the above mentioned problems of administrative units

FURs (GEMACA II)	This studies neighbouring municipalities with an employment density of more than seven jobs per hectare (core), plus the ring of contiguous municipalities for which more than 10 % of their commuters travel towards the above-identified core	Census data on population, employment, and commuting at a municipal level	Dynamic unit of analysis in time Easy and clear methodology that could be applied to almost every European country. There should be some problems for those countries that have large municipalities units	There is ambiguity on which kind of land to use (urban land, municipal land, etc.) It is very sensitive to the urbanisation pattern In polycentric or contiguous metropolitan areas, the direction of the expansion of densities doesn't have to follow the direction of economic interaction	Excellent performance. Despite the integration of the core with an urban ring, the methodology does not always take into account that neighbour FURs constitute a single city, especially when the identification is carried out for planning or transport policy purposes
Urban areas (Serra et al. 2002; Carreras et al. 2009)	This studies urban cores with at least 100,000 inhabitants and a density higher than 1,500 inhab/km^2. All contiguous municipalities with a density higher than 250 inhab./km^2 must be added to the core	Population and municipal surface data Morphological approach	Unit of analysis dynamic on time Basic requirements of information and simplicity of application Possibility of a European comparison between units identified in this way	It does not take into account the relationships between different parts of the metropolitan area; therefore it is uncertain that this unit of analysis coincides with actual economically integrated areas	Application is simple with little data needed. However, due to its pure morphological approach, it appears inadequate for economically integrated areas
LUZ (Urban Audit 2006)	This studies urban core, plus all the municipalities that present more than 15 % of total commuters travelling towards the core	Census data on flows of work commuters, employed residents, jobs, and residential population	Dynamic method in both time and space	The identified urban areas are usually too small, often limited to the central city of a bigger metropolitan area	After having applied this methodology to some countries, it emerges that the identified units are even smaller than Local Labour Market Areas

(continued)

Table 3.1 (continued)

Identification method	Description of the method	Source and kind of data	Advantages	Disadvantages	On the whole
	NUTS3 can be used as a proxy when there is no available statistical information	If there is enough statistical information this is a functional approach, while NUTS3 is only an administrative approach	It takes into account socioeconomic relationships between municipalities Easy method Possibility of a European comparability of the units of analysis	Due to the dimensions of the identified units, the methodology cannot capture the polycentric spatial organisation of cities	(LLMAs) (ISTAT 1997). The problem with these units is that they tend to separate sub-centres of the same metropolitan area
Metropolitan areas of USA's Census Bureau	The central core is made up of a municipality of more than 50,000 inhabitants, and of other municipalities that provide the municipality with at least 15 % of its employed residential population	Census data on commuting to work flows, employed residential population, jobs, and residential population	Dynamic method both spatially and temporally It takes into account socioeconomic relationships Use of high quality (census) data	Metropolitan areas with this method may be too small to be suitable for planning, transportation purposes, or to catch polycentricity. However, they are usually bigger than LUZ	This method seems to work well, but it still doesn't solve the problem of the study of polycentricity, and doesn't seem to be suitable for the planning of infrastructure and mobility

	The urban ring has to be built with the municipalities in which more than 15 % of employed residents work in the central core, and have a density of at least 62 inhab/km^2. Alternatively, the conditions for adding ring municipalities are that they must have a density of 37 inhab/km^2, and at least 30 % of the employed residential population must work in the central core. In this way, both contiguity and consolidation criteria are applied	Surface area at the municipal level	Possible European comparability Use of consolidation criteria Possibility to classify areas in different levels	Studies only one interaction between the central core and the urban belt, as the aim of the method is to build statistical areas rather than identify the real city

representation of the metropolitan area can be reduced to a dense network of cities. This approach has two basic implications.

The first implication is that the basic model is polycentric and can have several first-order centres. Thus, the monocentric model is only a constrained form of the polycentric one. This also leads to a change of paradigm, from central places to network models (Camagni and Salone 1993).

The second implication is that the methodologies of identification of metropolitan areas will evolve towards graph-based methods, found on the systems theory, better prepared to deal with polycentricity, networking, and increasing complexity.

In fact, we can consider morphological and functional approaches as restricted versions of the network approach, where there are important constraints in the available information, as well as a notable reduction in conceptual complexity.

While the administrative approach is clearly inadequate in identifying economically and socially-integrated urban areas, the morphological approach presents the further problem of finding cities which are too small and inappropriately calling them metropolitan areas. The functional approach appears to be a suitable method as it takes into account socioeconomic relations between several units which form the metropolitan area. However, it is limited in its complexity and is implicitly based on monocentricity. If the aim of the analysis is the study of urban polycentricity or, in general, of the urban spatial structure, the network approach appears to be the most suitable. In the absence of symmetrical information, it is possible to combine several criteria in order to apply the best option when available, or as an alternative otherwise, for instance in the definition of Larger Urban Zones (LUZ) by Urban Audit (Table 3.1).

3.3 Metropolitan Areas in Spain and Italy: A Review of the Literature

3.3.1 Spain

The Spanish Constitution (art. 141.3 and 152.3) grants the possibility to associate sets of contiguous municipalities in territorial entities that are different from the region or the province they belong. The law of local corporations (LRBRL, art. 43) asserts that metropolitan areas are local entities composed of the municipalities of large urban agglomerations with social and economic linkages, where joint coordination and planning is necessary.

First attempts aimed to identify metropolitan areas in Spain involved the Dirección General de Urbanismo of the Ministry of Housing (1965, 1967). The morphological criterion, inspired by Davis (1959), consisted of the identification of a central core of at least 50,000 inhabitants, and a strong socioeconomic relationship between the core and surrounding municipalities. The whole metropolitan area should have a population of at least 100,000 inhabitants, a density larger than 100 inhabitants/km^2, high rates of growth, and contiguity. Following these criteria, 26 areas were identified in 1960 (34 % of the national population), and 24 in 1967 (36 % of the national population).[3]

A second approach, also from an institutional source, is found in the "III Plan de Desarrollo Económico y Social" (1972). The document proposes three criteria to identify metropolitan areas: statistics, economic development, and planning. The application of the statistical criterion to 1965, 1969, and 1985 provided 25, 30 and 32 statistical metropolitan areas, respectively (De Esteban 1981).

The Ministry of Housing (Ministerio de Vivienda 2000, 2005, 2007) has recently provided other maps, though these are more centred on the identification of urban rather than metropolitan areas. The procedure follows a morphological approach that departs from data of population, housing, territorial structure, urban dynamics, and the transportation network. The Spanish territory has 82 LUAs (with at least one municipality containing more than 50,000 inhabitants), and 269 Small Urban Areas. The first has 9 % of Spanish municipalities and 71 % of the country's total population, and can be considered a proxy of the metropolitan phenomenon.

Serrano (2006) adopts a morphological approach to identify urban areas and agglomerations in Spain. This category contains those continuous areas formed by a central municipality of more than 75,000 inhabitants surrounded by a belt of municipalities, so the entire area has at least 100,000 inhabitants. The belt is determined using distance-based criterion; namely 40 km from the central city for the large areas, and 15 km for the small areas. For the year 2001, Serrano identified 45 urban agglomerations which have 9 % of the Spanish municipalities and 61 % of the total population. The largest agglomerations are Madrid (41 municipalities and five million inhabitants), Barcelona (74 municipalities and 3.8 million inhabitants) and Valencia (63 municipalities and 1.56 million inhabitants). This methodology is quite simple and only population and distance data are required. On the other hand, no justification is raised for the election of the distance thresholds and why they are the same for the range of large and small urban areas. In fact, the small number of municipalities surrounding Madrid suggests the inability of this morphological criterion to take into account the socioeconomic structure of complex metropolitan areas.

Clusa and Roca (1997) provide an algorithm in two stages for the identification of the metropolitan area of Barcelona, based on the former USA Federal Register (Office of Management and Budget 1990) procedure for the identification of metropolitan areas in New England. In the first step, they identify a central core as a municipality of more than 50,000 inhabitants, plus those municipalities in which at least 15 % of their resident employees commute to this municipality. The hinterland is formed by those municipalities in which at least 15 % of their resident employees commute to the central core. In comparison with the USA procedure, Clusa and Roca iterate the criterion four times to form the hinterland, each time

[3] A review of the identification of metropolitan areas in Spain from 1960 to 1980 is provided by De Esteban (1981).

using the result of the previous iteration as the core. Contiguity criteria are used after the last iteration. As labour markets tend to be self-contained, the choice of four iterations is based on the empirical fact that after the third iteration the number of municipalities included is very small, and in subsequent iterations tend to be nil. The area identified using this procedure for the year 1991 has 145 municipalities and 4.2 million inhabitants.

This criterion has been applied to the entire region of Catalonia by Trullén and Boix (2000) and Boix and Galletto (2004) who identify five metropolitan areas and their evolution since 1986. Roca et al. (2005) extended the procedure to identify the metropolitan areas of the seven largest cities in Spain in 1991 and 2001. The 2001 results remark the size of Madrid (608 municipalities and a population of 5.6 million), and Barcelona (227 municipalities and a population of 4.5 million). With more than one million inhabitants they also identify Valencia (152 municipalities and a population of 1.7 million), Seville (60 municipalities and a population of 1.4 million), and Bilbao (104 municipalities and a population of 1.1 million) as metropolitan areas.

Other attempts to identify metropolitan areas in Spain have been carried out at a regional level. The administrative approach prevails when Public Administrations approach the metropolitan area (e.g. Madrid is usually assimilated to the province and Valencia to the county). The morphological approach prevails in Martínez de Lejarza and Martínez de Lejarza (2002) for Valencia, and Sánchez (1998) for Zaragoza. Functional approaches have been applied to Barcelona by Salvador et al. (1997), and to Andalusia by Feria and Susino (2005). Rubert (2005) applied a pool of methodologies to Castellon.

Focusing on functional approaches in more detail, Esteban (1995) and Salvador et al. (1997) applied the FUR methodology (Cheshire and Hay 1989; GEMACA 1996) to identify the boundaries of the metropolitan area of Barcelona. The latter is similar to the FUR procedure proposed in the next section, and using 1991 data produces a FUR composed of 131 municipalities and 4.1 million inhabitants.

Feria and Susino (2005) employ a functional approach based on absolute and relative cut-offs of population and commuting flows. Following this approach, each metropolitan area must have a central city of at least 100,000 inhabitants. The hinterland is composed of those municipalities which send at least 15 % of their resident employees to the central city, or where the commuters received from the central city exceed 15 % of the local jobs. In both cases, the minimum flow must reach 100 commuters. As this procedure performs better on centralised structures, the authors propose that the relative threshold could also be reached by iterating, although in this case they require a minimum value of 500 commuters. Contiguity criteria are applied to obtain the final shape of the metropolitan areas. The procedure identifies eight metropolitan areas in Andalusia, where the most important are Seville (40 municipalities and 1.29 inhabitants) and Malaga-Marbella (29 municipalities and one million inhabitants).

From an international point of view, the OECD identifies three metropolitan regions above 1.5 million inhabitants in Spain (Madrid, Barcelona and Valencia). Urban Audit (2006) finds 18 LUZs, where Madrid (5.4 million inhabitants) and

Barcelona (4 million inhabitants) are the largest metropolitan units. Rozenblat and Cicille (2003) differentiate 22 Spanish large European agglomerations. ESPON (2006) identifies 100 FUAs, where Madrid is the only above five million inhabitants, and Barcelona, Valencia and Seville have over one million inhabitants each one.

3.3.2 *Italy*

Italian Metropolitan Areas is an institution legally recognised since the early 1990s by national law n. 142. The law provides a general criterion to guide the identification of metropolitan areas, where each pivotal municipality must be strongly integrated from an economic, social, or cultural point of view. The act fixes nine metropolitan areas, while the other five have been introduced by regional laws. Despite the importance of the urban and metropolitan in Italy, there are very few works aimed at identifying metropolitan areas.

Two decades before the metropolitan question was formally recognised in Italy, scholars attempted to identify the boundaries of the Italian metropolitan areas. For example, Cafiero and Busca (1970) adopted a morphological approach based on a threshold of density and spatial contiguity in their research for SVIMEZ (Associazione per lo sviluppo dell'industria nel Mezzogiorno). These criteria have been used in other reports by SVIMEZ, and in Cecchini (1988), who identified 39 metropolitan areas. The main limitations of these procedures are the choice of the size and density thresholds, and density and dimension. As a result, the findings do not seem to fit well to different territorial situations (for example, the metropolitan area of Milan seemed too big compared with the small area obtained for Rome). A few years later, Vitali (1990) identified urban areas using a morphological approach, in this case departing from the basis that each province's capital is the centre of a larger "area of attraction". Around each centre, a circle is drawn to delimit the area of attraction using a radius of 10, 15 or 20 km, depending on the dimensions of the centre. The three identified groups of urban areas had the same geographical extension and shape (circular).

Soon after the 1990s law n. 142 was established, several works were published on the issue of metropolitan areas in Italy. Among those worth mentioning are Bertuglia and La Bella's (1991), and Costa and Toniolo's (1992), where Italy's metropolitan question was discussed in a comprehensive way, from the role of metropolitan areas in the global competition to the methodologies for the delimitation of boundaries. Focusing more on the identification of metropolitan units, Marchese (1997) obtained 32 metropolitan areas following a morphological procedure in two steps. First, he selected all contiguous municipalities which showed a certain threshold of employment density, then divided these continuums into four groups on the basis of their dimensions. In the second step, he selected sets of contiguous municipalities that could be considered metropolitan areas on the basis of the existence of centrality factors, such as high rank services for families and firms. More recently, Basta et al. (2009) replicated the analysis of Marchese using 2001 Census data.

Among recent works, Bartaletti's (2009) is worth mentioning. The study identifies 33 metropolitan areas by taking into account commuting flows towards metropolitan nodes, demographic dynamics, and the physical extent of built environment. The author considers municipalities with a certain number of jobs in the private sectors to be part of the metropolitan area. In another work by Tortorella and Andreani (2009), 15 metropolitan areas are identified following a method adopted in the US for the identification of the Metropolitan Statistical Areas (MSAs). The algorithm combines three factors – the physical expansion of settlements, functional relationships, and economic performances – in order to identify the extent to which various rings of peripheral municipalities are integrated with the central core. Finally, a very recent work by City Railways (2011) identifies 22 metropolitan areas, where the latter are conceptualised as areas composed by a metropolis and its functionally dependent hinterland. The basis of this work suggests in order to identify a metropolitan area it is fundamental that the diverse parts of the area are well connected through an appropriate transport network, as well as through strong socioeconomic integration.

Regarding the institutional point of view, as far as the work of national and regional institutions is concerned, ISTAT-IRPET (1989) provides the most significant attempt to identify large urban units using a functional approach. It departs from the previously identified local labour markets which are subsequently aggregated in Functional Labour Regions. More specifically, by using 1981 Census data, ISTAT-IRPET (1989) identified 995 local labour markets which merged in 177 Functional Labour Regions. Recently, the Italian government has proposed to apply a population threshold to the 2001 local labour markets to identify these Local Metropolitan Systems (Consiglio dei Ministri 2006 – p. 228). Although this provides a feasible approach for small and medium metropolitan areas, local labour markets are clearly inappropriate for the largest metropolitan areas, such as Milan or Rome, which are formed by several local labour markets. The notion of Local Metropolitan Systems was introduced by ISTAT (1997), for local labour markets having at least a city of more than 250,000 inhabitants.

The annual ISTAT report (2007 and 2008) offers additional contributions for the identification of "urban areas" and "functional regions", starting from the 2001 local labour markets. The "Rapporto Annuale 2006" (ISTAT 2007, p. 137–147) provides 32 labour markets with characteristics of LUZs coming from the third Urban Audit report. Moreover, there are another 46 local labour markets defined as urbanised, but are not considered in the Urban Audit 3 project. Finally, the "Rapporto Annuale 2007" (ISTAT 2008, p. 149–153) identifies 41 metropolitan regions as local labour markets which combine morphologically urban characteristics and urban functions. These metropolitan regions cover 34.7 % of the national population.

3.4 FUR and DMA: Funtional and Network Methodologies

The general approaches mentioned in Sect. 3.2 of this chapter suggest the use of functional and network methodologies when data are available. Two methodologies are proposed: FUR by GEMACA II (1996) which is a functional methodology applicable to most UE countries; and DMA, an improvement on Clusa and Roca's (1997) iterative methodology which is a network approach elaborated from the basis of a functional methodology.

3.4.1 A Previous Thought About the Basic Territorial Unit of Observation and the Use of Commuting Data

Perhaps the first problem of comparative research is the choice of the basic territorial unit. In Anglo-Saxon countries it is usually the council, in Mediterranean countries the municipality, whereas in others the typical election could be the parish, districts, and so on. Even when using similar units we must deal with different sizes and characteristics of these units. From this point of view, the use of a homogeneous unit independent from the particularities of each country or region appears reasonable. Recent working papers by OECD (2010) propose the use of "building blocks" in the form of 1 km^2 grids which are subsequently aggregated to form the core of the metropolitan area.

This proposal should be taken into consideration as it has the advantage of homogeneity of the basic units of departure. However, some points should be made. The first is that this is only viable when detailed micro data, including the location (postal address) of the population, are available in all countries, which is quite unusual. The absence of these data means the lineal homogenous assignation of the population of the county or municipality to the grids do not have any evident advantage regarding the use of the aggregated unit. The second point is if the use of grids makes more social or economical sense than the use of counties or municipalities, or, if we are looking for smaller units of departure, quarters of postal districts. Again, it is not evident if the use of grids overcomes the problem of heterogeneity, and/or rather than overcoming it introduces an artificial ecological fallacy problem.[4] In any case, correct method of analysis continues to be an open question, for which there is no sole solution. Careful reflection is necessary in the first stage of the analysis. In the next sections, we follow the most extended tradition by departing from the idea of the municipality as the basic territorial unit. Our justification is that it better reflects the administrative, social, and economic spatial structure.

[4] A third solution is suggested by Coombes (2000), who proposes the re-definition of localities using synthetic data. Even though this relies on the availability of information, it provides an interesting solution to the homogeneity of the units and their social and economic significance.

A second relevant point is the use of commuting flow data as the basic measure of interaction. Davoudi (2003) reports some objections to its use as a sole indicator of interrelationship. Basically, the advantage of travel-to-work commuting is that it incorporates a synthetic mix of social, economic, infrastructural, housing, and administrative interactions. On the other hand, in an extended network approach, as discussed in Sect. 3.2, the use of multi-layer information will provide a richer set of measures in the future. In any case, even being critical, our experiences gleaned over a long period suggests that travel-to-work commuting is an excellent synthetic measure of interaction and deals perfectly with networking between cities.

3.4.2 Functional Urban Regions (FUR)

The concept of FUR was used for the first time by Berry (1967) for the USA. In Europe, it was introduced by Cheshire and Hay (1989). The main reason for the use of this concept was to identify comparable urban units across Europe, as Hall and Hay (1980) had done several years before by introducing the similar concept of Daily Urban System (DUS). Despite their name evoking the concept of a region, FURs are metropolitan areas (Cheshire and Hay 1989) and the methodology for their identification follows a functional approach, as their boundaries are determined on the basis of economic relationships (Davoudi 2008). The procedure employed follows the works by GEMACA (1996, 2001) for the North-West Europe Urban System[5]:

1. A core, composed of one or more contiguous municipalities with a density of at least seven jobs per hectare and with no less than 20,000 jobs; and
2. A hinterland, which consists of all contiguous municipalities where at least 10 % of the resident employees commute to the core. Municipalities that are completely surrounded by the FUR are also included.

3.4.3 Dynamic Metropolitan Areas (DMAs)

Network methodologies to identify metropolitan areas could depart from aggregative (bottom-up) or partitive (top-down) techniques. In the first case, we group cities in successive iterations to form the metropolitan area, whereas in the second the objective is to divide the cities into blocks or clusters and subsequently decide which of them have metropolitan characteristics. Most procedures included in our proposal thus far have relied on bottom-up approaches.

---

[5] Cheshire and Magrini (2008) use a variation of this procedure where the density of job per hectare increases to 12.35. In the case of Spain and Italy both thresholds produce the same empirical results.

The network methodology we propose to map metropolitan areas is based on adaptations by Clusa and Roca (1997) and Roca et al. (2005) of the USA Federal Register's methodology (Office of Management and Budget 1990). Similar to the FUR, we can interpret that the metropolitan area is composed of a central core and a hinterland. The main differences are that the initial relative threshold of commuting between the core and the hinterland is exigent, although it is iterated to take advantage of the trend of labour markets to be self-contained and the recursive networking between cities.

However, we propose to conceptually overcome the traditional approach and introduce the assumption that the core of the metropolitan areas can be polycentric. We propose to lay aside the traditional logic behind core-hinterland, and to introduce a logic based on the existence of a constellation of first order centres organising a network of cities. This involves introducing a previous step in the procedure to better differentiate between first and second-order centres, and to take into account the polycentric nature of some of these areas. The complete procedure is named Dynamic Metropolitan Area (DMA):

1. The first stage of the DMA algorithm aims to determine the "core network", formed by the "first-order centres" of the metropolitan area and their closer network of cities. A first-order centre must have at least 50,000 inhabitants. The core network is formed by one or more first-order centres and those municipalities that commute at least 15 % of their resident employees with these first-order centres.[6]
2. In the second stage, as compared with the USA procedure, the rest of the network of cities (similar to the hinterland in the traditional nomenclature) is aggregated in four iterations. It is in this part of the procedure where the existence of networking between cities is exploited. In the first iteration we include those municipalities for which at least 15 % of resident employees commute to the core network. This criterion is applied another three times, with "core" being the result of the previous iteration. In other words: network 1 = core network + municipalities commuting 15 % of their resident employment to the core; network 2 = network 1 + municipalities commuting 15 % of their resident employment to network 1, etc. Contiguity criteria are used after the last iteration, such that all isolated municipalities which are completely surrounded by others that belong to a DMA are included, while those that are not contiguous are excluded. In the network logic, contiguity is not strictly necessary, but it does help to produce a more compact metropolitan unit.

[6] After 1991, the Federal Register introduced several changes in the identification of the core and has increased the commuting threshold to 25 % in order to prevent growth of the statistical units. It is noted that its primary assignment is not to identify metropolitan areas, but rather to provide manageable statistical units. However, as our purpose is different, we prefer to base our procedure on the former 1990s methodology due to the fact that: (1) the 2001 version eradicates cities and towns in favour of counties and reduces its applicability to Spain and Italy; and (2) based on previous research, the 15 % threshold is considered to produce good results.

However, in the large metropolitan areas it is usual to find several contiguous and non-contiguous cities with more than 50,000 inhabitants, making it difficult to differentiate a first-order centre from a second-order sub-centre, or to avoid the assignation of the sub-centres of a polycentric metropolitan area to different areas. To separate first-order centres (central cities in the traditional nomenclature) from other large municipalities, we propose a pre-application of the procedure so that:

1. The percentages of commuters between all potential first-order centres are calculated. If one of these cities sends more than 15 % of its total commuting to another, the first is considered a sub-centre of the latter. If both cities each share more than 15 % of their total commuters, then both have to be considered to be a unique central network of the same metropolitan area.
2. A recursive pre-application of the procedure is proposed in order to differentiate first-order centres from the remaining second-order sub-centres. Thus, if in some of the four iterations a potential first-order centre reveals a city of the core network or the peripheric network of another metropolitan area, this city is removed from the list of first-order centres and the pre-application starts again until it separates all first-order centres from second order sub-centres holding more than 50,000 inhabitants.

3.4.4 Consolidation of FURs and DMAs

Following the Federal Register (Office of Management and Budget 1990), contiguous FURs or DMAs can be aggregated in a single area if some conditions are respected. To simplify these conditions, we consider that two areas must be aggregated in only one metropolitan area if some of them have a flow of commuters from one to another of more than 10 % of their total resident employees. If the percentage is close although lower to the 10 %, the integration is done if there is other robust evidence that the areas are economically and socially-integrated.

3.4.5 Names of the FURs and DMAs and Classification by Intervals

For simplicity, the name of the FUR or DMA corresponds to the name of the largest city.

Following the suggestion by the Federal Register (Office of Management and Budget 1990) and GEMACA (2001), we propose to divide the FURs and DMAs in four intervals or classes regarding the total size of the areas:

1. Level A, formed by metropolitan areas larger than one million inhabitants;
2. Level B, formed by metropolitan areas between 250,000 and one million inhabitants;
3. Level C, formed by metropolitan areas between 100,000 and 250,000 inhabitants; and
4. Level D, formed by metropolitan areas with less than 100,000 inhabitants.

3.5 Application and Results

Most data for the identification of metropolitan areas in Spain and Italy (population, employment, and commuting) come from the 2001 national censuses undertaken by the Spanish Institute of Statistics (INE) and the Italian Institute of Statistics (ISTAT). Land data has been obtained from national property registers. Cartographical basic layers used for GIS (municipalities and regions) come from INE and ISTAT.

3.5.1 Functional Urban Regions

The FUR procedure identified 65 FURs in Spain. They contain 51 % of municipalities (4,200), 76 % of the population (31 million), and 77 % of employment (16.3 million jobs). There are five Level A FURs (above one million employees) which have 13 % of Spanish municipalities, 35 % of the national population, and 38 % of employment (Table 3.2). Madrid is the largest FUR, with 575 municipalities, 5.9 million inhabitants, and 2.6 million employees. Barcelona has 174 municipalities, 4.3 million inhabitants, and 1.9 million employees. Valencia has 150 municipalities, 1.7 million inhabitants, and 700,000 employees. Seville has 57 municipalities, 1.3 million inhabitants, and 480,000 employees. Bilbao has 87 municipalities, 1.06 million inhabitants, and 420.000 employees.

There are 23 Level B FURs (between 250,000 and one million inhabitants). They have 20.5 % of Spanish municipalities, 28 % of the population, and 27 % of employment. There are 26 Level C FURs (between 100,000 and 250,000 inhabitants) which have 14.6 % of Spanish municipalities, and 10.5 % of population and employment. Finally, the 11 Level D FURs have 3.8 % of municipalities, and 2 % of the population and employment.

Regarding their spatial distribution, FURs are distributed across the country. However, the more extensive FURs tend to be localised in the centre-north of the country, whereas the most populated tend to concentrate in the centre and north-east (Fig. 3.1).

In Italy, 81 metropolitan areas have been identified through the FUR procedure. They contain 43 % of municipalities (3,475), 67.6 % of the total population, and 71.5 % of employment. There are six Level A FURs, which have 14.4 % of Italian

Table 3.2 Metropolitan areas in Spain and Italy. Main results. Total values

(a) Functional Urban Regions

FURs	No. of areas	Municipalities	Population	Employment
Spain				
Level A (>1,000,000)	5	1,043	14,436,219	6,180,480
Level B (250,000–1,000,000)	23	1,666	11,412,405	4,438,068
Level C (100,000–250,000)	26	1,185	4,251,746	1,676,858
Level D (<100,000)	11	306	869,903	348,329
Total Spanish FURs	65	4,200	30,970,273	12,643,735
Total Spain		8,108	40,847,371	16,329,713
Italy				
Level A (> 1,000,000)	6	1,172	17,361,480	6,417,324
Level B (250,000–1,000,000)	34	1,217	14,794,555	5,559,483
Level C (100,000–250,000)	38	1,036	6,124,900	2,336,696
Level D (<100,000)	3	50	250,452	104,770
Total Italian FURs	81	3,475	38,531,387	14,418,273
Total Italy		8,101	56,995,744	20,993,732

(b) Dynamic Metropolitan Areas

DMAs	No. of areas	Municipalities	Population	Employment
Spain				
Level A (>1,000,000)	5	1,049	14,506,823	6,219,367
Level B (250,000–1,000,000)	24	1,672	11,326,179	4,409,462
Level C (100,000–250,000)	24	990	3,951,546	1,568,868
Level D (<100,000)	14	258	1,091,995	402,086
Total Spanish DMAs	67	3,969	30,876,543	12,599,783
Total Spain		8,108	40,847,371	16,329,713
Italy				
Level A (>1,000,000)	6	1,355	17,479,230	6,510,073
Level B (250,000–1,000,000)	31	1,614	14,956,574	5,779,957
Level C (100,000–250,000)	40	905	6,358,585	2,308,902
Level D (<100,000)	9	88	766,873	281,186
Total Italian DMAs	86	3,962	39,561,262	14,880,118
Total Italy		8,101	56,995,744	20,993,732

Source: Elaboration from INE (Spain) and ISTAT (Italy) Census Data, 2001

municipalities, 30.5 % of the national population, and 32.4 % of the total employment. The largest FUR is Milan, with 499 municipalities, 5.2 million inhabitants, and 2.4 million employees. Rome is the second biggest, with 239 municipalities, 4.3 million inhabitants, and 1.5 million employees. Naples, Turin, Florence, and Palermo have 125, 215, 51 and 43 municipalities, as well as 3.5, 2, 1.2 and 1 million inhabitants, respectively. Naples has 778,000 employees, Turin 826,000, Florence 528,000, and Palermo 224,000.

There are 34 Level B FURs that represent 15 % of Italian municipalities, 26 % of the population, and 27.4 % of the total Italian employment. The 38 Level C FURs have 12.8 % of municipalities, 10.7 % of the population, and 11.1 % of national

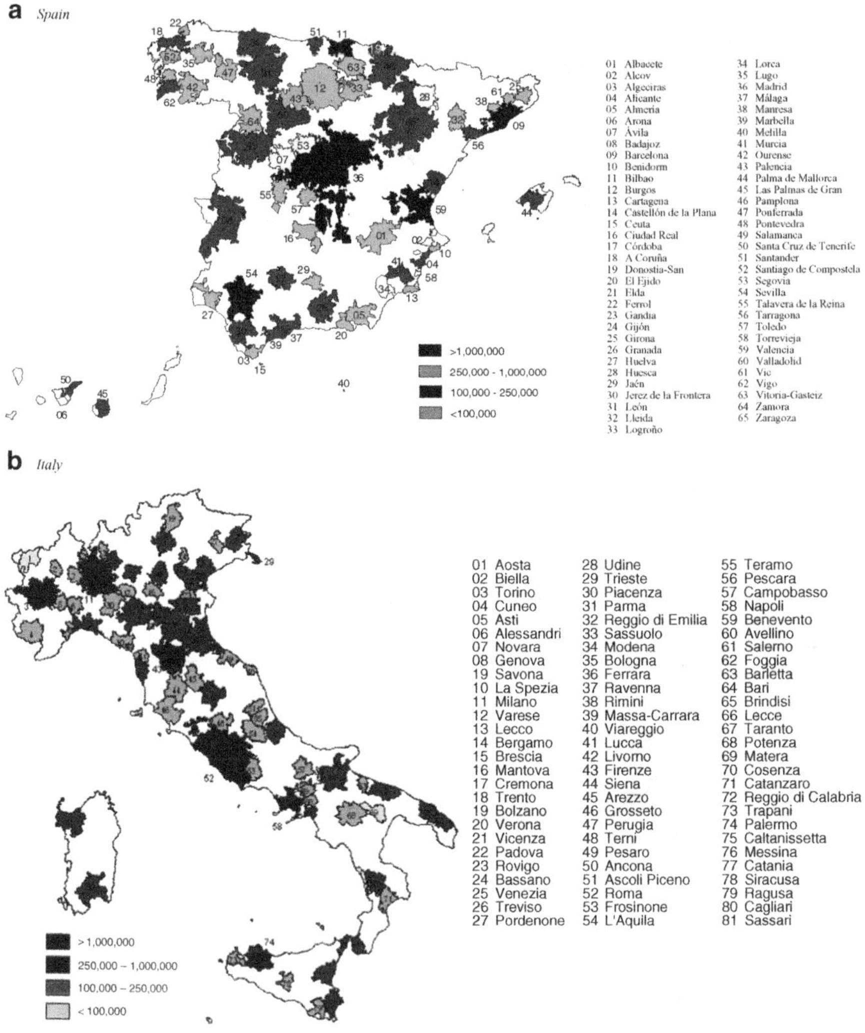

Fig. 3.1 Functional Urban Regions

employment. The three Level D FURs have 0.6 % of municipalities, 0.4 % of the Italian population, and 0.6 % of employment.

Italian FURs are distributed quite uniformly across the Italian territory, even if in the north east part of Italy a higher density of FURs can be observed. Many metropolitan areas are identified along the "Via Emilia" and the Po Valley, while in the south the FURs tend to be more spatially separated (Fig. 3.1).

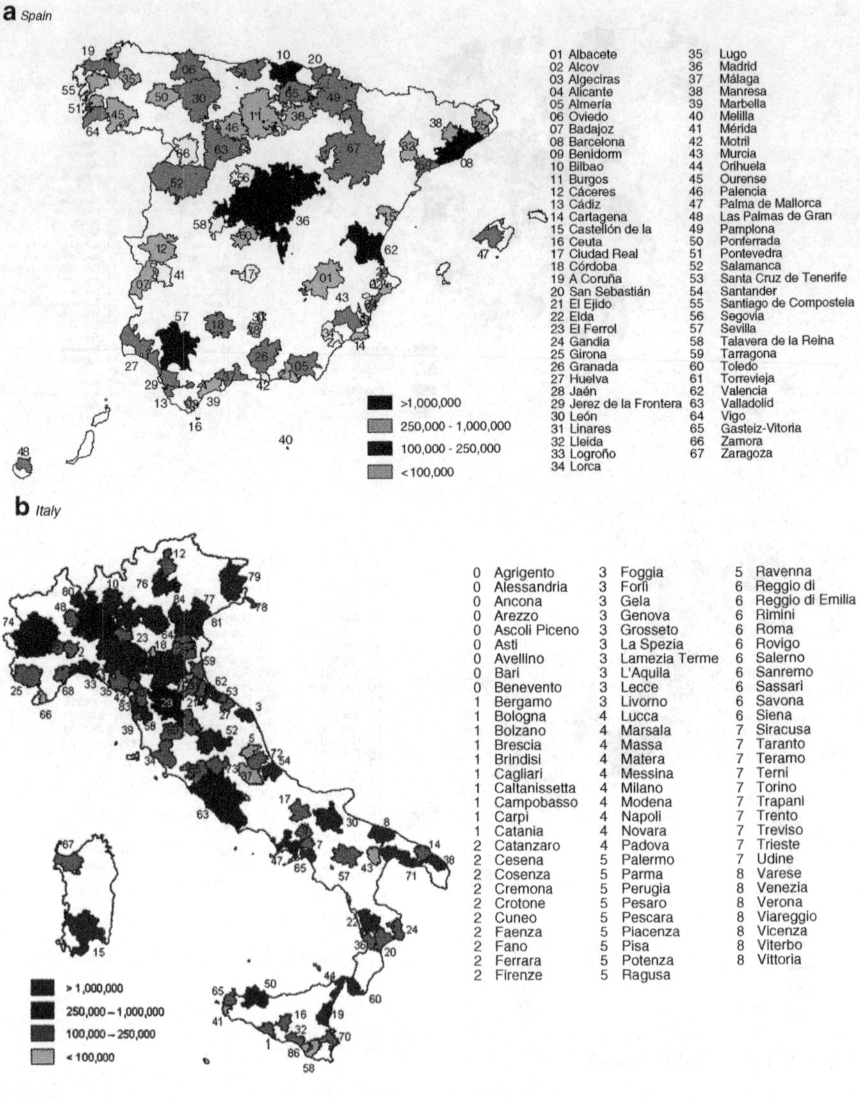

Fig. 3.2 Dynamic Metropolitan Areas

3.5.2 Dynamic Metropolitan Areas

The DMA procedure identifies 67 DMAs in Spain. They have 49 % of Spanish municipalities (4,000), 76 % of the population (31 million), and 77 % of employment (16.3 million jobs). There are five Level A DMAs, which have 13 % of Spanish municipalities, 35 % of the national population, and 38 % of employment (Fig. 3.2; Table 3.2 and 3.3). Madrid is the largest DMA, with 548 municipalities,

5.8 million inhabitants, and 2.6 million employees. Barcelona has 209 municipalities, 4.5 million inhabitants, and two million employees. Valencia has 129 municipalities, 1.7 million inhabitants, and 700,000 employees. Seville has 60 municipalities, 1.4 million inhabitants, and 480,000 employees. Bilbao has 108 municipalities, 1.1 million inhabitants, and 430.000 employees.

There are 24 Level B FURs which have 20.6 % of Spanish municipalities, 28 % of the population, and 27 % of employment. There are 24 Level C FURs which have 12 % of Spanish municipalities, as well as 9.7 % of the national population and employment. Finally, the 14 Level D FURs have 3.2 % of municipalities and 2.5 % of the population and employment.

The application of the DMA procedure to Italy identifies 86 urban areas. They have 48.9 % of Italian municipalities (3,962), 69.4 % of the total national population (39.6 million), and 73.4 % of employment (14.2 million jobs). There are six Level A DMAs, which have 16.7 % of the Italian municipalities (1,355), 30.7 % of the population, and 32.7 % of employment (Fig. 3.2; Table 3.2 and 3.3). The rank of the first DMAs is the same as in the FUR case. Thus, Milan is the biggest metropolitan area, with 597 municipalities, 5.3 inhabitants, and 2.4 million employees. Rome is the second, with 200 municipalities, 4.2 million inhabitants, and 1.5 million employees. Naples has 119 municipalities, 3.4 million inhabitants, and 757,000 employees. Turin has 341 municipalities, 2.2 million inhabitants, and 896,000 employees. Florence has 59 municipalities, 1.3 million inhabitants, and 580,000 employees. Finally, Palermo has 39 municipalities, one million inhabitants, and 222,000 employees.

Regarding the other dimensional classes of metropolitan areas identified with the dynamic procedure, there are 31 Level B DMAs which have 19.9 % of the Italian municipalities, 26.2 % of the population, and 28.6 % of the total national employment. The 40 Level C DMAs have 11.2 % of municipalities and population, and 10.8 % of total employment. Finally, the nine Level D DMAs have 1.1 % of total municipalities and 1.3 % of the national population and employment (Table 3.3).

3.5.3 *FUR, DMA and NUTS3*

It is noticeable that FUR and DMA methodologies produce very similar results regarding the total figures and their distribution among levels in both countries. The spatial patterns of distribution are also very similar. In Spain, the different criteria for the identification of the cores provide the basis for the inclusion in FUR of some smaller local labour markets, such as Vic, Arona or Avila, whereas these cities do not comply with the DMA initial criterion. On the other hand, due to the iterative procedure, DMA produces more clearly defined boundaries in both countries, and facilitates the consolidation in more compact metropolitan areas of Jerez-Cadiz and Badajoz-Caceres-Merida in Spain, and Sassuolo and Modena in Italy.

However, there is a strong difference between metropolitan areas (FURs and DMAs) and NUTS3 (provinces) in both countries. NUTS3 is too small to characterise Madrid, Rome and Milan, but usually too large to catch the rest of the

Table 3.3 Metropolitan areas in Spain and Italy. Main results. Percentages

(a) Functional Urban Regions

FURs	Municipalities (%)	Population (%)	Employment (%)
Spain			
Level A (>1,000,000)	12.9	35.3	37.8
Level B (250,000–1,000,000)	20.5	27.9	27.2
Level C (100,000–250,000)	14.6	10.4	10.3
Level D (<100,000)	3.8	2.1	2.1
Total Spanish FURs	51.8	75.8	77.4
Total Spain	100	100	100
Italy			
Level A (>1,000,000)	14.5	30.5	30.6
Level B (250,000–1,000,000)	15.0	26.0	26.5
Level C (100,000–250,000)	12.8	10.7	11.1
Level D (<100,000)	0.6	0.4	0.5
Total Italian FURs	42.9	67.6	68.7
Total Italy	100	100	100

(b) Dynamic Metropolitan Areas

DMAs	Municipalities (%)	Population (%)	Employment (%)
Spain			
Level A (>1,000,000)	12.9	35.5	38.1
Level B (250,000–1,000,000)	20.6	27.7	27.0
Level C (100,000–250,000)	12.2	9.7	9.6
Level D (<100,000)	3.2	2.7 %	2.5 %
Total Spanish DMAs	49.0	75.6	77.2
Total Spain	100	100	100
Italy			
Level A (>1,000,000)	16.7	30.7	31.0
Level B (250,000–1,000,000)	19.9	26.2	27.5
Level C (100,000–250,000)	11.2	11.2	11.0
Level D (<100,000)	1.1	1.3	1.3
Total Italian DMAs	48.9	69.4	70.9
Total Italy	100	100	100

Source: Elaboration from INE (Spain) and ISTAT (Italy) Census Data, 2001

metropolitan areas. In the case of Madrid and Milan, the metropolitan area expands to another six provinces, where Madrid basically absorbs the neighbouring province of Guadalajara and Milan absorbs Lodi. Only in rare cases (Álava and Valladolid in Spain, and Taranto and Pescara in Italy) are the metropolitan areas close to administrative boundaries.

The results show an additional issue in the consideration of FUR and DMAs below 100,000 inhabitants as true metropolitan areas, as well as in Spain the case of the autonomous cities of Ceuta and Melilla (isolated in the north of Africa). As the classification in levels allows the filtering, our position has been to provide the results according to the proposed criteria. However, we do warn against this fact and let potential users of these results make the final decision (Fig. 3.3).

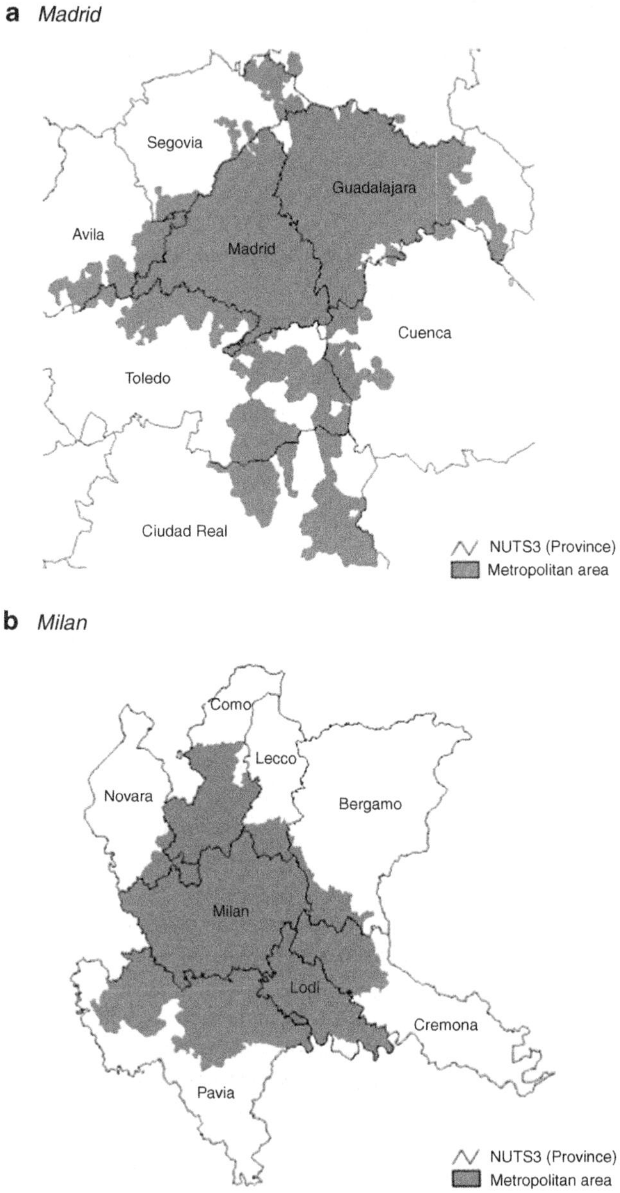

Fig. 3.3 FUR and NUTS3 (provinces). Details for Madrid and Milan

3.6 Conclusions: Towards an Agenda for a Unified Proposal of Delimitation of Metropolitan Areas in Europe

The aim of the research is to identify metropolitan areas in Spain and Italy using comparable methodologies in order to: give evidence about the metropolitan processes in each country; provide a comparison between the metropolitan configurations of both countries; and generate metropolitan units to be used in other research. For these purposes, FUR (functional approach) and DMA (network approach) methodologies have been used. Several conclusions have been made.

First, both methodologies produce very similar results. This can be explained because the lower commuting shares of the FUR procedure tend to converge to the iterative results of the DMA algorithm. This unexpected coincidence reinforces the feasibility of the commuting thresholds in both procedures and the validity of the metropolitan units identified to be used in further research.

Second, metropolitan areas (both FURs and DMAs) clearly diverge from the administrative boundaries (regions or provinces). As a matter of fact, the points highlighted in this section should help focus on the discrepancy between the administrative level of governance and the functional urban organisation of the territory.

Third, there is a high level of metropolitanisation in both countries analysed. These results support the relevance of metropolitan areas as socioeconomic units of analysis, and their importance in the design and implementation of policy strategies. In particular, we identified 65 FURs and 67 DMAs in Spain, which have about 50 % of municipalities, 76 % of the population, and 77 % of employment; and 81 FURs and 86 DMAs in Italy, which have between 43 % and 49 % of municipalities, 70 % of the national population, and about 72 % of national employment.

Fourth, almost half of the metropolitan population and employment is concentrated in the largest metropolitan areas of the country; those which hold over one million inhabitants. In terms of FUR or DMA, there are five large metropolitan areas in Spain (Madrid, Barcelona, Valencia, Seville and Bilbao) which hold about 35 % of the national population and 38 % of employment. In Italy, there are six large metropolitan areas (Milan, Rome, Naples, Turin, Florence, and Palermo) which have about 30 % of the national population and 32 % of employment. These results suggest that the metropolitan areas are keystones to be considered for the implementation of economic policies, and will face globalisation and competitiveness.

On the basis of these results, we propose to advance an agenda towards a unified delimitation of metropolitan areas in Europe. This could be implemented through successive stages. First, the proposed procedures should be applied to other countries where commuting data are available (e.g. France, Germany, Portugal, the UK, etc.). This allows for corrections and improvements in the methodology if necessary. Herein lies the problem with countries where commuting data are not available. However, in this case a provisional solution can be provided by means of the dynamisation of stock data; this is, through estimation of the interaction between cities by means of gravity models. The second stage of the agenda involves coordination of the National Statistic Offices to homogenise commuting statistics without losing the detail of cross-country interactions. This also allows cross-country metropolitan areas to be taken into account, and to extend the use of functional and network methodologies to identify mega-regions, currently identified using morphological approaches.

The third stage is the use of full-network procedures based on the integration of a full range of measures of interaction in multiple layers. However, this stage appears not to be possible until the distant future.

Acknowledgment The opinions expressed and arguments employed here are the responsibility of the authors and do not necessarily reflect those of the OECD.

References

Bartaletti F (2009) Le aree metropolitane in Italia e nel mondo. Bollati Boringhieri, Torino

Basta M, Morchio E, Sanguineti S (2009) Aree metropolitane in Italia: indagine empirica alla luce del censimento 2001. Alinea, Firenze

Berry BJL (1967) Geography of market centers and retail distribution. Prentice-Hall, Englewood Cliffs, NJ

Bertuglia C S, La Bella A (1991) I sistemi urbani, vol.1°: Le teorie - Il sistema - Le reti, vol. 2°: I modelli - Gli strumenti di supporto alle decisioni - I processi di governo. Franco Angeli, Milano

Boix R, Galletto V (2004) Anàlisi econòmica comparada del Cens de 2001. Principals resultats per a Catalunya. Nota d'Economia 79:65–85

Cafiero S, Busca A (1970) Lo sviluppo metropolitano in Italia. SVIMEZ, Giuffré, Milano-Roma

Camagni R, Salone C (1993) Network urban structures in Northern Italy: elements for a theoretical framework. Urban Studies 30(6):1053–1064

Carreras JM, Otero M, Ruiz E (2009) European metropolitan agglomerations. Papers Regió Metropolitana de Barcelona, nº 50, pp 226–243

Cecchini D (1988) Le aree urbane in Italia: scopi, metodi e primi risultati di una ricerca. Rivista economica del Mezzogiorno 2(1):39–70

Cheshire P, Hay DG (1989) Urban problems in Western Europe: an economic analysis. Hyman, London

Cheshire P, Magrini S (2008) Urban growth drivers in a Europe of sticky people and implicit boundaries. SERC discussion paper, 10

Cityrailways (2011) Urbanismi, Cluster urbani e aree metropolitan. Cityrailways, Roma

Clusa J, Roca J (1997) El canvi d'escala de la ciutat metropolitana de Barcelona. Revista Econòmica de Catalunya, nº 33, pp 44–53

Coombes M (2000) Defining locality boundaries with synthetic data. Environ Plann A 32 (8):1499–1518

Costa P, Toniolo M (eds) (1992) Città metropolitane e sviluppo regionale. Franco Angeli, Milano

Comisaría del Plan de Desarrollo Económico y Social (1972) III Plan de Desarrollo, monografía de "Desarrollo Regional", Presidencia del Gobierno. Madrid

Consiglio dei Ministri (2006) La statistica per le aree metropolitane e sulle aree metropolitane: Proposta per un sistema informativo integrato. Rapporto di Indagine 06.2. Roma, Presidenza del Consiglio dei Ministri

Davis K (1959) The world's metropolitan areas, by International Urban Research, Institute of International Studies, University of California. University of California Press, Berkeley

Davoudi S (2003) Polycentricity in European spatial planning: from an analytical tool to a normative agenda. Eur Plann Stud 11(8):979–999

Davoudi, S (2008) "Conceptions of the city-region: a critical review", Journal of Urban Design and Planning, 161(2):51–60.

De Esteban A (1981) Las áreas metropolitanas: Un análisis ecológico. Centro de Investigaciones Sociológicas, Madrid
de Vivienda M (2007) Atlas Estadístico de las Áreas Urbanas de España 2006. Ministerio de Vivienda, Madrid
ESPON (2006) ESPON atlas: mapping the structure of the European territory. Federal Office for Building and Regional Planning, Bonn
Feria JM, Susino J (2005) Movilidad por razón de trabajo en Andalucía. Dimensiones básicas y organización espacial. Instituto de Estadística de Andalucía
Freeman A (2005) Towards a common standard. Comparing European and American cities. Economics working paper 13, GLA, London
GEMACA II (1996) http://www.iaurif.org/fr/savoirfaire/reseaux/gemaca.htm
GEMACA (1996) North-West European metropolitan regions. Geographical boundaries and economic structures. GEMACA, Paris
GEMACA (2002) The metropolises of North-West Europe in figures. IAU, Paris
Hall P, Hay D (1980) Growth centres in the European urban system. Heinemann Educational Books, London
ISTAT (1997) I sistemi locali del lavoro 1991. ISTAT, Roma
ISTAT (2007) Raporto Annuale: La situazione del paese 2006. ISTAT, Roma
ISTAT (2008) Raporto Annuale: La situazione del paese 2007. ISTAT, Roma
ISTAT-IRPET (1989) I mercati locali del lavoro in Italia. Franco Angeli, Milán
Marchese U (1997) Aree metropolitane in Italia alle soglie del duemila. ECIG, Naples
Martínez de Lejarza I, Martínez de Lejarza J (2002) Delimitación de áreas metropolitanas mediante un modelo anisótropo de decrecimiento exponencial. Una aplicación al caso del Área Metropolitana de Valencia. Estudios de Economía Aplicada 20(2):471–486
Ministerio de Vivienda (1967) Las áreas metropolitanas en general en España. Secretaria General Técnica. Madrid
Ministerio de Vivienda (2000) Atlas estadístico de las áreas urbanas en España. Madrid
Ministerio de Vivienda (2005) Atlas estadístico de las áreas urbanas en España 2004. Madrid
OECD (2006) Competitive cities in the global economy. OECD Territorial Reviews, OECD, Paris
OECD (2010) Defining metro-regions for comparison of economic and environmental performance. GOV/TDPC/TI (2010) 9. OECD, Paris
Office of Management and Budget (1990) Revised standards for defining metropolitan areas in the 1990s. Fed Regist 55:12154–12160
ÖIR (2006) The role of small and medium-sized town. Final Report of the ESPON Project 1.4.1. www.espon.eu
Roca J, Burns M, Moix M (2005)Las áreas metropolitanas españolas. Evolución 1991-2001, en Centro de Política del Suelo y Valoraciones (mimeo)
Rozenblant C, Cicille P (2003) Les villes éuropeénnes: analyse comparative. Datar
Rubert JJ (2005) "El distrito industrial de la cerámica y la estructura urbana de Castelló", en Ana María Fuertes (Dir.) "El distrito industrial de la cerámica". Fundación Davalos-Fletcher
Salvador N, Mora C, Salvat E (1997) La regió urbana funcional de Barcelona en el context europeu. Revista Econòmica de Catalunya, nº 33
Sánchez Y (1998) Interpretación del hecho metropolitano en Zaragoza. Geographicalia (1996), 33
Serra J, Otero M, Ruiz R (2002) Grans aglomeracions metropolitanes Europees. IRMB
Serrano Martínez JM (2006) Aglomeraciones y áreas urbanas en España, dimensión y tendencias: Breves precisiones. Lurralde: Investigación y Espacio, nº 29
Tortorella W, Andreani V (2009) (eds) Rapporto Cittalia 2009. Cittá mobile. Cittalia, Fondazione Anci Ricerche
Trullén J, Boix R (2000) La ampliación del área metropolitana de Barcelona y su creciente interacción con las áreas metropolitanas de Tarragona, Lleida y Girona. Avance de resultados. Ajuntament de Barcelona (mimeo)
Urban Audit (2006) http://www.urbanaudit.org/
Vitali O (1990) Mutamenti nelle aree urbane. Popolazione e occupazione dal 1951 al 1987. Franco Angeli, Milano

Chapter 4
Measuring Metropolitan Areas: A Comparative Approach in OECD Countries

Monica Brezzi, Mario Piacentini, and Daniel Sanchez-Serra

4.1 Introduction

Metropolitan areas play a crucial role on the economic performance of countries. They tend to concentrate important shares of the national population and economic activity, but also important shares of innovation, highly educated workers and infrastructures. The 90 largest metropolitan areas in OECD countries, for example, account for around 40 % of OECD population and almost 50 % of its economic activity (OECD 2011).

The links between urbanisation and productivity growth – so-called "agglomeration economies", addressed in the economics literature since long time, are critical to the higher levels of productivity and income per capita we typically find in urban areas. Metropolitan areas might confer economic advantages to the firms located within them as they allow increased interactions of economic agents. The concentration of workers and firms located in close proximity contribute to increase innovation and technology transfer, and reduce operational cost which have an impact on productivity increments and growth promotion. On average, OECD metropolitan regions have been growing faster than the total OECD in terms of population and GDP per capita (OECD 2011).

The opinions expressed and arguments employed here are the responsibility of the authors and do not necessarily reflect those of the OECD. We are grateful to Konstantin Rosina for research assistance and suggestions, Lewis Dijkstra, Hugo Poelman and Tomas de Leus at the European Commission DG for Regional Policies and César de Diego Diez and Oliver Heiden at Eurostat for their contribution to the computations on European countries. Comments from Richard Baker, Rafael Boix Domènech, Mike Coombes and Joaquim Oliveira Martins are gratefully acknowledged.

M. Brezzi (✉) • D. Sanchez-Serra
Regional Development Policy, OECD, 2, rue André Pascal, Paris 75775 Cedex 16, Francia
e-mail: Monica.BREZZI@oecd.org

M. Piacentini
Trade and Business Statistics, OECD, 2, rue André Pascal, Paris 75775 Cedex 16, Francia

E. Fernández Vázquez and F. Rubiera Morollón (eds.), *Defining the Spatial Scale in Modern Regional Analysis*, Advances in Spatial Science,
DOI 10.1007/978-3-642-31994-5_4, © Springer-Verlag Berlin Heidelberg 2012

The economic and social role of cities in the national and regional performance has increased the consciousness for policy makers of the importance of metropolitan areas as strategic places. Currently, urban areas are experiencing profound economic, environmental and social changes. They are key players of national and transnational flows at a global scale and magnets of national and international migration (Frey 2005). Urban areas not only concentrate people they also create hot spots for energy and natural resource consumption, emissions of pollutants and greenhouse gases, and are the places where high income disparities are observed (Brender et al. 2007). As such, urban areas should be treated as interdependent systems on whose dynamics depend the quality of life of large part of the population and the strength of the economy of countries.

Additionally, the form and speed of urban development has been, purposely or not, affected by a variety of regional policy interventions, from spatial planning to cluster policies. The changing spatial organisation of cities (evolving from mono-centric agglomerations to more complex systems made of integrated cores and sub-centres) directly affects the quality of life of urban dwellers, the demand for transport infrastructures, the surrounding landscape, the directions of human and capital flows and the global environmental footprint of urbanisation. Key goals of regional policies, such as increased social cohesion, would critically depend on how cities grow, interact among themselves and with their urban/rural hinterland.

Regional policies need sound information on the efficient use of resources (land, energy, skilled labour, technology etc.) in urban areas, as well as a better account for the fact that urbanisation can take many forms (suburbanisation of metropolitan areas, networked medium cities, and growing towns in rural areas…). Despite the recognised effect of metropolitan areas on the economy, on living quality and on the environment, urban development is still poorly monitored. Moreover, statistically robust comparisons of urban areas across countries are lacking. This knowledge gap is partly due to the absence of an international agreement on what we wish to measure, i.e. what is 'urban' and of what is the real area of a city's labour market (its functional area). A harmonised definition of functional urban areas can help assess the links between the scale and type of urban growth and the sustainable development of a country.

This chapter presents some recent work carried out at the OECD to develop an international methodology for measuring the socio-economic and environmental performance of urban areas (OECD 2012). This work is developing at a time when maximising the sustainable growth potential of urban areas is at the hearth of policy agendas in many OECD countries.[1] A harmonised definition of functional urban areas has the potential to improve the analysis of urban growth and performance,

[1] Please see EU (2011a) "Cities of Tomorrow: challenges, visions and ways forward"; EU (2011b) "Territorial Agenda of the European Union 2020. Towards and Inclusive, Smart and Sustainable Europe of Diverse Regions"; HM Government (2011) "Unlocking growth in cities" and HIS Global Insight (2011) "U.S. Metro Economies", Prepared for The United States Conference of Mayors and the Council for the New American City.

enabling comparative evidence about drivers and constraints. The contributions to research and policy discussion made in this chapter are:

- First, an international methodology for the definition of urban areas applied to 27 OECD countries that classify more than 1,000 functional urban areas.[2] The methodology identifies urban areas as functional economic units, characterised by densely inhabited 'urban cores' and 'hinterlands' whose labour market is highly integrated with the 'cores', by commuting flows. The development of a harmonised functional economic definition overcome previous limitation linked to administrative definitions, by increasing cross-country comparison.
- Second, the definition of urban areas takes into account the possibility of polycentric development, since more cores physically separated can be included in the same urban area. This enables a better illustration of the economic and spatial organization of urban areas and can open up to consequent analysis on governance challenges and economic development.
- Third, the methodology integrates information from geographical sources with population data to get a better understanding of urban forms and urbanization processes.
- Fourth, the methodology identifies for each country all urban systems with at least 50, 000 population, enabling analysis of medium-sized urban areas and not only of large metropolitan areas. Several studies confirm the important role that medium-sized urban areas play in the national economic development (OECD 2010b; Mayfield et al. 2005).

This work is a first step to determining an international dataset through which monitor urban areas performance across countries.

The chapter is organized as follows. Section 4.2 discusses advantages and limitations of previous methodologies applied in the literature to define metropolitan regions. Section 4.3 presents the harmonised methodology to define functional urban areas in OECD countries. Section 4.4 shows some results on the characteristics of functional urban systems and Sect. 4.5 concludes.

4.2 Comparing Cities as 'Metropolitan Regions': Evidence and Limitations

Despite the clear relevance of cities for the aggregation of growth, individual well-being and sustainability of natural resource use, there have been relatively few attempts of building comprehensive datasets at the urban level for international comparative analysis of cities' performance. This is mainly a consequence of the difficulties of developing an internationally harmonized definition of 'city' that can be used as unit of analysis.

[2] The methodology has been applied in cooperation with the European Commission.

Several methodologies to identify metropolitan areas have been developed at national and international level. The level of comparison of metropolitan areas is directly influenced by the approach used to its identification. In fact, the demarcation of a metropolitan area will differ notably depending on the methodology used. In this sense, three approaches have been used generally to identify metropolitan areas (ÖIR 2006):

1. The administrative approach defines metropolitan areas on the basis of legal boundaries and of additional criteria such as population size or population density. Metropolitan areas identified with this first approach are generally used by public administrations in terms of governance issues.
2. The morphological approach defines metropolitan areas based on the aggregation of continuous built-up areas that fit certain criteria of population, density or proportion of the municipalities covered by urban settlements. This approach is better suited for environmental issues such as land use change or greenhouse gas emission or housing development and transportation policies. Currently GIS techniques based on aerial or satellite imagery have been used to identify metropolitan areas worldwide.
3. The functional approach defines metropolitan areas on the basis of flows between a core area and the surrounding territories. Travel-to-work commuting flows represent the flow information generally used for this approach. Small administrative units such as municipalities or census tracts are the building blocks generally used to construct the core and the hinterland of metropolitan areas. The functional approach better captures urban areas interactions and identifies self contained socio-economic urban units. It is therefore preferred to investigate the economic performance of cities as it allows understanding the extension of metropolitan areas over time, while the administrative approach captures static urban forms (Boix 2007).

Most countries have adopted a definition of metropolitan areas or urban systems, beyond the administrative approach. For example the U.S. Office of Management and Budget (2000) and Statistics Canada (2002) use a functional approach similar to the one here adopted to identify the metropolitan areas, respectively, in the United States and in Canada. Similarly, several independent research institutions and National Statistical Offices have identified metropolitan regions using different types of functionality in Italy, Spain, Mexico and United Kingdom based on the functional approach (Boix and Veneri 2009; CONAPO, SEDESOL, INEGI 2004; The Northern Way 2009). The novelty of our approach is to create a pan-OECD methodology which increases the comparability of metropolitan areas across countries. In order to do so, common thresholds and similar geographical units across countries are defined. These units and thresholds may not correspond to the ones chosen in the national definitions. Therefore, the resulting functional urban areas may differ from the ones derived from national definitions.

A previous methodology to identify and analyse metropolitan areas across member countries was developed by the OECD in 2006 and refined in 2009 (OECD 2006, 2009). According to this first definition, large metropolitan regions with at least 1.5 million population were identified on the basis of a mixed-functional approach composed of administrative boundaries, continuity of built up areas as well as commuting flows. This definition used TL3 regions as territorial unit of analysis since in several OECD countries these units are defined around a large urban centre. In this line, a single administrative unit (TL3 regions), or some aggregation of them, were considered a reasonable approximation for metropolitan regions.[3] At the same time, the large metropolitan regions in Canada, Mexico and United States were included according to the national functional economic definition. Around 90 monocentric metropolitan regions were included in the database and several socio-economic and innovation indicators have been collected. Analysis based on this 'regional' definition of urban areas has allowed the OECD to build a first body of evidence on the importance of agglomeration economies for regional development.

Despite the obvious advantage of data availability, this methodology also proved to have certain disadvantages. Firstly, the use of TL3 regions as territorial units of analysis can identify metropolitan areas either too large or too small to capture the socio-economic area of influence of the core cities. As a result, an imprecise delimitation of the area can give place to biased indicators, and consequently to wrong conclusions from the international benchmarking. Secondly, these errors in measurements get larger when one focuses on medium-sized cities, so that one is forced to focus exclusively on larger agglomerations. Thirdly, a definition using regions as building blocks is inherently unable to identify and characterize polycentric urban development. Fourthly, densely built-up cores and their hinterlands can have very different growth dynamics, and a definition of cities using regions as building block is unable to capture them.

4.3 An Improved Approach to Define Functional Urban Areas Worldwide

The process of urbanisation and process of economic transformation in which cities have been involved in the past years have brought about a phenomenon of territorial reorganisation. New patterns of location of population and economic activity have a significant impact on the structure of current metropolitan areas (Becerril 2000) as well as sustainable development implications (Wu 1998). In Europe, polycentricity has become one of the key components of the spatial development strategy promoted by the European Spatial Development Perspective (ESDP) to develop the potentials of all regions (Davoudi 2003). Based on the important emergence of polycentric urban structures well connected and after having analysed different

[3] A similar methodology was used in Europe. Cheshire and Magrini (2008, 2009) identified spatial units called Functional Urban Regions (FUR) by using population, density and commuting criteria.

methodologies for the delimitation of metropolitan areas, Liang et al. (2010) underlines that methods that combine commuting flows and statistical analysis based on GIS are the best choices for the delimitation of metropolitan areas worldwide

Recent analysis at the OECD has argued that policymakers concerned with understanding the impact of cities on the environment and sustainable development should focus more on the form and quality of the urbanisation process, rather than simply on the volume and speed of urbanisation (OECD 2010a).

For these reasons we suggest a functional definition of metropolitan areas that identifies a densely inhabited core on the basis of small administrative units and use commuting data to aggregate adjacent communities with a high degree of integration with the core (OECD 2012).

4.3.1 Data Inputs and Selection of Geographical Units

Given that data are generally disseminated according to administrative jurisdictions or statistical geographic units, functional urban areas are best defined as aggregations of these nationally defined subdivisions. The first key issue for a functional definition of urban areas is thus the choice of an appropriate geographic building block. Here the obvious trade-off is between the precision in the delineation of metro areas and the availability of data for smaller administrative units. For all European countries, the definition uses municipalities (LAU2 in Eurostat terminology).[4] In non-European countries, the selected building block is generally the smaller administrative units for which national commuting data are available. In the following description of the methodology for delineating urban areas, we will use the general term municipalities for indicating the building block used in our analysis.

4.3.2 Defining Urban Cores Through Gridded Population Data

The OECD has traditionally used thresholds based on population density (the ratio between population and the total area of the administrative unit) to classify regions in either urban or rural. While this approach has the obvious benefit of simplicity and performs well for several applications, it has clear limitations when applied to the analysis of urbanisation patterns and their effects on the environment.

One clear problem when using population density as the unique criterion for defining urban cores is the fact that administrative units are unevenly sized and highly heterogeneous within and between countries. It is fairly common to observe

[4] The only exception is Portugal, for which commuting data are available only for LAU1 regions.

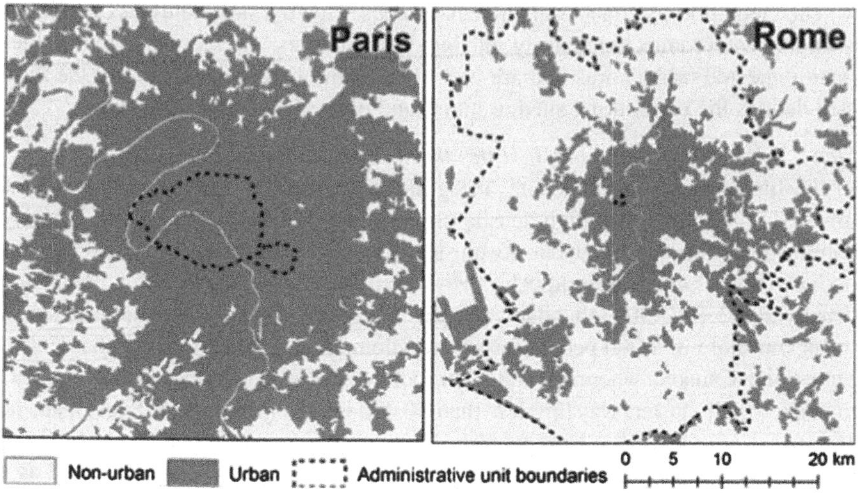

Fig. 4.1 Urban and non-urban population density in Paris and Rome (Source: OECD calculations based on population density disaggregated with Corine Land Cover)

municipalities that, for historic or economic reasons, cover surfaces that are much larger than those of the other municipalities of a country. These municipalities often host a relevant urban centre, but their administrative borders extend also over large mountainous areas, or include vast water surfaces, woodland and shrub. Large administrative borders are a key reason why we can observe low density values even for municipalities that contain non negligible urban agglomerations (in Europe, more than 250 communes above 20,000 inhabitants have a density lower than 150 and the majority of them host fairly large urban core). At the other extreme, considering simply the ratio population/area of the municipality, it is easy to end-up classifying as 'urban cores' some municipalities that have in reality a marked rural connotation.[5] The problem is non negligible also when we focus only on large metropolitan areas, such as Paris or Rome. In Fig. 4.1, it can be seen that population density difference between the two cities depends mainly on the boundary definition; the actual population distribution in the cities plays a secondary role.

The methodology we suggest uses population grid data at 1 km^2 to define urban cores in a way that is robust to cross-country differences in administrative borders. The source of the population grid data for European countries is the population density disaggregated with Corine Land Cover dataset, produced by the Joint Research Centre for the European Environmental Agency (EEA). For all the other countries, harmonised gridded population data from the Landscan project are used.

[5] An example is the municipality of Aldea de Trujillo, a small rural town of 439 inhabitants in 2000 which has very high density because its communal territory measures only 0.3 km. See other examples by Gallego at http://www.ec-is.org/docs/F11116/RURAL%20URBAN%20%20POPDENS.PDF).

The methodology consists in three main steps: The first step identifies contiguous or highly interconnected densely inhabited urban cores, the second step identifies interconnected urban cores that are part of the same functional areas and the third step defines the commuting shed or hinterland of the functional urban area.

Step 1. Identifying core municipalities through gridded population data. In the first step of the procedure, the gridded population data are used to define urbanised areas or 'urban high-density clusters' over the national territory, ignoring administrative borders. High density clusters are defined as aggregation of contiguous high density 1 km square grid cells.[6] High density cells are those with a population density of at least 1,500 inhabitants per km^2 in Europe, Japan, Korea and Mexico. A lower threshold of 1,000 people for square kilometre is applied to Australia, United States and Canada, where several metropolitan areas develop in a less compact manner. Small clusters (hosting less than 50,000 people in Europe, US and Canada, 100,000 people in Japan, Korea and in Mexico) are dropped, as they are likely to capture small agglomerations of built-up areas which cannot be characterised as a urban area. As Box 4.1 shows, a municipality is defined as being part of an urban core by calculating the fraction of its population living within an urban cluster. If the percentage of the population of a municipality living within the urban cluster is higher than 50 %, then the municipality is considered 'densely inhabited'. The final part of the procedure consists simply in aggregating contiguous densely inhabited municipalities in an 'urban core'.

Step 2. Connecting non-contiguous cores belonging to the same functional area. The urban cores defined through this procedure are found to be good approximations of contiguous, highly built-up surfaces. As already said, not all the urban areas in the OECD are characterised by contiguity in built-up development. Many of them are developing in a polycentric way, hosting high densely inhabited cores that are physically separated, but economically integrated. An important innovation of this work identifies which urban areas have such a polycentric structure. This is done by simply looking at the relationships among the urban cores, using the information contained in the commuting data.[7] Two urban cores are considered integrated, and thus part of the same polycentric metropolitan area, if more than 15 % of the residence population of any of the cores commutes to work in the other core. This

---

[6] Contiguity for high-density clusters does not include the diagonal (i.e. cells with only the corners touching). Gaps in the high-density cluster are filled using the majority rule iteratively. The majority rule means that if at least five out of the eight cells surrounding a cell belong to the same high-density cluster, the cell will be added. This is repeated until no more cells are added.

[7] The integration of different clusters of urbanized areas in a unique functional urban area considers only the information provided by travel-to-work data. In some countries, additional sources of information on functional linkages between different areas could be used to better identify polycentric patterns of development. For example, the Northern Way has used information on relative concentrations of employment by four-digit sector across neighboring urban centers to proxy sectoral business linkages, and thus the likelihood that different centers form part of the same economic area (The Northern Way 2009). In general different choices on how to measure the economic linkages economic among areas would of course result in different boundaries and size for the functional urban areas.

intermediate step allows a correction for possible discontinuities in population density within the same urban centre (e.g. natural surfaces larger than 1 km^2 splitting one city in two parts).

Using this simple functional criterion, it is possible to identify several polycentric metropolitan areas.[8] These polycentric metropolitan areas are generally constituted by one central city with a large population nucleus and a set of smaller sub-centres which have a high degree of integration with the nucleus. The direction of the relationship is not necessarily from the small sub-centres to the large central cores, as in many cases the sub-centres develop as dynamic industrial and service hubs, rather than as dormitory spaces for the workers in the big cities. For large metropolitan areas and in countries where commuting distance is steadily increasing, it is easy to find sub-centres situated far apart from the central city core. This is for example the case of London, whose increased connectivity with urban sub-centres has been the result of the combined effect of infrastructural improvements and increasing spatial re-organisation of production activities (firms keeping their administrative headquarters in the central core, and relocating production facilities to well connected agglomerations outside the central core).

Step 3. The identification of the urban hinterlands.

Once the densely inhabited municipalities are aggregated to form urban cores, and polycentric metro areas with tied cores are identified, the final step of the methodology consists in delineating the hinterland of the metro areas. The 'hinterland' can be defined as the "worker catchment area" of the urban labour market, outside the densely inhabited core. The size of the hinterland, relative to the size of the core, gives clear indications of the influence of cities over surrounding areas. Getting distinct information for cores and for hinterlands is also very important to understand where change is taking place.

We assign to each core as hinterland municipalities all those municipalities which send to the core a percentage of their workers above a given threshold. After extensive sensitivity analysis, the threshold has been fixed at 15 % of the resident employed for municipalities.

We consider the multiple cores within a polycentric metropolitan area as a single destination. In this way, a hinterland municipality is assigned to a polycentric municipal area if the level of its commuting to the tied cores exceeds the threshold. This adjustment is needed to take into account the fact that workers within the catchment areas of a polycentric system tend to commute towards multiple employment centres.[9] For the cases in which a municipality has commuting levels over 15 % to cores in different metropolitan areas, it is linked to the core to which it sends the highest share of its employed population. In addition,

---

[8] For example, the application of the criterion leads to the pairing of 94 urban cores in 20 countries in Europe.

[9] Without the adjustment, a hinterland municipality with 14 % commuting to three tied urban cores (thus strongly integrated into the urban agglomeration, with 42 % (14 times 3) of its resident population moving to work to the urban centres), would be excluded from the metropolitan area.

municipalities surrounded by a single functional area are included as part of the functional urban areas and non-contiguous municipalities are dropped (Box 4.2).

As a result of this methodology, it is possible to obtain an accurate representation of each country's 'urban system.[10] These systems are constituted by all the functional urban areas taking shape around high-density clusters with population higher than 50,000 people (100,000 in Mexico, Korea and Japan). The simple visualisation of the results is already informative about the concentration of urban people in particular regions of a country, and about the size distribution ("hierarchy") among the different urban centres. Figure 4.2 shows the Korean functional urban system; the complete set of maps of the functional urban systems can be found in OECD 2012.

Box 4.1 Defining the Urban Cores, and Illustration from Toulouse (France)

An urban core consists of a high-density cluster of contiguous[11] grid cells of 1 km^2 with a density of at least 1,500 inhabitants per km^2 and the filled gaps.[12] A lower threshold of 1,000 people for km^2 is applied to Canada and United States, where several metropolitan areas develop in a less compact manner. Small clusters (hosting less than 50,000 people in Europe, US and Canada, 100,000 people in Japan, Korea and in Mexico) are dropped.

A municipality is defined as being part of an urban core if at least 50 % of the population of the municipality lives within the urban cluster

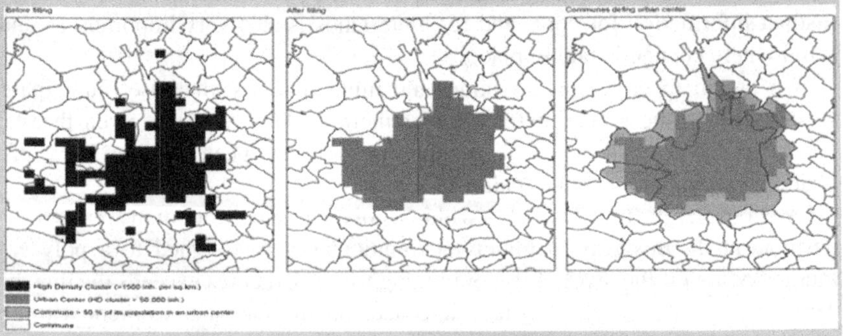

[10] It must be noted that few functional urban areas spread over national borders.

[11] Contiguity for high-density clusters does not include the diagonal (i.e. cells with only the corners touching).

[12] Gaps in the high-density cluster are filled using the majority rule iteratively. The majority rule means that if at least five out of the eight cells surrounding a cell belong to the same high-density cluster it will be added. This is repeated until no more cells are added.

Box 4.2 Defining the Hinterlands, and Illustration from Toulouse (France)

All municipalities with at least 15 % of their employed residents working in a certain urban core are assigned to that functional urban area. Municipalities surrounded by a single functional area included and non-contiguous municipalities are dropped

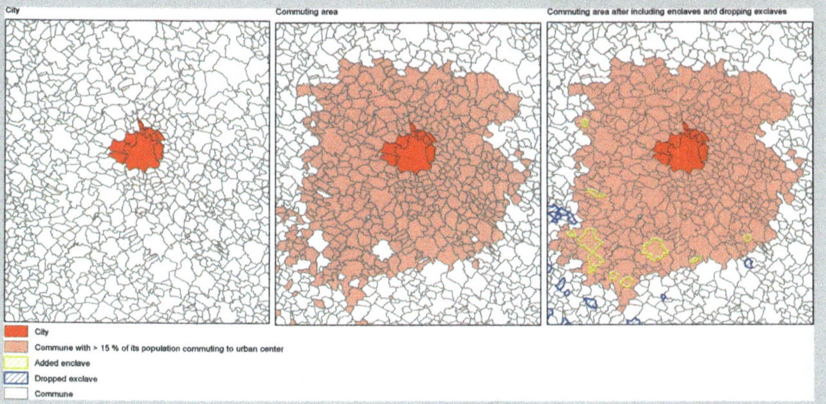

4.4 A Description of Urban Systems in OECD Countries Based on the New Methodology

The above described methodology is applied to 27 OECD countries, where a total of 1,140 functional urban areas have been identified. According to this definition, the proportion of population living in urban areas with cores larger than 50 (100) thousand inhabitants in OECD countries is around 66 % ranging from almost 90 % in Luxembourg, to less than 40 % in the Slovak Republic (Fig. 4.3).

For the remaining OECD countries Australia, Chile, Iceland, Israel, New Zealand and Turkey the work has not started yet.

The functional urban areas are classified according to their population size in four classes:

- Small urban areas, with population below 200,000 people.[13]
- Medium-sized urban areas, with population between 200 and 500 thousand people

[13] Given that cities are identified on the basis of high-density clusters with minimum size of 50,000 people, there is a lower bound in the population of the functional urban areas. The smallest cities identified (Thousand and Palm Desert in the United States, Granollers in Spain) have a total population of around 45,000 people.

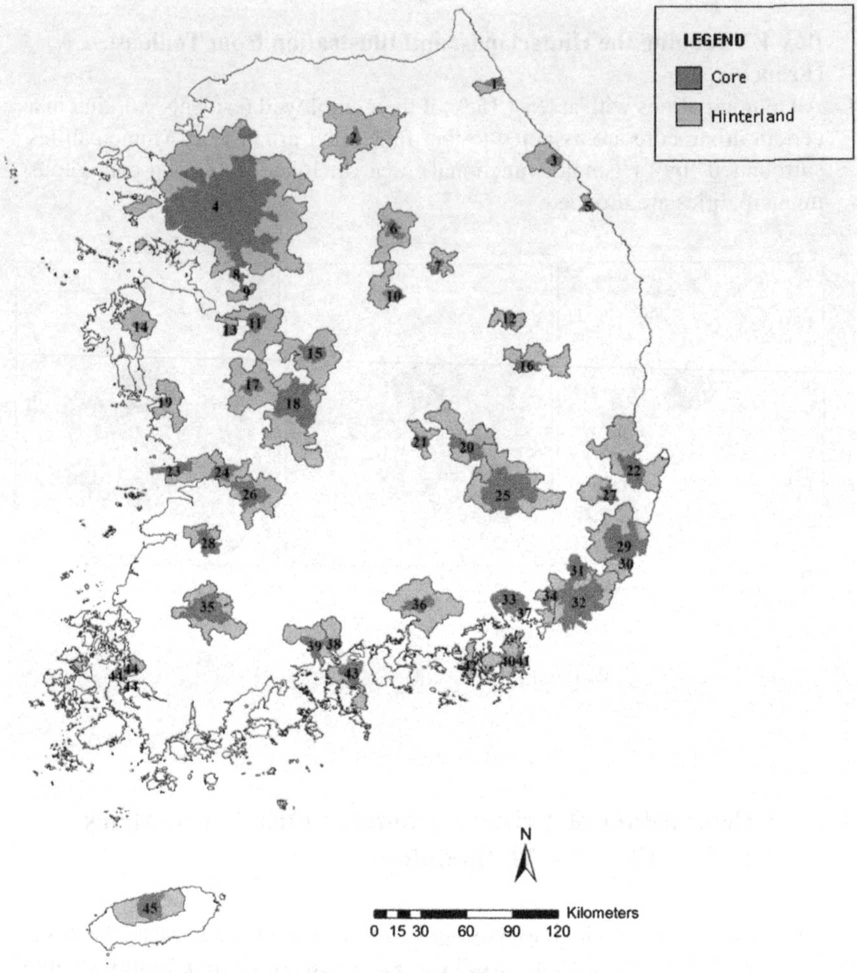

Fig. 4.2 Korean functional urban system (Note: Due to the unavailability of commuting data among municipalities for the whole country, here it is used a representative sample of the commuting in 2009 provided by the (MLTM). The sample data covers around 700,000 commuters referring to the "home to offices" category Source: OECD calculations)

- Metropolitan areas, with population between 500 thousand and 1.5 million people
- Large metropolitan areas, with population of 1.5 millions or higher

On the basis of this classification, it is possible to study the relative importance of medium-sized urban areas with respect to large metropolitan areas in each country.[14] The eight countries on the bottom of Fig. 4.4 have no large metropolitan

[14] Several studies confirm the important role that middle-size cities play in the national economic development. In fact, middle-size cities are often seen as a vehicle of diffusion of opportunities of growth and as a more sustainable form of urbanization, with lower footprints on the natural environment (Mayfield et al. 2005).

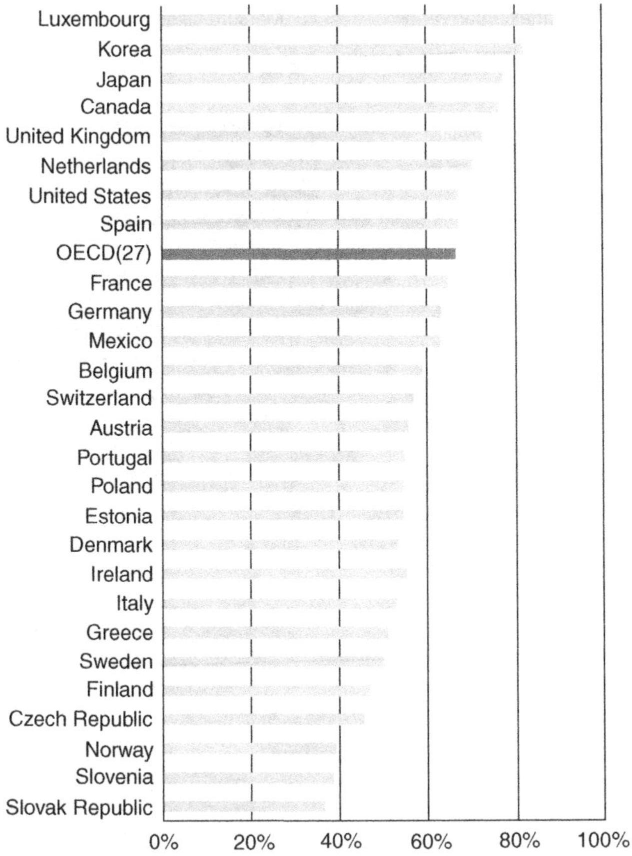

Fig. 4.3 Percentage of national population living in a urban area (year 2000) (Note: The national population living in a urban area is defined as the population living within an identified functional urban area, thus any city whose functional urban cluster hosts more than 50,000 people (100,000 in Mexico and Japan and Korea)

areas, while in all the other countries the urban centres with 1.5 million people or higher host at least 20 % of the urban population. The primacy of large metropolitan areas is particularly clear in Korea, Greece, Denmark, Japan and the United States where at least 60 % of the urban population live in cities of this class.

Among the 1,140 functional urban areas identified in the 27 OECD countries, 73 are large metropolitan areas with more than 1.5 million people, 194 are metropolitan areas, 391 are medium-sized urban areas and 482 are small urban areas. A larger share of urban population lives in large metropolitan areas in North America, Japan and Korea than in Europe and the average size of the large metropolitan areas is much bigger in Japan (more than ten million inhabitants), Korea (almost nine millions) and North America (around four millions) than in Europe (around three millions). On the other hand the weight of population in small and medium sized

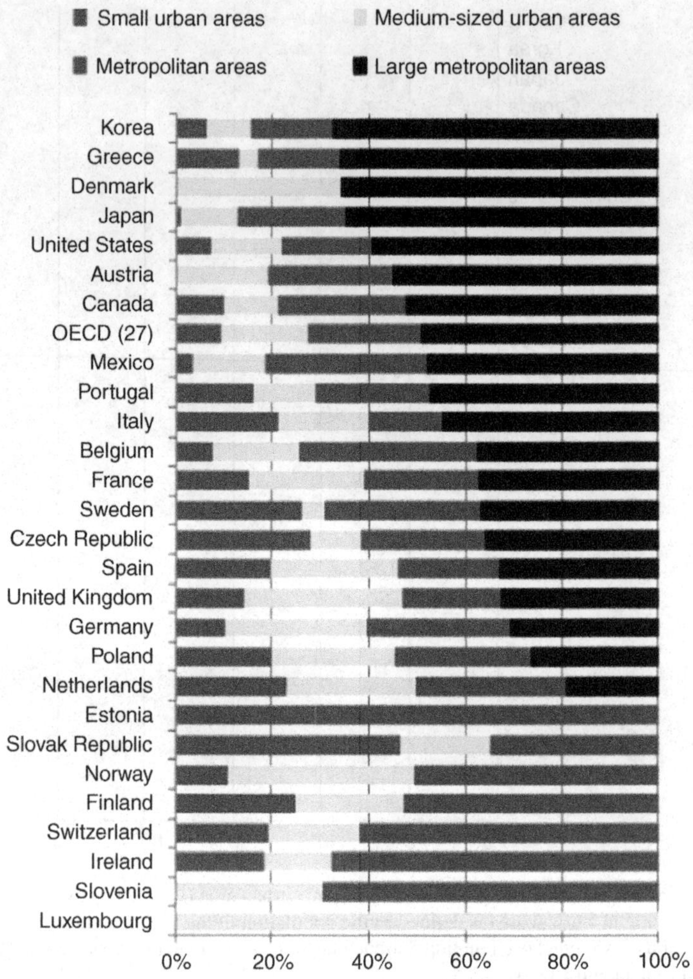

Fig. 4.4 Distribution of OECD urban population across city types (Note: Population data refer to year 2000)

urban areas is bigger in Europe than in Japan, Korea and North America, even though the average size of these two city types is comparable across OECD countries. Additionally, small urban areas are the category more represented in Korea and Europe (they account for almost 50 % of all functional urban areas), while 45 % of the functional urban areas in Japan are classified as medium-sized urban areas (Fig. 4.5).

Population growth between 2000 and 2006 was especially marked in the metropolitan areas and small urban areas. In small urban areas, the acceleration of population growth after 2000 is particularly marked in the city cores. Across all the four types of functional urban areas, the population of the hinterland has been

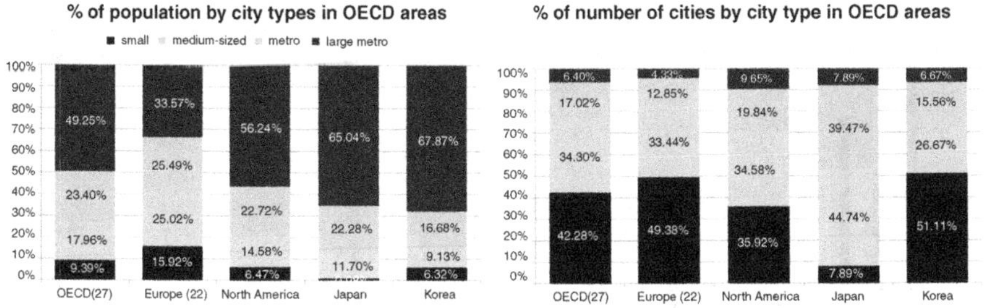

Fig. 4.5 Share of number of cities and population by city types in OECD countries (Note: Population data refer to year 2000)

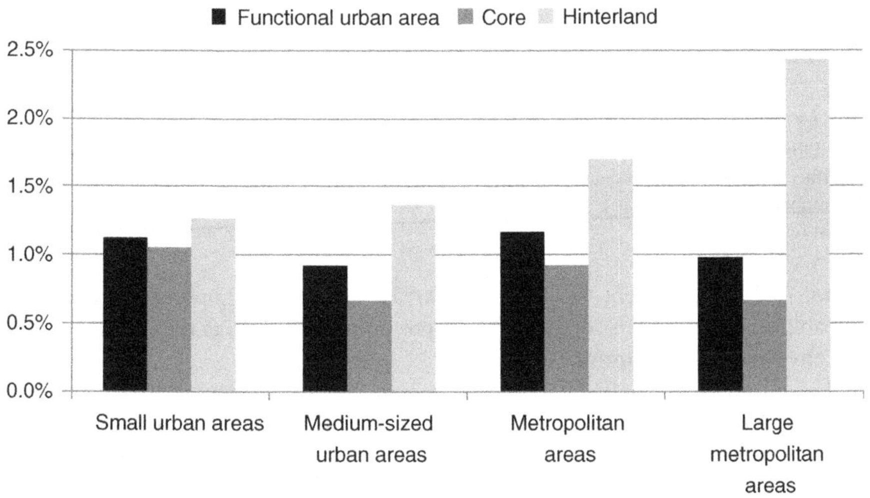

Fig. 4.6 Population growth by city types and core/hinterland (Note: The period of growth in the case of Korea is 2000–2010, and Portugal 1991–2011)

growing at a faster rate than the population of the core, suggesting a common trend of 'suburbanisation' or densification of peri-urban areas. The largest increases in population are observed in the hinterlands of the large metropolitan areas, with a yearly population growth of 2.5 % in the period between 2000 and 2006 (Fig. 4.6). This evidence on the fast growth of the hinterlands of metropolitan cities, based on a sample of 14 countries[15] warrants further analysis on the consequences of such a

[15] These are Belgium, Canada, Denmark, Estonia, France, Italy, Japan, Korea, Luxembourg, Mexico, Norway, Portugal, Spain and the United States. The population in urban systems of these 14 countries represents around 80 % of the population in urban system of the 27 OECD countries included in this paper.

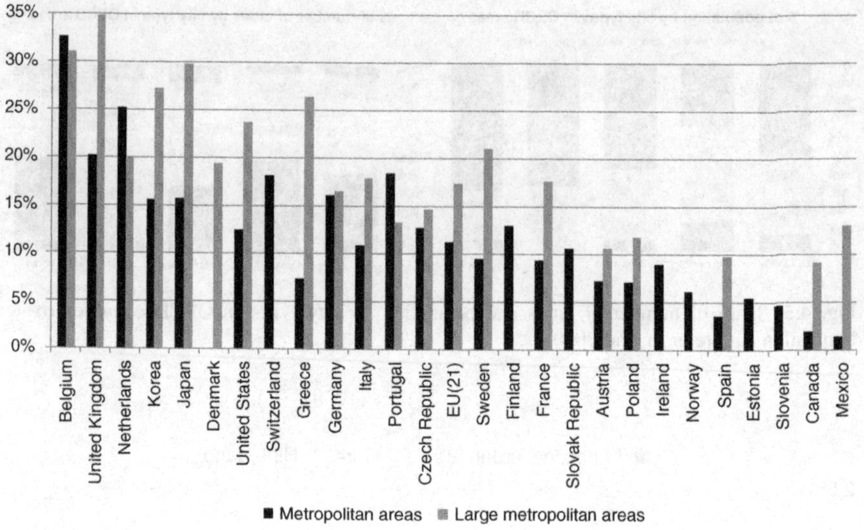

Fig. 4.7 Share of urbanised land over total area in metropolitan areas, by country (Note: It must be noted that for Canada, Korea and Mexico data are derived from medium spatial resolution (500 m) satellite imagery (MODIS) and should be taken as rough estimates. The functional city of Luxembourg is classified as a medium-sized city so it is not included in this figure. The data for Japan refer to 1997)

trend. The development of peri-urban areas has in fact important impacts on liveability and equity in access to job-opportunities, as well as relevant effects on the environmental footprint of cities.

Urbanisation not only concentrates people but also trigger a variety of land changes processes in natural environments. Detailed spatial information on the changes in land cover can help identify which areas have been exposed to larger urban pressure, guiding targeted policy interventions where this expansion threatens the quality of the landscape or bio-diversity.

Making use of global land cover datasets at high geographical resolution, we can derive a measure of the share of "urbanised land" (artificial land with built-up cover or urban use such as parks and sport facilities) within the functional urban areas and its change over time. The percentage of urbanised land over total area in metropolitan areas varies from less than 4 % in Mexico and Canada to around 30 % in Belgium, the Netherlands and the United Kingdom. This percentage is generally higher in large metropolitan areas than in metropolitan areas, especially in Japan, Korea and the United Kingdom, with the only exception of Portugal (Fig. 4.7).

Urbanised land in the metropolitan areas and large metropolitan areas in the United States have grown at almost 1 % per year, while 0.7 % in Japan and 0.4 % in Europe. Among European metropolitan areas a very steep increase in urbanised land is observed in Dublin (Ireland), La Palmas, Madrid, Murcia and Zaragoza (Spain), Tallin (Estonia) and Lisbon (Portugal) (Fig. 4.8).

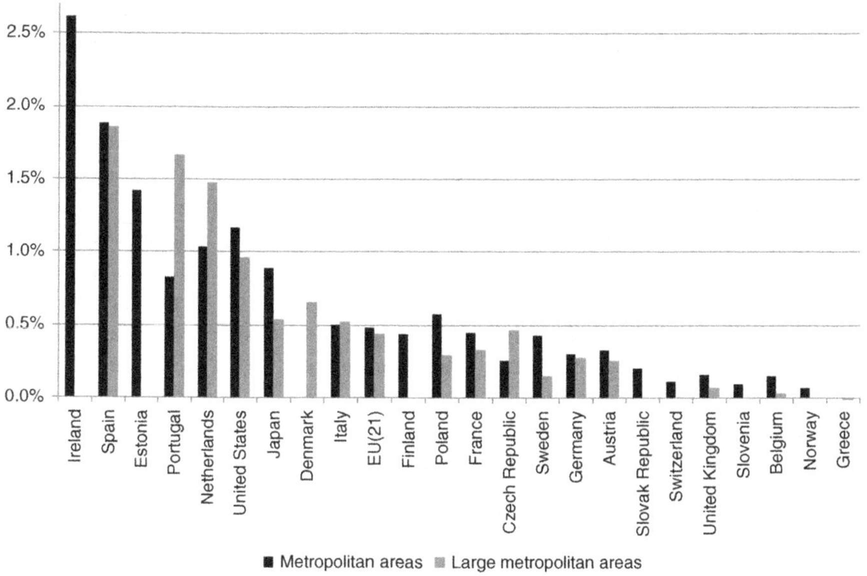

Fig. 4.8 Growth of urbanised land in metropolitan areas, by country (Note: The functional city of Luxembourg is classified as a medium-sized city so it is not included in this figure. The data for Japan refer to 1997. In Canada and Mexico data are available only for 1 year, so changes can't be computed)

As a result of the population and urbanised land dynamics, we observe a sub-urbanisation of the metropolitan areas, corresponding to a low or negative growth of population density in the cores of urban areas. In particular, in Estonia, Italy, Japan, Portugal and Spain the rate of population growth in the cores of metropolitan areas was lower than the rate of growth of urbanised land.

4.5 Conclusions

Metropolitan areas have intensified their competition to attract and retain specific economic activities, investors and labour force. This has increased the need for public policies to evaluate current situations as well as quantify results achieved by the implementation of policy efforts. The common methodology to define urban systems presented in this chapter can be a useful tool to increase comparable evidence of metropolitan areas between and within countries.

This chapter starts from the premise that we need new comparative analysis of urban areas, based on an economic definition of the spatial extent of metropolitan areas. Building on previous OECD work, this chapter presents the results of a joint effort of the OECD and the European Commission to provide a functional definition of metropolitan areas, which proved to be effective in delineating both the densely inhabited urban cores and the hinterlands of the cities.

The definition has been successfully applied so far to 27 OECD countries. In order to ensure comparability of the statistics obtained for the functional areas, a particular effort was made to use administrative building blocks of comparable size and to reduce to the minimum country-specific adjustments in the methodology. Only limited variations in the population thresholds to define the urban cores were allowed to adjust for the large cross-country differences in the form of urban settlements. This search for international comparability might come at the cost of a loss of accuracy in the delimitation of the urban borders. A validation work with national experts has een carried out to ensure that the results are good representations of OECD cities. The relative simple steps of the methodology make the result replicable by interested countries and possible to update, as new data from censuses become available and administrative units are modified.

A wide application of this methodology can generate the basis for building new comparable indicators of urbanisation trends and quality of life in cities. The main constraint to further extend the geographical coverage is the availability of travel-to-work (commuting) data to define the hinterlands of the functional urban areas. Further methodological work is in progress to identify suitable substitute for the commuting data and to adapt such definition to urban areas in emerging and developing economies (OECD 2012).

References

Austrian Institute for Regional Studies and Spatial Planning (ÖIR) (ed) (2006) ESPON Project 1.4.1. The role of small and medium sized towns

Becerril M (2000) Policentrismo en las ciudades latinoamericanas. Theomai: estudios sobre sociedad, naturaleza y desarrollo N°. 1

Boix R (2007) Concepto y delimitación de áreas metropolitanas: una aplicación a las áreas metropolitanas de España. Paper presented in workshop Las grandes áreas metropolitanas en una perspectiva comparada, Sevilla, 11 de Abril de 2007

Boix R, Veneri P(2009) Metropolitan areas in Spain and Italy. IERMB working papers in economics n.09.01

Brender N, Cappe M, Golden A (2007) Mission possible: successful Canadian cities. In: 2007 the conference board of Canada, Canada

Cheshire PC, Magrini S (2008) Urban growth drivers in a Europe of sticky people and implicit boundaries. SERC discussion paper 10

Cheshire PC, Magrini S (2009) Urban growth drivers and spatial inequalities: Europe – a case with geographically sticky people. LSE 'Europe in question' discussion paper series, 11

CONAPO, SEDESOL, INEGI (2004) Delimitación de las zonas metropolitanas de México, Mexico

Davoudi S (2003) EUROPEAN BRIEFING: polycentricity in European spatial planning: from an analytical tool to a normative agenda. Eur Plann Stud 11(8):979–999

European Union (2011a) Cities of tomorrow: challenges, visions, ways forward. European Union, Brussels

European Union (2011b) Territorial agenda of the European Union 2020. Towards an inclusive, smart and sustainable Europe of diverse regions. European Union, Brussels

Frey W H (2005) Inmigration and domestic migration in US Metro areas: 2000 and 1990 census findings by education and race. Population Studies Center Research Report 05–472

HIS Global Insight (2011) U.S. metro economies. In: Prepared for the United States conference of mayors and the council for the New American City
HM Government (2011) Unlocking growth in cities. Cabinet Office, London
Liang J, Li F, Mao L (2010) Review of the methods of delimitation for the spatial scope of urban agglomeration. Geoinformatics, Beijin
Mayfield L, Courtney P, Tranter R, Jones P (2005) The role of small and medium-sized towns in rural development. Final report. Centre for Agricultural Strategy, The University of Reading
OECD (2006) Competitive cities in the global economy. OECD, Paris
OECD (2009) A description of OECD metropolitan regions. www.oecd.org/gov/regional/statisticsindicators
OECD (2010a) Cities and climate change. OECD, Paris
OECD (2010b) Trends in urbanisation and urban policies in OECD countries. What lessons for China. OECD-China Development Research Fundation, Paris
OECD (2011) Regions at a glance, 2011. OECD, Paris
OECD (2012) Redefining "urban": a new way to measure metropolitan areas. OECD, Paris
Statistics Canada (2002) Census metropolitan area and census agglomerations. Statistics Canada, Ottawa
The Northern Way (2009) "City relationships: economic linkages in northern city regions", Methodology Paper, November
U.S. Office of Management and Budget (2000) Standards fir defining metropolitan and micropolitan statistical areas, notice. Office of Management and Budger, Federal Register 65 (249)
Wu F (1998) An experiment on the generic polycentricity of urban growth in a cellular automatic city. Environ Plann B: Plann Des 25(5):731–752

Part II
Estimation of Spatial Disagregated Data

Chapter 5
Ecological Inference with Entropy Econometrics: Using the Mexican Census as a Benchmark

Esteban Fernández-Vazquez and Rafael Garduño-Rivera

5.1 Introduction

One relatively frequent limitation for empirical economics is the lack of data available at an appropriate spatial scale. Although the target, in principle, would be to work at a smaller geographical scale, the non-availability of geographically disaggregated information usually limits the conclusions of the analysis at an aggregate level. To overcome this problem, a process of Ecological Inference (EI) is required in order to recover the information at the required spatial scale.

Generally speaking, EI is the process of estimating disaggregated information from data reported at aggregate level. Research in this area has grown enormously in recent years, given its usefulness in many academic disciplines of social science as well as in policy analysis. The foundations of EI were introduced in the seminal works by Duncan and Davis (1953) and Goodman (1953), whose techniques were the most prominent in the field for more than 40 years, although the work of King (1997) supposed a substantial development by proposing a methodology that reconciled and extended previously adopted approaches. An extensive survey of recent contributions to the field can be found in King et al. (2004).

The objective of this chapter is to explore the potential of an estimation procedure based on entropy econometrics to recover disaggregated information from more aggregated data. The chapter is divided into five further sections. Section 5.2 presents the basic Cross Entropy solution to the estimation problem, and this approach is extended in Sect. 5.3 in order to consider the case of having

E. Fernández-Vazquez (✉)
REGIOlab - University of Oviedo, Oviedo, Spain
e-mail: evazquez@uniovi.es

R. Garduño-Rivera
Centro de Investigación y Docencia Económicas, A.C. (CIDE), Mexico

E. Fernández Vázquez and F. Rubiera Morollón (eds.), *Defining the Spatial Scale in Modern Regional Analysis*, Advances in Spatial Science,
DOI 10.1007/978-3-642-31994-5_5, © Springer-Verlag Berlin Heidelberg 2012

non-reliable aggregate information. Section 5.4 provides a picture of the availability of information in Mexico at small scale (municipalities). In Sect. 5.5, the performance of the technique is evaluated by applying to a real-world problem where the disaggregated data are observable. Specifically we take the Mexican 2009 *Censo Económico* as a benchmark. This database is published every 5 years and contains information about gross value added, wages and labor by industry disaggregated at the level of municipalities. The data contained in this census will be confronted with estimates obtained assuming that only aggregated information is observable. The main conclusions and possible further lines of research complete the chapter in Sect. 6.

5.2 Ecological Inference as a Matrix Balancing Problem

The problem of estimating spatially disaggregated data can be described in the same terms as on the matrix-balancing problem depicted in Golan (2006, page 105), where the goal is to fill the (unknown) cells of a matrix using the information that is contained in the aggregate data of the row and column sums (Fig. 5.1).

Graphically, the point of departure of our problem is a matrix where the cells z_{ij} are the unknown elements we would like to estimate and we define the following sums $\sum_{j=1}^{T} z_{ij} = z_{i\cdot}$, $\sum_{i=1}^{K} z_{ij} = z_{\cdot j}$, and $\sum_{i=1}^{K}\sum_{j=1}^{T} z_{ij} = z$. The z_{ij} elements can be expressed as a bi-dimensional probability distribution simply dividing the quantities of the matrix by the sum $\sum_{i=1}^{K}\sum_{j=1}^{T} z_{ij} = z$. In such a case, the previous matrix can be rewritten in terms of a new matrix P where the p_{ij}s are defined as the proportions $\frac{z_{ij}}{z}$, with new row and column margins defined as $R_i = \frac{Z_{i\cdot}}{Z}$ and $C_j = \frac{Z_{\cdot j}}{Z}$ respectively. Consequently, the followings equalities are fulfilled by the p_{ij} elements

$$\sum_{j=1}^{T} p_{ij} = R_{i\cdot}; \forall i = 1, \ldots, K \tag{5.1}$$

$$\sum_{i=1}^{K} p_{ij} = C_{\cdot j}; \forall j = 1, \ldots, T \tag{5.2}$$

These two sets of equations reflect all we know about the elements of matrix P. Equation 5.1 shows the cross-relationship between the (unknown) p_{ij}'s in the matrix and the (known) sums of each row and column. Additionally, Eq. 5.2 indicates that the p_{ij}'s can be viewed as (column) probability distributions. In such a situation, the Cross Entropy (CE) principle can be applied to recover the unknown p_{ij} probabilities if we have available a prior distribution Q that reflects our initial assumptions about the target matrix P. In other words, we want to transform an a priori probability matrix X into a posterior matrix P that is consistent with the vectors R and C.

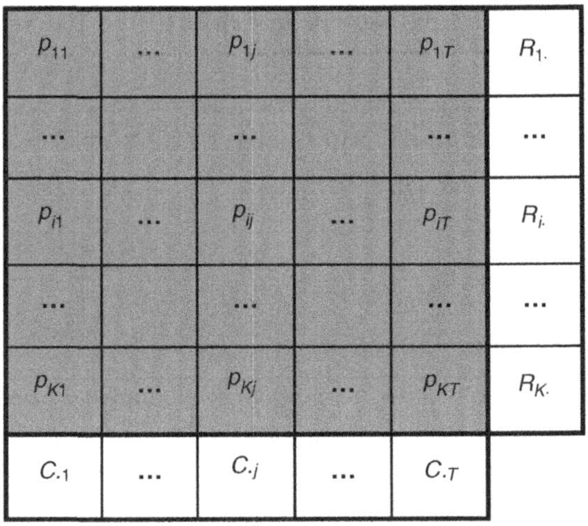

Fig. 5.1 The matrix balancing problem

The solution to this type of problems is obtained by minimizing a divergence measure with the prior probability matrix Q subject to the set of constraints Eqs. 5.1 and 5.2, which can be written in the following terms:

$$\underset{P}{\mathrm{Min}}\, D(P \parallel Q) = \sum_{i=1}^{K} \sum_{j=1}^{T} p_{ij} \ln\left(\frac{p_{ij}}{q_{ij}}\right) \tag{5.3}$$

The divergence measure $D(P \parallel Q)$ is the Kullback-Liebler entropy divergence between the posterior and prior distributions. The entropy-based estimation techniques outlined above can be directly applied to the field of Ecological Inference (EI). Judge et al. (2004) suggested the use of information-based estimation techniques for EI problems, although in a different context (the estimation of individual voters' behavior from aggregate election data). In this chapter we suggest an application of CE following the line of Judge et al. (2004) and posing the EI as a particular case of the more general matrix balancing problem.

Consider a geographical area that can be divided in T smaller spatial units. Besides this first geographical partition, suppose that there is another dimension on which we would like to observe some variable. Consider that this second dimension is the classification into K different industries on which the economic activity can be divided. The objective of the estimation exercise would be to recover the values of the variable disaggregated by regions and industries from aggregate information at the industrial and regional scale. Graphically, this estimation problem can be represented by Fig. 5.1.

Each one of the p_{ij} elements is defined as the proportion of the variable allocated in region i and industry j, forming a $(K \times T)$ matrix P of unknown values. The $(K \times 1)$ row vector R and the $(1 \times T)$ column vector C contain respectively the regional and sectoral shares of the variable across the country. If an a priori probability distribution X is also available, the Cross Entropy procedure outlined previously can be directly applied.

5.3 A Flexible CE Estimation with Non-reliable Margins

The above-sketched procedure assumes that we have perfectly reliable information of the margins R and C, which can be considered as an unrealistic assumption. Suppose that we observe row and column margins as $\tilde{R}$ and $\tilde{C}$, where:

$$\tilde{R}_{i\cdot} = R_{i\cdot} + \varepsilon_i; \forall i \tag{5.4}$$

$$\tilde{C}_{\cdot j} = C_{\cdot j} + \epsilon_j; \forall j \tag{5.5}$$

Where ε_i and ϵ_j are random errors that make the observed margins diverge from the real margins of the target matrix. In this situation is still possible to adjust our prior Q with row and column margins not perfectly reliable by means of a Generalized Cross Entropy approach (GCE), following a similar approach to the ideas suggested in Golan and Vogel (2000) or Robinson et al. (2001).

The basic idea is to re-parameterize the errors ε_i and ϵ_j in terms of unknown probability distributions. The uncertainty about the realizations of these errors is introduced in the problem by considering each element ε_i and ϵ_j as discrete random variables with $L \geq 2$ possible outcomes (for the sake of simplicity L is assumed common for both). These values will be contained in two convex sets $v' = \{v_1, \ldots, 0, \ldots, v_L\}$ and $u' = \{u_1, \ldots, 0, \ldots, u_L\}$ respectively. We also assume that these possible realizations are symmetric ($-v_1 = v_L;\ -u_1 = u_L$) and centered on zero. The unknown probability distributions for the support vectors will be denoted as w_ε and w_ϵ and, consequently, the random errors are defined as:

$$\varepsilon_i = v w_{\varepsilon i} = \sum_{l=1}^{L} w_{\varepsilon il} v_l; \forall i$$

$$\epsilon_j = u w_{\epsilon j} = \sum_{l=1}^{L} w_{\epsilon jl} u_l; \forall j$$

Consequently, the GCE problem can be written in the following terms:

$$\begin{aligned}\underset{X, w_\varepsilon, w_\epsilon}{\text{Min}}\ D\left(P, w_\varepsilon, w_\epsilon \parallel Q, w_\varepsilon^0, w_\epsilon^0\right) &= \sum_{i=1}^{K}\sum_{j=1}^{T} p_{ij} \ln\left(\frac{p_{ij}}{q_{ij}}\right) \\ &+ \sum_{i=1}^{K}\sum_{l=1}^{L} w_{\varepsilon il} \ln\left(\frac{w_{\varepsilon il}}{w_{\varepsilon il}^0}\right) \\ &+ \sum_{j=1}^{T}\sum_{l=1}^{L} w_{\epsilon jl} \ln\left(\frac{w_{\epsilon jl}}{w_{\epsilon jl}^0}\right)\end{aligned} \tag{5.6a}$$

Subject to:

$$\sum_{j=1}^{T} p_{ij} = R_{i\cdot} + \sum_{l=1}^{L} w_{\varepsilon il} v_l; \forall i \tag{5.6b}$$

$$\sum_{i=1}^{K} p_{ij} = C_{\cdot j} + \sum_{l=1}^{L} w_{\epsilon jl} u_l; \forall j \tag{5.6c}$$

$$\sum_{l=1}^{L} w_{\varepsilon il} = 1; \forall i \tag{5.6d}$$

$$\sum_{l=1}^{L} w_{\epsilon jl} = 1; \forall j \tag{5.6e}$$

Note that both the bounds specified in the support vectors as well as the a priori probability distributions (w^0_ε and w^0_ϵ) reflect our assumptions on the way the errors are affecting the observed margins. Larger bounds in v and u would allow, obviously, for larger errors. In the context of GCE problems, the values of the supporting vectors for the errors are usually fixed following the three-sigma rule (Pukelsheim 1994), which in this case implies to take as upper and lower bound $\pm$ three times the standard deviation of $\tilde{R}$ and $\tilde{C}$ respectively; whereas the a priori distributions $\left(w^0_\varepsilon \text{ and } w^0_\epsilon\right)$ are set as uniform.

5.4 The Case of Mexico: Available Data for Economic Analysis at Small-Scale

Availability of economic data at small-scale varies very much depending on the country studied. In the specific case of Mexico both aggregated (at state level) and disaggregated (at municipal level) economic information is available in several surveys published by the National Institute of Statistical and Geographical Information(INEGI).

The INEGI performs population census every 10 years with population conteos (counts) 5 years after each population census. It performs economic census every 5 years and Anuarios Estadisticos (Statistics Yearbooks) every year, but with less detail as the censuses. They also perform the agricultural, livestock and forestry census (Censo Agropecuario) as well as the Censo Ejidal; which are done every 10 years (except from the 2007 censuses, which was done 6 years after the last censuses). Other surveys that provide economic data are the Household Surveys (Encuesta en Hogares). These surveys provide detailed information from households across Mexico, which study their characteristics, habits, living conditions, welfare and economic development. The most popular household surveys are the Encuesta

Nacional de Ingresos y Gastos de los Hogares (ENIGH) and the Encuesta Nacional de Ocupación y Empleo (ENOE). The ENIGH survey is done biannual since 1992; whereas the ENOE is done quarterly, starting from 2005. Apart from the Anuarios Estadisticos, all the surveys have information at a municipal level, but their periodicity its of 5 years or more; living researchers without any annual observable information to carry out regional studies.

If the researcher is interested in information observable at a municipal level, but not classified into industries on a yearly basis, the most appropriate source of information is the SistemaEstatal y Municipal de Bases de Datos (SIMBAD). This database puts together information from several different surveys elaborated by the INEGI and provides information at a municipal scale of different variables every 5 years. This information, however, does not allow for an industry classification, which is a problem if we are interested in analyzing the figures for some specific industry on any specific year. The only source of information that provides a yearly analysis by industry is the Anuario Estadistico. But this source only provides information at the State level.

If the research to be conducted requires the observation of data at a municipal level that refers to production, gross value added, wages, etc., the main source of data is the Censo Economico. The National Economic Census in Mexico is the main source of economic information for the INEGI, and gives the basis for the development of many other economic indicators. The information contained in the economic census refers to every economic unit in the country and it provides data for every level of geographic disaggregation and for each one of the more than 950 NAICS activities. Given the enormous amount of resources required to elaborate it, INEGI publishes one economic census every 5 years, corresponding the most recent one to 2009. The information on each census corresponds to the previous year; for example, the 2004 economic census presents information from the 2003 fiscal year, which runs from January 1st 2003 to December 31st 2003.

5.5 Estimates Versus "Real Data": Estimating Wages by Industry at Municipal Level in Four Mexican States

As commented in the previous section, state and industry aggregates are much easier to observe than the disaggregated data by industry and municipality contained in the Economic Census. In this context, it would be interesting the application of an estimation procedure that produce disaggregated values quicker than the official ones. For practical purposes, we have focused on the following four Mexican states located in the central area of the country and close to Mexico City: Aguascalientes, Guanajuato, Jalisco and San Luis Potosí (see Map 5.1). We have chosen this specific set of states because they constitute a quite heterogeneous group. For example, Aguascalientes is a small geographical

Map 5.1 Group of Mexican states analyzed

area with a population in 2010 of approximately 1.2 million people; most of them live in the municipality of Aguascalientes and the rest on the other ten municipalities that compose the state. On the contrary, Jalisco is composed by more than 100 municipalities and a population larger than 7.3 millions in 2010. The other two states lie in an intermediate position between these two cases. Map 5.2 shows this population distribution among all the municipalities of these four states. Table 5.1 summarizes the main demographic and economic indicators for the four states:

In our estimation problem we assume that the target to be estimated is the distribution of wages paid at the municipal level and classified by industry, considering a breakdown into five industries (we exclude agriculture). For this estimation exercise only some aggregated information is supposed to be observable. Specifically, for every state of our study, these aggregates are: (1) vector R with dimension ($T \times 1$, being T different in each case) that contains the municipal shares of wages paid across the state (equivalent to the row sums of Table 5.1); and, (2) vector $C(1 \times 5)$ given by the proportion of wages by industry at the state level (equivalent to the column sums of Table 5.1). From these aggregates, we will apply the entropy-based estimation strategy explained in previous sections to recover the distribution of wages by municipality and industry.

From this point of departure, the estimation problem could be written as a CE matrix adjustment as follows:

$$\underset{P}{\operatorname{Min}} D(P \parallel X) = \sum_{i=1}^{11} \sum_{j=1}^{5} p_{ij} \ln\left(\frac{p_{ij}}{q_{ij}}\right) \tag{5.7}$$

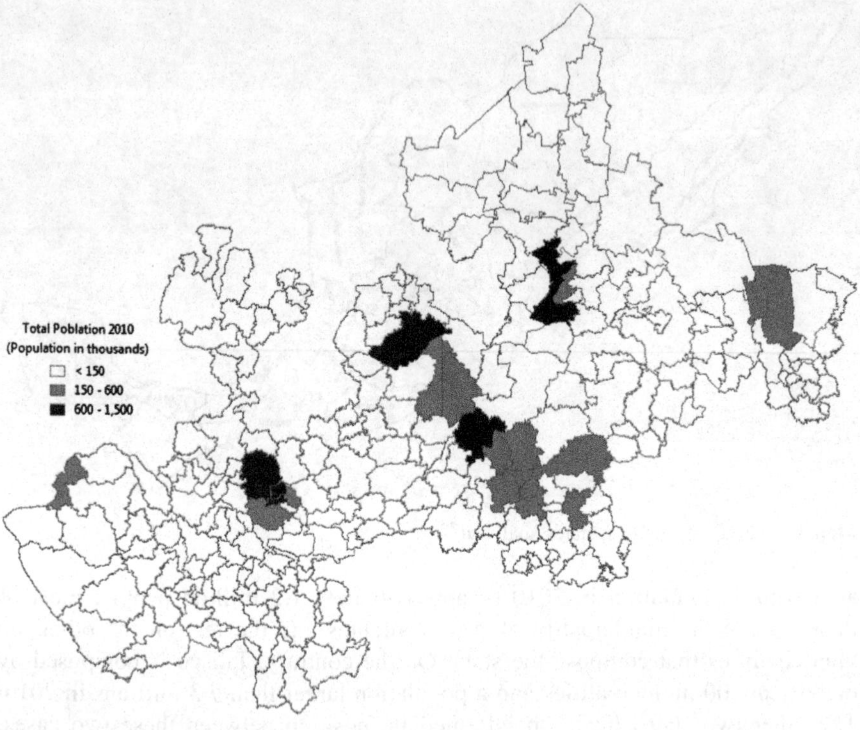

Map 5.2 Total population in 2010 by municipality

Table 5.1 Descriptive statistics of the states analyzed

State	Population in 2010	Surface (km^2)	Number of municipalities	Total wages in 2009 (million Pesos)
Aguascalientes	1,181,390	5,616	11	13,287.56
Guanajuato	5,456,936	30,607	47	43,756.55
Jalisco	7,266,952	78,588	125	68,637.83
San Luis Potosí	2,563,012	60,983	58	18,559.81

Source: INEGI, Economic Census 2009 and Population Census 2010

Subject to:

$$R_{i\cdot} = \sum_{j=1}^{5} p_{ij};\ \forall i = 1, \ldots, 11 \tag{5.8}$$

$$C_{\cdot j} = \sum_{i=1}^{11} p_{ij};\ \forall j = 1, \ldots, 5 \tag{5.9}$$

The above-sketched adjustment method assumes that we have perfectly observable information on margins C and R. Whereas assuming that vector C could be observable without error (the Anuario Estadistico provides official data yearly on wages by industry at state level), it is difficult to assume that we have perfectly reliable information on R, given that this information (distribution of wages by municipality) is only estimated in the Economic Census.

Some proxy variable, however, could be used instead. For example, in the Anuario Estadistico we can find data on the distribution of total labor by municipality, which can be supposed to be strongly correlated with the wages paid. In this context, the information of this variable is taken as vector $\tilde{R}$, being $\tilde{R} = R + \varepsilon$ and ε_i where each is defined as $\varepsilon_i = v'w_{\varepsilon i} = \sum_{l=1} w_{\varepsilon il} v_l$. In this context the following GCE problem can be solved:

$$\underset{X, w_\varepsilon}{\mathrm{Min}}\, D\left(P, w_\varepsilon \parallel Q, w_\varepsilon^0\right) = \sum_{i=1}^{11} \sum_{j=1}^{5} p_{ij} \ln\left(\frac{p_{ij}}{q_{ij}}\right) + \sum_{i=1}^{11} \sum_{l=1}^{L} w_{\varepsilon il} \ln\left(\frac{w_{\varepsilon il}}{w_{\varepsilon il}^0}\right) \tag{5.10}$$

Subject to:

$$R_{i\cdot} + \sum_{l=1}^{L} w_{\varepsilon il} v_l = \sum_{j=1}^{5} p_{ij}; \forall i = 1, \ldots, 11 \tag{5.11}$$

$$C_{\cdot j} = \sum_{i=1}^{11} p_{ij}; \forall j = 1, \ldots, 5 \tag{5.12}$$

$$\sum_{l=1}^{L} w_{\varepsilon il} = 1; \forall i \tag{5.13}$$

Note that the last term in Eq. 5.6a is removed in Eqs. 5.10 and 5.12 is identical to Eq. 5.9, since we assume that C is observed without error. For the sake of simplicity, only three points ($L = 3$) are included in the support vectors of the errors in R, which have been fixed using the three sigma rule and being always the central point equal to zero.

Although several options are possible for the specification of an a priori distribution X, we opted for taking as reference the values published in the previous Economic Census in 2004. This has the disadvantage that the estimates obtained for those cases (one specific industry in one municipality) where in 2004 there was no economic activity will be equal to zero. In other words, since the zeros in the Q matrix are kept, this means that the CE estimation cannot predict the "birth" of new industries in one municipality.[1]

[1] This problem can be solved, on the other hand, by specifying a prior different from the example used in our empirical analysis.

Table 5.2 summarizes the estimates of this GCE adjustment for each one of the Mexican states studied. The figures have been obtained as the respective estimate of p_{ij} multiplied by the total volume of wages paid at the state level as a whole (z). In order to have indicators of the accuracy of our estimates by municipality and industry, we computed the weighted absolute error percentage -WAPE-, a measure that is frequently used in the studies that evaluate the performance of matrix adjustment techniques. The municipal ($WAPE_i$), industry ($WAPE_j$) and total (WAP E_T) error indicators are defined as:

$$WAPE_i = \sum_{j=1}^{5} 100 \frac{|z_{ij} - \hat{z}_{ij}|}{\sum_{j=1}^{5} z_{ij}} \tag{5.14a}$$

$$WAPE_j = \sum_{i=1}^{T} 100 \frac{|z_{ij} - \hat{z}_{ij}|}{\sum_{i=1}^{11} z_{ij}} \tag{5.14b}$$

$$WAPE_T = \sum_{i=1}^{T} \sum_{j=1}^{5} 100 \frac{|z_{ij} - \hat{z}_{ij}|}{\sum_{i=1}^{11} \sum_{j=1}^{5} z_{ij}} \tag{5.14c}$$

where the $\hat{z}_{ij}$ elements denote the estimated wages.

Table 5.2 reports the respective $WAPE_j$ for each industry and state. Additionally, it shows the deviations at the municipal level for the largest and smallest municipalities in every state studied.

From the figures in Table 5.2 we can see a large variability in the accuracy of the estimates, but under the general rule that the largest cells produce comparatively smaller errors. At an industry dimension, the economic sectors with largest weights are those activities with the smallest errors. Oppositely, "Energy" with a relative low weight on the economic activity across the states studied, was the industry with the largest errors on average.

At a geographical dimension, this same pattern is observed as well, even to a greater extent. The errors range from more than 100 % for the smallest municipalities to less than 5 %. It is remarkable the low error (4.99 %) yield for Guadalajara, the capital city of Jalisco, which is the largest city of all the states analyzed concentrating almost 50 % of the wages and the employment in the state according to the 2009 Economic Census. Map 5.3 expands the summarized results reported in Table 5.2 showing the detailed distribution of absolute deviations.

In our results we found a negative correlation between the size (population) of the municipalities and the deviation with the actual values, as we can see in Fig. 5.2. In other words, the estimation performs comparatively better when applied for larger areas, where the errors are in general moderate, whereas they can be much bigger for the case of small villages and towns.

Table 5.2 Summary of deviations, GCE estimates (WAPE, %)

State		Energy	Manufacturing	Commerce	Services to companies	Other services	$WAPE_{mun}$
Aguascalientes	San José de Gracia	256.08	54.49	91.71	100.00	57.23	**72.42**
	Aguascalientes	15.56	9.03	7.02	6.24	4.16	**7.89**
	Total	**29.60**	**15.14**	**12.71**	**11.47**	**7.37**	**13.97**
Guanajuato	Santa Catarina	189.35	38.72	30.36	80.98	121.97	**66.16**
	León	175.35	6.11	8.64	35.24	3.70	**18.18**
	Total	**42.47**	**15.69**	**18.09**	**51.31**	**15.76**	**27.52**
Jalisco	Ejutla	99.99	26.84	44.93	23.46	59.86	**47.73**
	Guadalajara	6.31	2.62	7.07	5.69	4.76	**4.99**
	Total	**13.81**	**14.48**	**11.53**	**15.47**	**12.42**	**13.84**
San Luis Potosí	Cerro de San Pedro	98.82	N.C.	99.99	1371.49	N.C.	**101.28**
	San Luis Potosí	10.79	8.71	5.66	6.23	0.93	**7.32**
	Total	**21.52**	**17.54**	**12.71**	**12.65**	**6.54**	**15.13**

N.C.: non-computed because the wage in 2009 was zero in this industry and municipality

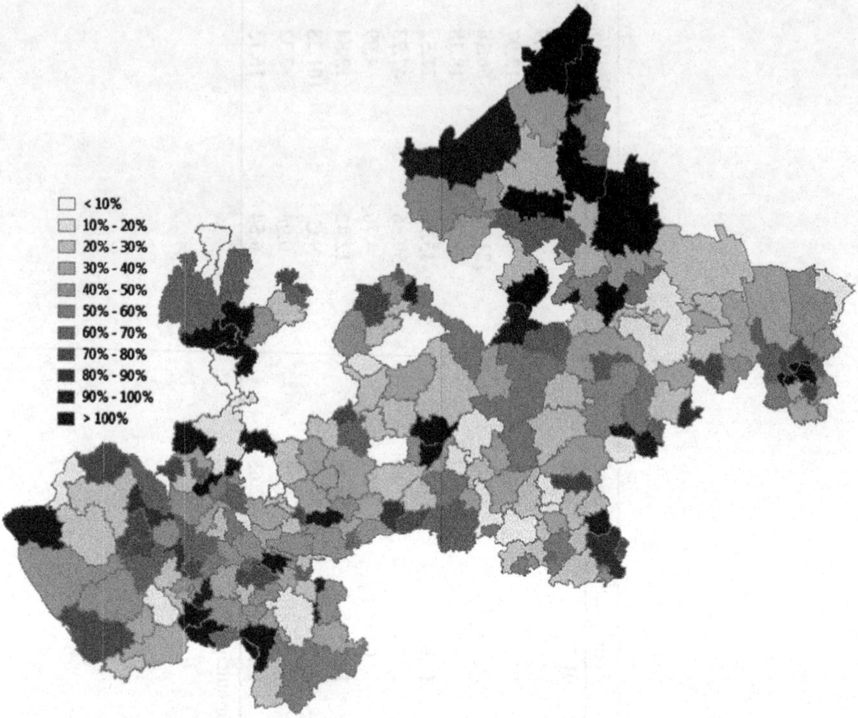

Map 5.3 Distribution of errors by municipality

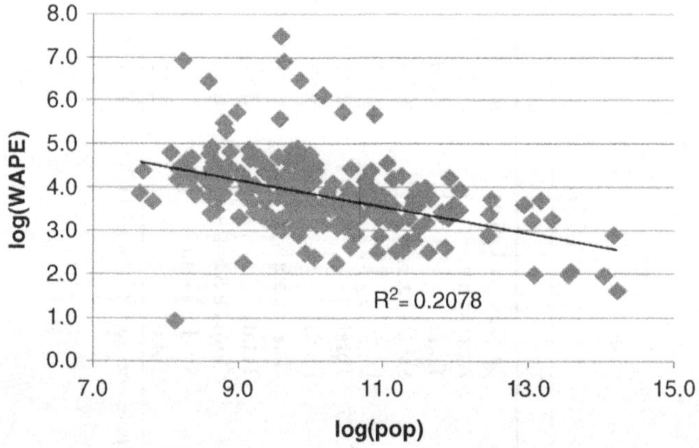

Fig. 5.2 Correlation between size (log of population) and error (log of WAPE)

5.6 Conclusions and Future Research

This chapter applies a methodology based on Entropy Econometrics to estimate data at a municipal scale and classified into industries. Our point of departure is observable information (a prior matrix) that reflects our beliefs about a possible distribution across municipalities and industries of the variable of interest. Besides, observable municipal and industry aggregates are required in order to put into practice the technique proposed. From these two pieces of information, a Cross Entropy (CE) adjustment is applied and a comparison between actual values and estimates is made for a case study. Specifically, the methodology proposed is evaluated with observable data from the Mexican Economic Census for 2009. The case of Mexico is an especially useful case, because official data are available with a high level of disaggregation, which is not usual in many other countries. Specifically, our objective was to estimate, by means of the CE procedure, the amount of wages paid by industry and municipality published by INEGI in the Economic Census of 2009 for a group of four Mexican states. Given that assuming perfect observable information on the municipal aggregates was quite unrealistic, we extend the standard CE adjustment in order to consider the possibility of errors in the margins and transform the problem into a Generalized Cross Entropy (GCE) estimation program. When we confront the estimates we obtained by applying this GCE technique with the actual values published in the Census we found that the general performance of the method seems to yield quite moderate errors, in spite of some cases where the errors were exceptionally large. Nevertheless, large errors are concentrated in smaller locations, which suggest that the proposed technique can be seen as a valid method for disaggregating spatial data, especially for the case of large cities.

The results obtained in our case study encourage further research in the area of ecological inference and spatial disaggregation by using entropy-based estimators. In the research agenda should be a systematic comparison of the GCE adjustment presented in this chapter with other estimation techniques frequently used for the same purpose, like the distributionally weighted regression techniques that require of disaggregated observations on the regressors included in the equations.

References

Duncan OD, Davis B (1953) An alternative to ecological correlation. Am Sociol Rev 18:665–666

Golan A, Vogel SJ (2000) Estimation of stationary and non-stationary accounting matrix coefficients with structural and supply-side information. Econ Syst Res 12:447–471

Golan A (2006) Information and entropy econometrics. A review and synthesis. Found Trends Econ 2:1–145

Goodman L (1953) Ecological regressions and the behavior of individuals. Am Sociol Rev 18:663–666

Judge G, Miller DJ, Cho WTK (2004) An information theoretic approach to ecological estimation and inference. In: King G, Rosen O, Tanner MA (eds) Ecological inference: new methodological strategies. Cambridge University Press, New York, pp 162–187
King G (1997) A solution to the ecological inference problem: reconstructing individual behavior from aggregate data. Princeton University Press, Princeton
King G, Rosen O, Tanner MA (2004) Ecological inference: new methodological strategies. Cambridge University Press, Cambridge
Pukelsheim F (1994) The three sigma rule. Am Stat 48:88–91
Robinson S, Cattaneo A, El-Said M (2001) Updating and estimating a social accounting matrix using cross entropy methods. Econ Syst Res 13(1):47–64

Chapter 6
GDP Estimates for Regions Within the Province of Quebec: The Changing Geography of Economic Activity

André Lemelin, Pierre Mainguy, Daniel Bilodeau, and Réjean Aubé

6.1 Introduction

The purpose of this chapter is twofold. First, we report on a method developed at the Institut de la statistique du Québec (ISQ) to estimate the GDP of regions within the Province of Quebec. And, second, we analyse the estimates to examine the recent evolution of the geographical pattern of economic activity in Quebec.

There are two families of methods for calculating regional GDP. So-called "bottom-up" methods consist in collecting economic data at the individual production-unit level (establishment), and then adding them up to obtain the corresponding regional data. Various adjustments are then performed in order to make the regional data consistent with national data, so that the sum of regional products is equal to total production over the national territory. So-called "top-down" methods consist in allocating an overall national figure across regions. They do not require knowledge of local establishment data. The national figure is distributed using an indicator which is as close as possible to the variable to be estimated. Practically speaking, most methods are mixed. For, on the one hand, the kind of data required for bottom-up estimation almost always has gaps that must be filled using a top-down method. And, on the other hand, top-down methods also make use of exhaustive data sources similar to those required by bottom-up methods.

A. Lemelin (✉)
INRS-UCS, Universite du Québec, 385, rue Sherbrooke est, Montréal, Québec H1M3W2, Canada
e-mail: andre_lemelin@ucs.inrs.ca

P. Mainguy
Consultant, Québec, Canada

D. Bilodeau • R. Aubé
Direction des statistiqueseconomiqueset du développement durable, Institut de la statistique du Québec, Québec, Canada

E. Fernández Vázquez and F. Rubiera Morollón (eds.), *Defining the Spatial Scale in Modern Regional Analysis*, Advances in Spatial Science,
DOI 10.1007/978-3-642-31994-5_6, © Springer-Verlag Berlin Heidelberg 2012

6.2 A Brief Survey of Methods[1]

There are few examples of regional GDP computed for entities smaller than States or provinces, except in the European Union (see below). In Canada, the Conference Board produces annual estimates of the GDP of metropolitan areas (MAs).[2, 3] The Conference Board first estimates gross value added at basic prices in real terms (at constant prices) for some 60 industries (depending on the information available for each MA), using monthly employment data from Statistics Canada's Labour Force Survey (LFS). Metropolitan GDP is then obtained as the sum of gross value added over industries. But the LFS generates data on a place-of-residence basis, rather than on a place-of-work basis, whereas GDP is defined, according to national accounting principles, in terms of where production takes place. So the Conference Board method adjusts its employment figures to take commuting into account, using population census data: employment in a MA is obtained by multiplying the LFS figure by the ratio of the number of workers whose place of employment is in the MA, over the number of workers whose residence is in the MA. Industry gross value added is obtained by multiplying estimated employment in each MA by labour productivity at the Provincial level. So the underlying hypothesis is that labour productivity is the same within any industry everywhere in a given Province. This may constitute a weakness in the case of industries encompassing a wide range of activities when the mix of activities varies from one region to another. Finally, the Conference Board method could hardly be applied to all regions, given that LFS data are unreliable for small regions or for small industries, because of high error margins in small sub-populations, not to mention Statistic Canada's confidentiality rules, which can result in masked data.[4]

In the United-States, until recently, the Bureau of Economic Analysis (BEA) did not estimate GDP at a lesser scale than the State level.[5] However, since September

[1] For a more detailed survey, see Lemelin and Mainguy (2009b).

[2] Technically, Census Metropolitan Areas (CMAs), following the statistical system in Canada. It is understood that, throughout this text, metropolitan areas are Census Metropolitan Areas. We shall use the full expression "Census Metropolitan Areas" when discussing the definition of MA boundaries by Statistics Canada.

[3] Those estimates are published in the spring issue of the Conference Board's quarterly Metropolitan Outlook/Note de conjoncturemetropolitaine.

[4] Statistics Canada defines a minimum threshold below which no information may be disseminated (Statistics Canada 2011, 71-543-G, p 31, "Release criteria"). For Quebec, that threshold is 1,500. It follows that for small CMAs, some industries "disappear" at times, only to reappear later on, just because the number of employees has temporarily fallen below the confidentiality threshold. This limits the level of detail at which the method is applicable and occasionally forces to make adjustments.

[5] The BEA method for computing Gross State Products (GSP), like Statistics Canada's method for provincial GDP, is a mixed, bottom-up/top-down method which makes use of fiscal and administrative data.

2007, the BEA publishes GDP estimates for metropolitan areas.[6] The BEA methodology uses a top-down approach, distributing state-level output by industry to metropolitan areas according to earnings (reported by place of work). Earnings – which consist of wage and salary disbursements, supplements to wages and salaries, and proprietors' income – are estimated on the basis of data from the Quarterly Census of Employment and Wages of the Bureau of Labor Statistics (BLS). The BEA also produces statistics on personal income for counties. Now, personal income is the sum of incomes of persons who live in a given area. Since county personal income is computed mostly from data collected on a place-of-work basis, the BEA applies a correction using decennial census data on home-to-work commuting. The adjustment is interpolated or extrapolated to non-census years, following a method described in BEA (2008).

Within the European Union, the rules for distributing so-called "structural" funds (in support of poorer regions) requires knowing regional GDP. Calculations follow common principles laid down by Eurostat, the EU statistical agency. But it must be kept in mind that the regions of Europe are of a much greater demographic and economic weight than most of Quebec's 17 administrative regions. Particular attention was paid to the techniques of the French Institut National de la Statistiqueet des EtudesEconomiques (INSEE), and of the UK Office for National Statistics (ONS).

In France, INSEE applies a method which rests on a complex system of enterprise data, and calls upon the expertise of "regional accountants", whose local presence and knowledge of the environment make it possible to better validate the information (Delisle et al. 2000).[7] The INSEE method is mixed, but predominantly bottom-up. It would seem to be more accurate than the ONS's (described below), but also more demanding. That is probably why it is fully applied only for certain benchmark years, on the basis of which other years are estimated by inter- or extrapolation.

In the United Kingdom, the ONS applies a method which is quite similar to that of the US BEA for estimating GSP, and to that of Statistics Canada for provincial GDP. It appears however that the top-down side is more important in the ONS method (Lacey 2000). The main shortcoming of that method is that wages and salaries are allocated on a place-of-residence basis, rather than on a place-of-work basis, as the concept of gross domestic product implies. On the other hand, the data requirements of the ONS method are moderate, especially when compared to INSEE's.

The method developed for estimating the GDP of Quebec's 17 administrative regions and 6 metropolitan areas is described in greater detail in the next section. It is

[6] Panek et al. 2007. The 2001–2009 estimates of GDP by metropolitan area in current and real (chained) dollars are available from the Regional Economic Accounts page of the BEA Web site at http://www.bea.gov/regional/index.htm.

[7] A brief description of the INSEE method can be found at http://www.insee.fr/fr/themes/detail.asp?reg_id=99&ref_id=pib-va-reg

a mixed method, closely akin to that of the ONS in the UK. Regional GDP at basic prices is computed by industry or group of industries, following the incomebased approach, defined in the OECD System of National Accounts as the sum of components of value added (OECD 2011). The key ingredients in the method are a compilation of fiscal data on incomes, and reliable home-to-work commuting tables by industry. We believe a similar approach could be used to estimate regional GDP anywhere these are available. Our method has the advantage of being less limited than the Conference Board method by small-region or small-industry sampling errors in labour survey data, and it implicitely takes into account inter-regional productivity differences within industries, since it relies on income, rather than employment data. Compared to the French INSEE method, ours is certainly less demanding. And compared to the British ONS method, it is consistent with the GDP place-of-production definition, and it has the advantage of using all of the detailed fiscal data, rather than the 1 % sample of tax records, compiled by Inland Revenue. We now proceed to give a more complete account of the method. In Sect. 6.2, we will examine the evolution of regional GDP from the core-periphery perspective.

6.3 ISQ Regional GDP Estimation Method

6.3.1 Broad Outline

The process of applying the method can be summarized as follows:

1. The starting point is the Quebec total, to be allocated between the regions: value added (VA) at basic prices, by industry and by component, in current dollars, according to Quebec Economic Accounts. They are the "target data".
2. Data on regional distribution are obtained from Revenu Québec, which extracts them from individual income tax returns:
   - Wages and salaries, by administrative region of residence and by Standard Industrial Classification (SIC) industry;
   - NIUB, by administrative region of residence and by SIC industry until 2000, and by North-American Industry Classification System (NAICS) industry afterwards.
3. Revenu Québec data undergoes two conversions before being used to form allocators:
   - Data by SIC industry are converted to the NAICS.
   - Place-of-residence fiscal data are converted to place-of-work data using home-to-work commuting patterns by industry (special tabulation by Statistics Canada, based on population census data).
4. The table of NIUB by NAICS industry and by region is adjusted, applying the minimum-information-gain method (also known as minimum cross-entropy), so as to exploit all of the information in Revenu Québec's fiscal data, accommodating for the relatively high rate of missing information relating to industry of

origin. That procedure makes it possible to use NIUB data for which the region of residence is known, but not the industry.

5. Wages and salaries and adjusted NIUB, by industry and by region, are used as allocators for the other components of VA:
   - Supplementary labour income is distributed in proportion to salaries;
   - Other components are distributed in proportion to the total of salaries, supplementary labour income, and NIUB.
6. VA by administrative region (that is, regional GDP) is obtained by first summing the components within each industry, and then summing the VA of industries.

In addition, eight industries are treated in a special way: Fishing, hunting and trapping; Metal ore mining; Non-metallic mineral mining and quarrying; Construction; Petroleum and coal products manufacturing; Primary metal manufacturing; Lessors of real estate property; Owner-occupied dwellings. We shall come back to those later. Finally, the top-down approach of the method ensures that estimated regional GDP is consistent with provincial economic accounts.

6.3.2 Geographical Divisions in the Estimation Process

The Province of Quebec is divided into 17 administrative regions (AR). There are also 21 regional conferences of elected officials (ConférencesRegionales des Élus – CRÉ), where the mayors and prefects of the region meet (since it is an institution unique to Quebec, we shall not attempt a translation, and will henceforth use the acronym CRÉ). Fifteen of the 17 administrative regions coincide with the territory of a single CRÉ, but two of them have three each. Besides the administrative regions, there are six metropolitan areas (MA) in Quebec. Two of the administrative regions are entirely included within the Montreal metropolitan area, while seven others are partly inside, partly outside a metropolitan area. Finally, one metropolitan area, Gatineau, is part of the larger Ottawa-Gatineau metropolitan area which extends into the neighboring Province of Ontario.

The objective in producing regional data was to provide local authorities with information on which to base their development strategies. So it was desirable to estimate GDP for all of the territorial divisions described above. To make that possible, the Quebec territory was subdivided into 30 areas which can be aggregated into administrative regions, CRÉ territories,[8] or metropolitan areas. It was possible to obtain the target data and the Revenu Québec fiscal data for each of the 30 territories. So the GDP estimates are perfectly consistent for any aggregation of the 30 territorial subdivisions. However, for the eight special industries mentioned above, two separate calculations must be made, one for the 17 administrative

[8] Except for the three regional conferences of the Nord-du-Québec administrative region, whose economic and demographic weight is too small to subdivide the region.

regions, with one of them divided into three regional conferences, and another for the six metropolitan areas and the non-metropolitan territories.

6.3.3 *Steps in the Estimation Process*

This section examines the estimation process in greater detail.[9] In what follows, we consider successively: source data used (target data on GDP by industry, and Revenu Québec fiscal data); prior processing applied to Revenu Québec data to construct the allocators; and application of the allocators to the target data.

6.3.3.1 Target Data on GDP by Industry

Data on GDP by industry and by component for Quebec as a whole are the "target data", which are to be distributed among regions by the estimation process. Indeed, the estimation results must be consistent with other official data. Such consistency is ensured because the totals distributed among regions with allocators correspond to the official figures. However, these target data are not drawn as such from a single source; rather, they are constructed from three main sources: (1) GDP at basic prices, in current dollars, by industry and by province, in current dollars (Statistics Canada, Provincial Gross Domestic Product by Industry, 15-203-XIE); (2) Statistics Canada's Provincial Input–output tables (IO) for Quebec; (3) GDP at basic prices, by component, for 18 industry groups, in current dollars, estimated by the Institut de la statistique du Québec (Compteseconomiques des revenus et dépenses du Québec).

Industry GDP at basic prices in current dollars is consistent with Economic Accounts. So it is a good starting point for estimating regional GDP by a top-down method. Indeed, the choice of estimating regional GDP in current, rather than constant, dollars was made with the purpose of obtaining a regional GDP which is consistent with Economic Accounts. Moreover, the regional GDP estimation method uses allocators based on Revenu Québec fiscal statistics, which are, obviously, in current dollars.

The level of aggregation chosen for the estimation of regional GDP is 63 NAICS industries. At that level of detail, however, the first source does not disaggregate industry GDP into its components. So if one were to rely on that single source, one would be forced to apply the same regional allocator to all of each industry's value added. For that reason, industry GDP data are used jointly with data from other sources. In Compteseconomiques des revenus et dépenses du Québec, the Institut de la statistique du Québec publishes the value of GDP at basic prices for 18 industry groups, with three components: total labour income (wages and salaries, plus supplementary labour income); gross operating surplus and miscellaneous

[9] For an even more detailed description, see Lemelin and Mainguy (2005, 2009a).

adjustments; and the total of accrued net income of farm operators from farm production and net income of non-farm unincorporated business, including rent (mixed income). These data are utilized jointly with the input–output tables, to overcome certain classification and Economic Account benchmarking problems, and to obtain the necessary information to distribute the GDP of the 63 industries between the three value added components mentioned earlier. The result of those adjustments constitutes what we call the "Economic Accounts target data".

Finally, note that the source data on GDP by industry is published with a certain time-lag, and in several stages: preliminary estimates, revised estimates, etc. The definitive figures are often published only after several years. Consequently, the target data for the estimation of regional GDP in recent years must be projected on the basis of available source data, and the GDP estimates are revised as better and more complete data become available.

6.3.3.2 Revenu Québec Fiscal Data

Revenu Québec fiscal data on salaries and individual business income are used to construct the allocators according to which the target values are distributed between regions.[10] So they are the two main allocators utilized. Other allocators are used for special industries.

6.3.3.3 Salaries

The data are extracted from R1 slips[11] (equivalent to Canadian federal T4 slips). By combining the amounts of worker incomes according to the R1 slips, employer economic activity codes, and employees' postal codes, Revenu Québec produces an estimate of the amount of salaries by activity and by territory of employee residence. Those data are quite complete. For instance, in 1997, Revenu Québec was able to determine taxpayer territory of residence and employer industry for 93 % of R1 slips, amounting to 95 % of the total value of salary income according to Revenu Québec fiscal data. It is interesting to note that the total value of salaries according to Revenu Québec data is very close to the Wages and salaries component of GDP according to the Economic Accounts: the total value reported by Revenu Québec in 1997 represents 99.9 % of wages and salaries (excluding supplementary labour income) according to the Economic Accounts. The way Revenu Québec fiscal data are utilized is quite similar to what the United Kingdom's ONS does. But the ONS

[10] Fiscal data are compiled by Revenu Québec, and passed on to the Institut de la statistique du Québec in an aggregated form that complies with the confidentiality rules which protect personal information.

[11] R1 slips are issued by employers to employees, with a copy forwarded to Revenu Québec. They contain information on the worker's labour income to be entered in his/her income declaration for income tax purposes.

builds its estimates on the basis of the 1 % sample of tax records, compiled by Inland Revenue. In contrast, Revenu Québec fiscal data are based, not on a sample, but on all of the R1 slips.

6.3.3.4 Net Income of Unincorporated Business (NIUB)

The net income of unincorporated business (NIUB), also called mixed income, corresponds to net individual business income (revenu net des particuliers en affaires) in personal income tax returns. It is taken from taxpayer income declaration form TP1 (equivalent to Canadian federal T1 form). It includes accrued net income of farm operators from farm production as well as the net income from other types of unincorporated business.

However, the completeness rate of the economic activity code is not entirely satisfactory: for instance, in 2003, the industry of origin could be identified for no more than 71.9 % of taxpayers declaring individual business income (70.6 % of the total value of individual business income declared). Moreover, when the two criteria, region of residence and activity code, were combined, data were complete for only 71.7 % of taxpayers declaring individual business income in 2003 (amounting to 70.6 % of the total value of individual business income declared). It should be pointed out that the completeness rate has been improving year after year since 1997; for that year, data was complete for only 41 % of taxpayers declaring individual business income. Let us mention that, for 2003 again, the total value of NIUB according to Revenu Québec fiscal data represents 96.8 % of mixed income according to the Economic Accounts (after subtracting net rent imputed to owners occupying their own dwelling). But if one retains only the NIUB for which the taxpayer's region of residence and activity code are known, only 69.3 % of the Economic Account NIUB remains. To try and compensate for the gaps in Revenu Québec NIUB fiscal data, an adjustment procedure, described below, is applied: it allows to use all the available information, including NIUB for which the region of residence is known, but not the economic activity code.

6.3.3.5 Conversions Applied to Revenu Québec Fiscal Data

Two conversions are applied to Revenu Québec fiscal data: (A) data by SIC industry are converted to the North-American Industry Classification System (NAICS); (B) place-of-residence fiscal data are converted to place-of-work data using home-to-work commuting tables by industry based on data from the population census. Now, while the 1996 census was classified according to the SIC, the 2001 census follows the NAICS. Therefore, in the period 1997–2000, conversion B must be performed before A, but for 2001 and the following years, A must precede B.

6.3.3.6 From the SIC to the NAICS

Revenu Québec fiscal data are converted from the 1980 SIC to the North American Industrial Classification System (NAICS), to make them comparable to Economic Accounts target data: that way, Revenu Québec data for 66 SIC industries are converted to 63 NAICS industries.[12] Conversion from one classification system to another is never perfect. In principle, it could be, at a very fine level of detail. However, given the level of aggregation at which Revenu Québec fiscal data are available, the correspondence is necessarily imperfect, particularly for service industries. Although every SIC industry corresponds mostly to one particular industry in the NAICS and vice-versa, parts of any SIC industry are usually distributed into several NAICS industries; conversely, each NAICS industry is generally made up of parts of several SIC industries. Thus, the conversion matrix can be considered a table of average distributions: the value added produced by a given SIC industry is distributed between NAICS industries following its average distribution. Most of the conversion matrix data come from a Statistics Canada table based on data from the Survey of Employment, Payrolls and Hours (SEPH). That table has been constructed from survey data in each province, for two periods of three consecutive months in 1998, and for another 3-month period in 1999, and also from information drawn from Statistics Canada's Business Register. Conversion data based on the SEPH were completed thanks to another one of Statistics Canada's conversion tables, concerning only manufacturing industries, and constructed from 1996 manufacturing shipments data.

6.3.3.7 From Place-of-Residence Data to Place-of-Production Data

The initial data on salaries and NIUB by industry and region of residence must also be converted to data by region of production, using home-to-work commuting tables by industry, for different categories of workers. Those tables are computed by Statistics Canada, from the 20 % sample of population filing the long census questionnaire.[13] The industrial classification applied in the 1996 census of population was the 1980 SIC, but it was the NAICS for the 2001 census. Commuting trips by salaried workers correspond to those of Paid workers, while commuting trips by

[12] Recall that, since 2001, Revenu Québec classifies NIUB data according to the NAICS, so that this conversion is now only necessary for fiscal data on salaries.

[13] For narrow ideological reasons, the governing Conservative Party of Canada has decided to make the long census questionnaire optional, beginning with the 2011 census. This was done against the advice of Statistics Canada, and in spite of widespread protest from numerous organizations across the country who need reliable data, including Provincial and local administrations. Statistics Canada will attempt to maintain the validity of the census, by distributing the long questionnaire to 40 % of the population, rather than 20 %. But it is doubtful that this will be successful in eliminating self-selection biases and ensuring an adequate representation of all categories of persons.

unincorporated business operators correspond to those of the Self-employed (unincorporated). The general principle of the conversion is simple: the home-to-work commuting tables show how the workplaces of the residents of a given territory who work in a given industry are distributed over the territories; the total income earned by those residents, according to Revenu Québec, is therefore distributed between regions of production in the same proportions. The underlying hypothesis is that, within a given industry, the average income per worker (wages and salaries, or NIUB) is the same for all residents of a territory, no matter which territory they work in.

6.3.3.8 Minimum Information-Gain Adjustment of NIUB Allocators

In the case of NIUB, a further adjustment is applied. As mentioned above, the completeness rate of the economic activity code is not entirely satisfactory, so that if only complete data were used, that would leave aside all the information contained in the significant fraction of tax returns for which territory of residence is known, but not activity code. In order to make full use of all the information contained in Revenu Québec fiscal data on NIUB, the conversions just described are applied, not only to NIUB by industry, but also to total NIUB (including NIUB for which industry is unknown). An adjustment based on information theoretic principles is then performed (Theil 1967). The first step in the adjustment is to hierarchize information, according to reliability: (1) It is imperative to respect Economic Accounts target data concerning NIUB by industry: they are considered the most reliable figures. (2) In second place comes the distribution among production territories of total NIUB compiled by Revenu Québec, and converted following the procedures described above (in 2003, that represented 99.7 % of the total amount of individual business income declared). (3) Last come the distributions among production territories of NIUB by industry, also compiled by Revenu Québec, and converted (in 2003, that represented 70.6 % of the total amount of individual business income declared). The hierarchy having been established, one proceeds with the adjustment itself, taking into account, however, that the adjustment process excludes values relating to the eight special industries. Second- and third-rank information is first made consistent with first-rank information by proportional adjustments. At that stage, second- and third-rank data have not yet been harmonized with one another. In the final step, first-rank data (industry totals) and previously adjusted second-rank data (regional totals) act as constraints controlling third-rank data adjustment.

That final adjustment is performed following the minimum information-gain principle (also known as minimum cross-entropy), which is an operational form of the rule of scientific neutrality (Jaynes 1957; Golan et al. 1996; Kapur and Kesavan 1992). The standard cross-entropy minimization technique considers the structure of the matrix to be adjusted as if it were a probability matrix. Cross-entropy is a measure of distance between the a priori probability distribution, i.e. the unadjusted matrix, and the posterior distribution, i.e. the adjusted matrix. The adjustment process then consists in adjusting the matrix to its marginal totals in such a way as to minimize that distance, which is readily interpreted in information theory as minimizing the quantity of extraneous information imposed upon the a priori matrix.

However, NIUB is negative for some industries in some regions, so the standard technique is not applicable. So, following Günlükenesen and Bates (1988), Junius and Oosterhaven (2003), and Lemelin (2010), the minimum information-gain approach is generalized to the case in which there may be negative entries. Essentially, we consider the initial matrix as the term-by-term product of two matrices: a matrix of +1 and −1 values, respectively corresponding to positive and negative values in the data, and the matrix of absolute values of the data. The first matrix is taken to be intangible (certain) information, while the second plays the part of a priori matrix, to which the minimum cross-entropy principle can be applied directly. The following optimization problem is solved in GAMS (Rosenthal 2010).

$$MIN \sum_i \sum_j |a_{ij}| \, z_{ij} \ln z_{ij},$$

subject to

$$\sum_j a_{ij} z_{ij} = u_i$$

$$\sum_i a_{ij} z_{ij} = v_i \tag{6.1}$$

where: $z_{ij} = x_{ij}/a_{ij}$; x_{ij} is the adjusted (posterior) amount of NIUB of industry i in region j; a_{ij} is the a priori amount of NIUB of industry i in region j, according to the converted Revenu Québec data; u_i is total NIUB of industry i in all regions, according to the Economic Accounts (target data); and v_j is the total amount of NIUB of all industries in region j, according to Revenu Québec data, after proportional adjustment to the target data total. Solving the above problem yields the table of NIUB by industry, for the 30 territorial divisions defined above.[14]

6.3.3.9 Applying the Allocators to GDP Components

The procedure began with two tables of income by SIC industry and by territory of residence obtained from Revenu Québec, one for salaries, and one for NIUB. The conversions and the adjustment just described yielded tables of salaries and NIUB by NAICS industry and territory of production. The rows of these tables are then

[14] For further details, see Annex 3 in Lemelin and Mainguy (2009a).

used to allocate total labour income (wages and salaries, plus supplementary labour income), and NIUB among regions, by industry, for all but the eight special industries mentioned earlier. The remainder of GDP at basic prices is aggregated under the label of "Other operating surplus" (OOS),[15] and is distributed among regions (by industry) using as an allocator the sum of total labour income and NIUB as estimated in the previous steps.

6.3.3.10 Distributing the GDP of Special Industries

Eight industries are dealt with in a special way. First, the estimation method applied to the other industries is inapplicable to Construction and to Owner-occupied dwellings, for reasons to be given below. As for the six other industries, it was decided to treat them, in part or totally, as special, on account of practical difficulties in applying the method.

Fishing, hunting and trapping (NAICS industry 114). Wages and salaries, and supplementary labour income from this industry are distributed between regions just like those of other industries. But the value of this industry's NIUB in 2003 Revenu Québec fiscal data is only 0.9 % of the same industry's NIUB in the Economic Accounts, while it makes up 27.5 % of the value added of the industry. Given that the bulk of that industry's value added comes from commercial fishing, its NIUB and OOS are distributed following the value of fish landed in each region (Fisheries and Oceans Canada, Quebec Marine Fisheries. Annual Statistics Review).

Metal ore mining (NAICS industry 2122) and Non-metallic mineral mining and quarrying. (NAICS industry 2123). Initial experiments in applying the regional GDP estimation method produced anomalous results for these industries (in particular, a significant presence of mining in a region that has no mines). Careful examination revealed that there were probably misclassifications in Revenu Québec's fiscal data, due to the fact that the activities of large corporations are classified by the reporting entities themselves. Misclassification is all the more likely between closely related industries, such as Metal ore mining and Primary metal manufacturing. So salaries in these two industries are distributed using microdata from the Census of mines, quarries and sand pits. These data are collected by the Institut de la statistique du Québec, in collaboration with Natural Resources Canada. In all other respects, these industries are treated like the others.

Construction (NAICS industry 23). By its very nature, that industry has a high percentage of workers without a fixed place of work, so that it would be inappropriate to distribute them proportionally to those whose place of work is known, all the more since, even if the regional distribution of construction workers were

[15] OOS therefore includes (Statistics Canada 13-213-PPB): Interest and miscellaneous investment income; inventory valuation adjustment; indirect taxes on production, net of subsidies on production (taxes and subsidies on production are not to be confused with taxes and subsidies on products: the latter are not taken into account in the calculation of GDP at basic prices); and, finally, capital consumption allowances (or depreciation).

known for census years 1996, 2001 and 2006, it would most likely have changed as old construction projects were completed and new ones began, in the years following census years. Thus, it was decided to distribute the value added of the construction industry using a special allocator: capital and repair expenditures, except on machinery and equipment, by administrative region or by metropolitan area.[16]

Petroleum and coal products manufacturing (NAICS industry 324). Petroleum refining represents 90 %–95 % of that industry in the Province of Quebec, and refining capacity is located in two regions: Chaudière-Appalaches and Montréal. Initial experiments in applying the regional GDP estimation method led to an underestimation of the share of Chaudière-Appalaches, and to locating a significant share of the industry in a region without refining capacity, namely Mauricie. So this industry is allocated among regions proportionately to value added according to the microdata of the Annual Survey of Manufactures and Logging.[17]

Primary metal manufacturing (NAICS industry 331). Initial experiments in applying the regional GDP estimation method produced anomalous results for this industry. Careful examination revealed that there were probably classification errors in Revenu Québec's fiscal data. Moreover, the anomalies seemed related to those of the Metal ore mining industry. So this industry was treated in the same way.

Lessors of real estate property (number 5A03 in Statistics Canada's NAICS-based input–output industry classification). In some regions, that industry's NIUB has an ups-and-downs pattern, even jumping from positive to negative values from one year to the next. And because of the peculiar structure of value added in that industry, such instability would result in rather wild fluctuations. So the wages and salaries of that industry are distributed in the same way as in other industries, but the NIUB and OOS are distributed according to total real estate tax base by administrative region or metropolitan area (the database "Évaluationsfoncières des municipalités du Québec" is provided to the Institut de la statistique du Québec by the ministère des Affaires municipales et des Régions).

Owner-occupied dwellings (number 5A04 in Statistics Canada's NAICS-based input–output industry classification). There is no Revenu Québec fiscal data whatsoever relating to that industry, because the imputed rents of owner-occupied dwellings are just that: imputed. That industry is bound to be a special one. Also, the stock of dwellings changes from year to year, at different rates in different regions. To take these factors into account, the value added of owner-

[16] *Immobilisations et réparations des secteursprivé et public, par région administrative*, Québec, years 1997 and following, Institut de la statistique du Québec, Direction des statistiques-économiques et du développement durable. Source: Statistics Canada, Investment and Capital Stock Division. Compilation: Institut de la statistique du Québec. These data may be downloaded from the Banque de données des statistiquesofficiellessur le Québec (BDSO): http://www.bdso.gouv.qc.ca/.

[17] More information on the *Annual Survey of Manufactures and Logging* at: http://www.statcan.gc.ca/cgi-bin/imdb/p2SV_f.pl?Function=getSurvey&SDDS=2103&lang=en&db=imdb&adm=8&dis=2

occupied dwellings is divided into two or more components, depending on the number of years since the last population census: (1) the industry's GDP for the census year (1996 or 2001) is distributed in proportion to the value of owner-occupied dwellings in each region, according to the 1996, 2001 or 2006 census of population (the special compilation by Statistics Canada, on the basis of the 20 % sample responding to the long form of the census questionnaire, may be consulted on the Institut de la statistique du Québec Web site); (2) the increase of that industry's GDP between the last census (1996, 2001 or 2006) and the current year is distributed, year by year, in proportion to the value of residential building permits for the corresponding year and the preceding one, so as to take into account the time-lag between permit emission and building completion (data published on the Institut de la statistique du Québec Web site). Finally, the regional value added of owner-occupied dwellings is simply the sum of its components.

6.3.3.11 Final Calibration

No final calibration is needed: from the outset, the method ensures that estimated regional GDP are consistent, both with Gross domestic product by industry in Quebec, and with Quebec Economic Accounts.

6.4 Core and Periphery in the Province of Quebec

Regional economic disparities are a long-standing policy issue in Canada, as in many other countries. Polèse and Shearmur (2002) have conducted a major project to assess the current situation and future prospects of peripheral regions in Canada.[18] Their analysis, based on 1971–1996 census data (Shearmur 2001) leads the authors to conclude that "The peripheral share of total (Canadian) employment and population will continue to fall in the foreseeable future. Employment and population will continue their gradual shift towards central locations, in and around major metropolitan areas." (Polèse and Shearmur 2002, p. 189).

In what follows, we use the 1997–2009 estimates of the GDP of Quebec's administrative regions and metropolitan areas, together with population and personal income data, to further contribute to a better understanding of the evolution of central and peripheral regions in Quebec.

[18] Six of the 11 study regions in Polèse and Shearmur (2002) are in Quebec: Abitibi-Témiscamingue, Côte-Nord, Gaspésie, Saguenay-Lac-Saint-Jean, and two subdivisions of the Bas-Saint-Laurent administrative region.

6.4.1 Geographical Divisions for Analysis

While metropolitan areas are an analytically meaningful category, the administrative regions have been defined for other purposes. So, for the objective of analyzing the changing geography of economic activity, we define 16 regions, which we shall call "analytical regions". The analytical regions are defined in such a way that their GDP can be obtained by summation or subtraction from the estimates published by the ISQ for administrative regions (AR) on one hand, and for metropolitan areas (MA) on the other. We shall also use ISQ population figures and estimates of personal income, published for the same geographical divisions as GDP estimates.

Every MA constitutes an analytical region, except for the Montreal metropolitan area, which is subdivided in three, a refinement made possible by the fact that two administrative regions, Montreal AR and Laval AR, are embedded within the metropolitan area. Next, six analytical regions consist of the non-metropolitan area surrounding a metropolitan area. These are made up of the non-metropolitan parts of one or more administrative regions having common territory with the MA. In some cases, the boundaries of administrative regions extend far away from the MA, and include areas which can hardly be qualified as "surrounding" the MA.[19] While this might be viewed as a shortcoming of the data, it is strongly mitigated by the fact that population density is very low, and economic activity generally weak in the remote parts of the ARs. Finally, we define two peripheral regions, Rest-of-the-North and East. Figure 6.1 displays the map of our analytical regions, and Appendix 1 gives their definition in terms of summation and subtraction.

It should be recalled that the metropolitan areas are Census Metropolitan Areas, and that their boundaries are adjusted in census years, according to new journey-to-work data. The Sherbrooke MA, in particular, underwent a significant enlargement in 2006. Since the ISQ's GDP and personal income estimates refer to the Census Metropolitan Areas, there are breaks in the series. Fortunately, demographic data is also available on the basis of the 2006 CMA boundaries for the whole 1997–2009 period. So the MA GDP and personal income for all years previous to 2006 were converted to the 2006 geography by multiplying the per capita figures computed from the original data by population according to the 2006 boundaries. The underlying hypothesis is that, in zones closely contiguous to an MA according to the 1996 or 2001 limits, per capita values are the same as inside the MA. This reconstruction of the data ironed out most of the kinks in the series. Remaining wrinkles are possibly due to the substitution of new to old commuting patterns in census years for the estimation of GDP (see above "From place-of-residence data to place-of-production data").

[19] Such is the case, for example, of the non-metropolitan area around the Saguenay MA, most of which could be attributed to the North peripheral region.

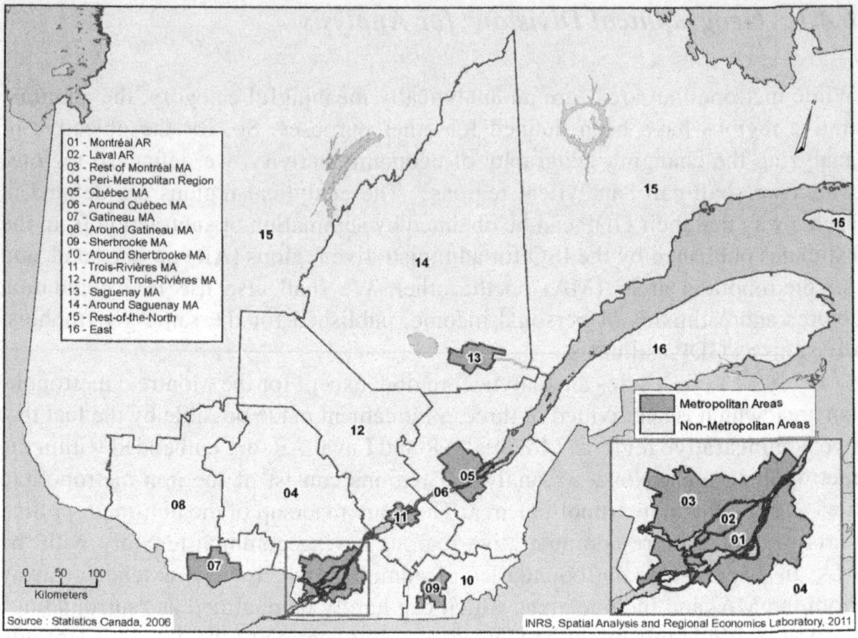

Fig. 6.1 Map of analytical regions

6.4.2 *Changes in the Distribution of Population, Production and Income, 1997–2009*

We want to examine the evolution of the geographic distribution of economic activity and income, that is, the evolution of the share of each region in the GDP or personal income of the Province of Quebec (what is said in the following paragraphs in reference to GDP should be understood to apply to personal income as well).

Let x_{it} be the GDP of region i in year t. Also let $X_t = \sum_i x_{it}$ be the GDP of the Province of Quebec in year t. Similarly, let n_{it} be the population of region i, and $N_t = \sum_i n_{it}$ be the population of the Province of Quebec in year t. Finally, let $y_{it} = x_{it}/n_{it}$ and $Y_t = X_t/N_t$ be the per capita GDP in region i and in the whole of Quebec respectively. Obviously, $x_{it} = y_{it}n_{it}$ and $X_t = Y_tN_t$, and the share of region i can be written as

$$(x_{it}/X_t) = (y_{it}/Y_t)\,(n_{it}/N_t) \tag{6.2}$$

The evolution of share (x_{it}/X_t) is summarized as

$$\frac{(x_{it}/X_t)}{(x_{i0}/X_0)} = \frac{(y_{it}/Y_t)}{(y_{i0}/Y_0)}\,\frac{(n_{it}/N_t)}{(n_{i0}/N_0)} \tag{6.3}$$

where (y_{it}/Y_t) is the relative per capita GDP of region i, and (n_{it}/N_t) is the population share of region i. From Eq. 6.3, the percent change between year 0 and year t of share (x_{it}/X_t) is

$$100 * \left(\frac{(x_{it}/X_t)}{(x_{i0}/X_0)} - 1\right) = \\ = 100 * \left[1 + \left(\frac{(y_{it}/Y_t)}{(y_{i0}/Y_0)} - 1\right)\right]\left[1 + \left(\frac{(n_{it}/N_t)}{(n_{i0}/N_0)} - 1\right)\right] - 100 \tag{6.4}$$

The three percentage changes in this simple decomposition are given in Table 6.1 for GDP.

It is striking that, without exception, the GDP shares of analytical regions (numbered 01–16 in Table 6.1) rise or fall according to whether their shares of population rise or fall: the correlation coefficient between the 1997–2009 change in the share of population and the change in the share of GDP is 71 % (F-statistic = 14.40).[20] A declining relative GDP per capita is also associated with a falling share of GDP: the correlation coefficient between 1997 and 2009 changes is 72 % (F-statistic = 15.45). But the change in relative GDP per capita and in the share of population are unrelated: the correlation coefficient is an insignificant 3.2 %.

In terms of geographical distribution, what is the picture that emerges? First, population (+3.3 %) and, to a lesser degree, production (+0.8 %) are concentrating in the Montreal metropolitan and peri-metropolitan area. But the most impressive feature locally is what is happening to the the Montreal AR relative to the rest of the metropolitan and peri-metropolitan area: its share of population has diminished slightly, while its share of GDP has fallen by a substantial 6.5 % over a 13-year period. Population and production are being decentralized within the MA and in its vicinity. Outside the Montreal metropolitan and peri-metropolitan area, however, the opposite is happening. All regions but three of the five MAs are losing shares. Of the three MAs gaining shares, two gain heavily in terms of production. Both are administrative cities: Quebec City (+10.1 %) is the capital of the Province, and Gatineau (+14.7 %) is part of the Ottawa-Gatineau MA, where Canada's federal capital is located. Everywhere else, population and GDP shares are declining, and they are delining faster in non-metropolitan areas.

Table 6.2 presents the same decomposition as Table 6.1 for personal income. Population data is repeated for readability.

---

[20] It is readily recognized that, to compare the closeness of relationships, given that the correlation coefficient is a measure of linear dependence, it would be mathematically more correct to take the logarithmic transform of Eq. 6.3 and compute correlations between the logarithm of the left-hand side variable and the logarithms of each of the two right-hand side variables. But that would make the exposition unnecessarily technical. Given the Maclaurin series $\ln(1 + z) = z - z^2/2 + z^3/3 - z^4/4 + \ldots$, we consider $[(x_t/x_0) - 1]$ to be a first order Taylor approximation of $\ln(x_t/x_0)$ when x_t is close to x_0, so the more correct mathematical approach would lead to the same observations.

Table 6.1 GDP at basic prices by analytical region, Quebec, 1997–2009

Analytical region	Relative GDP per cap. 1997	Relative GDP per cap. 2009	Relative GDP per cap. 1997–2009 proportional change (%)	Share of population 1997 (%)	Share of population 2009 (%)	Share of population 1997–2009 proportional change (%)	Share of GDP 1997 (%)	Share of GDP 2009 (%)	Share of GDP 1997–2009 proportional change (%)
	Province = 100								
Montreal and peri-metro area	106	104	−2.5	58.5	60.5	3.3	62.2	62.7	0.8
Montreal MA	113	109	−3.6	47.1	48.8	3.7	53.0	52.9	−0.1
01 Montreal AR	146	137	−5.9	24.7	24.6	−0.6	36.0	33.7	−6.5
02 Laval AR	78	86	9.5	4.6	5.0	7.7	3.6	4.3	17.9
03 Rest of Montreal MA	75	78	3.6	17.7	19.2	8.6	13.3	14.9	12.5
04 Peri-metro region	81	84	3.6	11.5	11.7	2.09	9.3	9.8	5.7
Other metropolitan areas	98	104	6.4	19.6	19.7	0.4	19.2	20.5	6.8
Other non-metro	83	84	0.9	33.4	31.6	−5.4	27.8	26.6	−4.5
Non-metro away from Montreal	85	85	−0.4	21.9	19.9	−9.2	18.6	16.8	−9.6
Peripheral regions	89	95	6.3	10.2	8.9	−12.7	9.1	8.5	−7.2
05 Quebec MA	105	115	9.6	9.5	9.5	0.5	10.0	11.0	10.1
06 Around Quebec MA	81	78	−4.1	4.7	4.4	−6.9	3.8	3.4	−10.7
07 Gatineau MA	80	84	5.3	3.5	3.8	9.0	2.8	3.2	14.7
08 Around Gatineau MA	57	60	5.7	0.8	0.7	−7.4	0.4	0.4	−2.1
09 Sherbrooke MA	90	88	−2.6	2.4	2.5	3.7	2.2	2.2	0.9
10 Around Sherbrooke MA	93	71	−24.2	1.5	1.4	−5.1	1.4	1.0	−28.1
11 Trois-Rivières MA	107	108	1.1	2.0	1.9	−5.1	2.1	2.0	−4.1
12 Around Trois-Riv. MA	81	79	−2.9	4.7	4.4	−5.8	3.8	3.5	−8.5
13 Saguenay MA	96	107	11.6	2.2	1.9	−12.6	2.1	2.1	−2.5
14 Around Saguenay MA	73	75	1.8	1.8	1.5	−12.9	1.3	1.1	−11.3
15 Rest-of-the-North	113	126	11.2	4.1	3.6	−12.5	4.6	4.5	−2.7
16 East	73	74	0.8	4.3	3.8	−12.8	3.2	2.8	−12.1
Province of Quebec	100	100		100.0	100.0		100.0	100.0	

Table 6.2 Personal income by analytical region, Quebec, 1997–2009

Analytical region	Relative pers. inc. per cap. 1997	2009	1997–2009 proportional change (%)	Share of population 1997 (%)	2009 (%)	1997–2009 proportional change (%)	Share of personal income 1997 (%)	2009 (%)	1997–2009 proportional change (%)
	Province = 100								
Montreal and peri-metro area	106	103	−2.4	58.5	60.5	3.3	61.9	62.4	0.9
Montreal MA	113	109	−3.6	47.1	48.8	3.7	51.5	51.9	0.7
01 Montreal AR	111	104	−6.0	24.7	24.6	−0.6	27.4	25.6	−6.6
02 Laval AR	107	103	−4.3	4.6	5.0	7.7	5.0	5.1	3.0
03 Rest of Montreal MA	108	110	1.8	17.7	19.2	8.6	19.1	21.1	10.6
04 Peri-metro. region	90	90	−0.5	11.5	11.7	2.0	10.4	10.5	1.6
Other metropolitan areas	100	104	4.5	19.6	19.7	0.4	19.6	20.5	4.8
Other non-metro	87	87	0.9	33.4	31.6	−5.4	28.9	27.6	−4.5
Non-metro away from Montreal	85	86	1.4	21.9	19.9	−9.2	18.6	17.1	−7.9
Peripheral regions	86	90	4.6	10.2	8.9	−12.7	8.7	8.0	−8.6
05 Quebec MA	105	110	5.0	9.5	9.5	0.5	10.0	10.5	5.4
06 Around Quebec MA	85	84	−1.3	4.7	4.4	−6.9	4.0	3.7	−8.1
07 Gatineau MA	97	109	12.5	3.5	3.8	9.0	3.4	4.2	22.6
08 Around Gatineau MA	68	76	11.8	0.8	0.7	−7.4	0.5	0.6	3.6
09 Sherbrooke MA	98	93	−5.0	2.4	2.5	3.7	2.3	2.3	−1.5
10 Around Sherbrooke MA	82	81	−1.6	1.5	1.4	−5.1	1.2	1.2	−6.6
11 Trois-Rivières MA	94	93	−0.8	2.0	1.9	−5.1	1.8	1.7	−5.8
12 Around Trois-Riv. MA	86	84	−2.6	4.7	4.4	−5.8	4.0	3.7	−8.2
13 Saguenay MA	92	94	1.9	2.2	1.9	−12.6	2.0	1.8	−10.9
14 Around Saguenay MA	83	86	3.7	1.8	1.5	−12.9	1.5	1.3	−9.7
15 Rest-of-the-North	94	98	4.9	4.1	3.6	−12.5	3.9	3.5	−8.2
16 East	79	83	4.7	4.3	3.8	−12.8	3.4	3.1	−8.7
Province of Quebec	100	100		100.0	100.0		100.0	100.0	

For personal income as for GDP, analytical regions (numbered 01–16 in Table 6.2) with declining shares of population also have declining shares of personal income: the correlation coefficient between the 1997–2009 change in the share of population and the change in the share of personal income is 80 % (F-statistic = 25.42). But there are a few exceptions (analytical regions 08 and 09). Not surprisingly, the same relationship is observed between the change in GDP shares and the change in personal income shares: the correlation coefficient between the last column of Table 6.1 and the last column of Table 6.2 is 74 % (F-statistic = 17.07). A declining relative personal income per capita is also associated with a falling share of personal income: the correlation coefficient between 1997 and 2009 changes is 48 % (F-statistic = 4.19). But the change in relative personal income per capita and in the share of population are unrelated: the correlation coefficient is a non-significant −14 %.

Let us now compare relative personal income per capita in Table 6.2 with relative GDP per capita in Table 6.1. The comparison illustrates the conceptual difference between personal income and domestic product. Indeed, personal income is the income that residents of a given territory receive, no matter where production took place; on the other hand, domestic product is the total value of what has been produced in a given territory, no matter where those who receive the income live. Therefore, per capita GDP is strongly influenced by home-to-work commuting. The high per capita GDP of the Montreal AR and, to a lesser degree, of most other MAs is explained by the large number of residents of the surrounding area who come to work there. Conversely, per capita GDP is systematically weaker in the areas around MAs than in the MAs themselves. Two exceptions among the MAs are Gatineau and Sherbrooke. In the case of Gatineau, the somewhat low relative GDP per capita is explained by the fact that Gatineau belongs to the Ottawa-Gatineau MA, and that large numbers of civil servants live in Gatineau and work in Ottawa; indeed, per capita GDP is even lower in the surrounding area. Finally, the high per capita GDP of the Rest-of-the-North region reflects the presence of capital intensive industries (such as mining and hydroelectric power), whose shareholders do not necessarily live there. There is much less variation in relative personal income per capita than in relative GDP per capita, and this results in a less contrasted picture. Nonetheless, it appears that personal income per capita is higher in the Montreal MA, and in the two administrative MAs of Quebec and Gatineau. It should be kept in mind that at least part of that greater personal income is swallowed up by a higher cost of living. Changes in the geographical distribution of personal income reflect both changes in population and changes in per capita personal income, and they are consistent with changes in the distribution of production. There is a slight movement of concentration towards the Montreal metropolitan and peri-metropolitan area, and decentralization within the Montreal MA, with the core Montreal AR losing share. Both administrative capital regions of Quebec and Gatineau increase their shares, and, in the latter case, the increase seems to be spilling over to the surrounding region. All other regions lose share. Changes in the geographical distribution of personal disposable income (not displayed here) are very similar to changes in the distribution of personal income, and the same comments apply.

6.4.3 Time-Paths

In the previous section, an examination of the first and last years of the time-series produced a relatively clear picture, which we synthesized as: stability in the dominant share of the Montreal metropolitan and peri-metropolitan area, accompanied by a movement of decentralization within the area, together with a rise in the shares of the administrative and political cities of Quebec and Gatineau MAs, and a concomitant fall in the shares of other regions, especially non-metropolitan and peripheral areas. In this section, we briefly look at the evolution in-between, to verify that the first and last periods are not outliers. Our approach is graphical, since 13 years is too short a time-series for serious econometric analyses.

Figure 6.2 presents the evolution of population shares in index form (1997 = 100). An index above 100 means that the region's share of population is higher than it was in base year 1997, and vice-versa for values below 100. In the graph, the curves of population shares are mostly monotonous (i.e. without ups-and-downs), although there are exceptions. The general impression is one of smooth evolution. From this point on, we shall distinguish four groups of analytical regions: (1) regions with increasing population shares: Gatineau MA, Rest of Montreal MA, Laval AR and Sherbrooke MA; (2) regions with stable population shares: Quebec MA and Montreal AR; (3) regions with falling population shares: non-metropolitan area around Sherbrooke MA, Trois-Rivières MA, non-metro area around Trois-Rivières MA, non-metro areas around Quebec and Gatineau MAs; (4) regions with collapsing population shares: Rest of the North, Saguenay MA, the East and non-metro area around Saguenay MA.

Panel (a) of Figs. 6.3, 6.4, 6.5, and 6.6 displays the evolution of an index of GDP shares for each of the four groups. Roughly speaking, regions with increasing or decreasing population shares also have increasing or decreasing GDP shares, but the evolution is not monotonous as it is for population shares. Now look at Panel (b) in Figs. 6.3, 6.4, 6.5, and 6.6, which shows the evolution of relative GDP per capita in index form (1997 = 100). Relative GDP per capita for 1997 and 2009 is given in the first two columns of Table 6.1. The index for 2009 is obtained by dividing the 2009 relative GDP per capita by the corresponding 1997 value. For example, the 2009 index of relative GDP per capita for the Montreal AR is $100 \times 137/146 = 94$ (after rounding). It is quite apparent that the shape of the relative GDP per capita index curves in Panel (b) of Figs. 6.3, 6.4, 6.5, and 6.6 mirrors that of GDP share index curves in Panel (a).

In the previous section, we have detailed the mathematical relationship between GDP shares, population shares and GDP per capita. Visually, it would seem that the relationship could be characterized by saying that population shares represent the trend, and GDP per capita the fluctuations around the trend. Changes in GDP per capita may reinforce the trend, as in the case of the Laval AR, or they may dampen or even reverse the trend, as in the case of the Saguenay MA. Not surprisingly, the relationship between the variables is supported by the correlations between

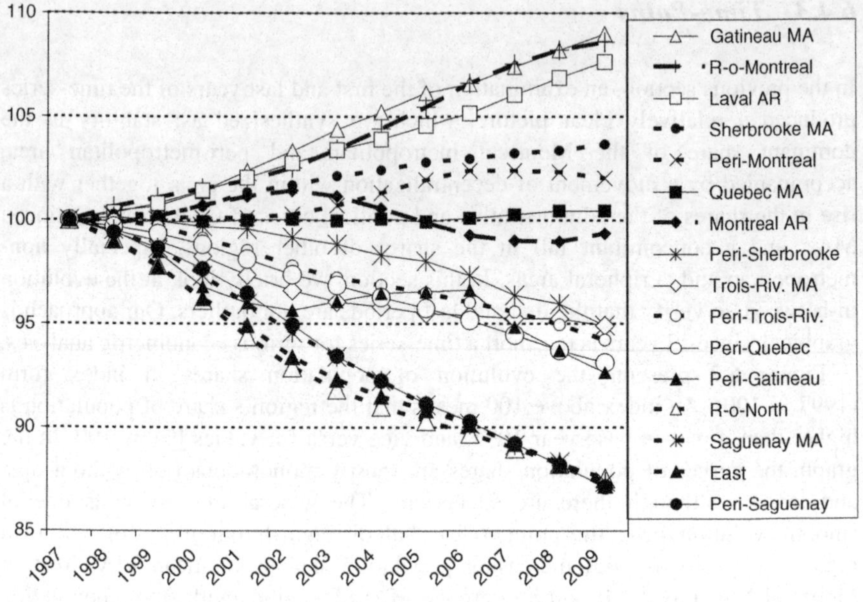

Fig. 6.2 Evolution of population shares, analytical regions, 1997–2009 (1997 = 100)

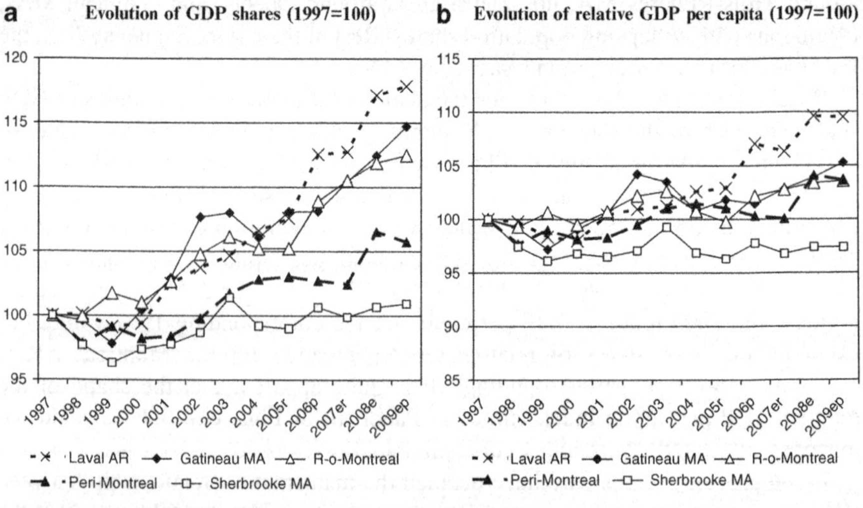

Fig. 6.3 Analytical regions with rising population shares, 1997–2009

them: the panel correlation coefficient[21] of year-to-year changes in GDP shares with year-to-year changes in population shares is 72 % (F-statistic = 199); with year-to-year changes in relative GDP per capita, it is 75 % (F-statistic = 246).

[21] What we call the panel correlation coefficient here is calculated for the panel of 16 regions and 12 year-to-year changes (from 1997–1998 to 2008–2009).

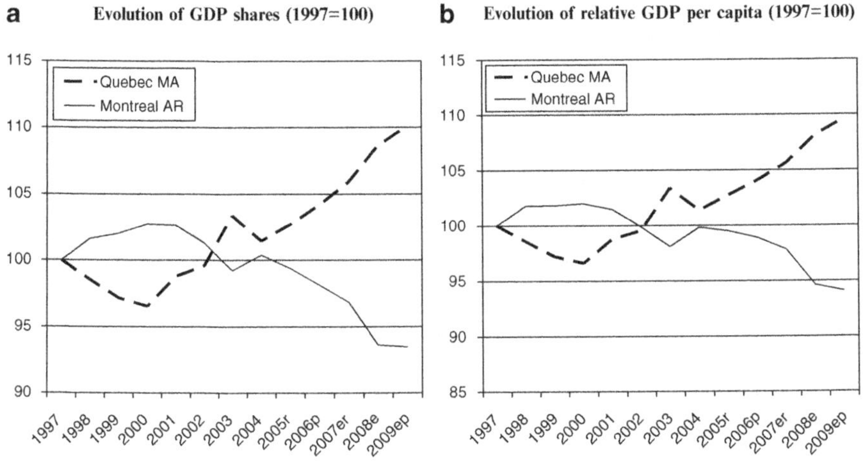

Fig. 6.4 Analytical regions with stable population shares, 1997–2009

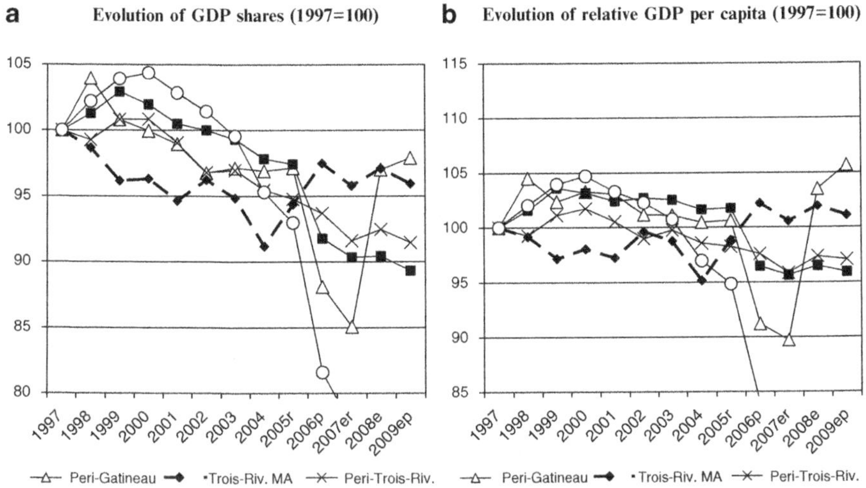

Fig. 6.5 Analytical regions with falling population shares, 1997–2009

Finally, the panel correlation coefficient of year-to-year changes in population shares with year-to-year changes in relative GDP per capita is not so strong, at 29 % (F-statistic = 17.8).

Compared with GDP shares, personal income shares vary somewhat less, but they display a similar relationship with population shares and with relative personal income per capita. The panel correlation coefficients of year-to-year changes in personal income shares are 29 % (F-statistic = 16.9) with year-to-year changes

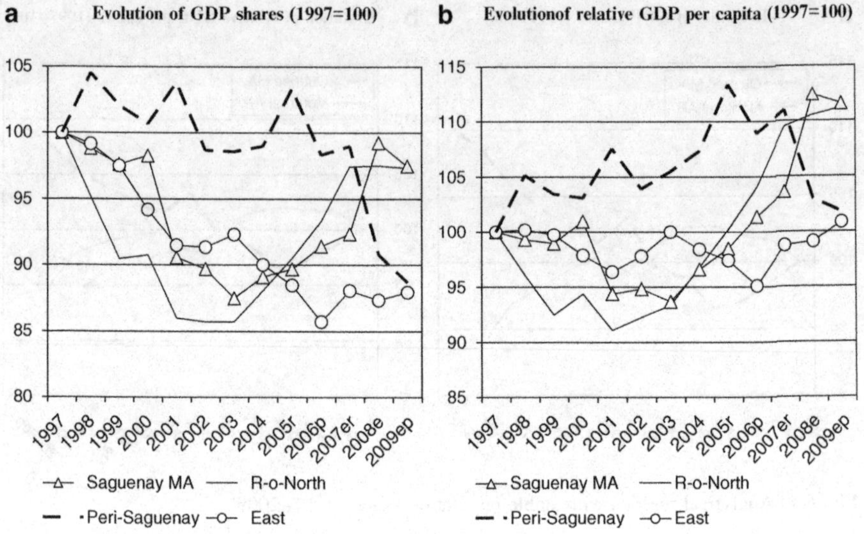

Fig. 6.6 Analytical regions with collapsing population shares, 1997–2009

in population shares, and 92 % (F-statistic = 1,016) with year-to-year changes in relative personal income per capita.

6.4.4 *Regional Gross Value Added by Industry*

Regional GDP is the sum over all industries of regional gross value added by industry. So we shall now use the estimates of gross value added by industry and by region to examine the evolution of regional GDP shares in more detail. The percent change between year 0 and year t of the share (x_{it}/X_t) of region i can be written as

$$100^*\left(\frac{(x_{it}/X_t)}{(x_{i0}/X_0)}-1\right)=$$

$$=100^*\left(\frac{(x_{it}/x_{i0})}{(X_t/X_0)}-1\right)=100^*\left(\frac{(x_{it}/x_{i0})-(X_t/X_0)}{(X_t/X_0)}\right) \tag{6.5}$$

The numerator of the ratio on the extreme right is the differential growth rate of the region, since $(x_{it}/x_{i0})-(X_t/X_0)=[(x_{it}/x_{i0})-1]-[(X_t/X_0)-1]$. We shall apply shift-share analysis to gross value added by industry in order to decompose differential GDP growth into two components, conventionally labelled "structural growth" and "residual growth". Structural growth is defined as the additional (possibly negative) growth that the region would have experienced if each of its industries had grown at the same rate regionally as at the national level (in the present context, at the provincial level).

It must be recognized at the outset that shift-share analysis has many shortcomings. Its principal limitation is that the amount of structural growth depends on the industry classification, and that the bias from using a less, rather than a more detailed classification is of unknown sign. Another point to keep in mind is that calculation of the structural component of growth depends on the choice of the base year, in much the same way as a Laspeyres price index; if, for any reason, the industry mix in the base year is not representative of a region's long-term economic structure, then the effect of that structure on regional growth is misrepresented. In addition, the shift-share decomposition of growth over a given period is not the sum of shift-share decompositions of growth over its sub-periods, since every sub-period has a different base year industry mix. In view of these limitations, there is no validity whatsoever in the interpretation of the residual effect as an indicator of the region's competitiveness. So, with all due caution, we apply shift-share analysis to answer the following question: at the available level of industrial detail, is it possible to identify industries or groups of industries whose relatively high or low provincial growth rates may have influenced regional growth?

Before tackling this task, two methodological challenges had to be solved. The first concerns changes in the metropolitan area boundaries, mentioned above (Sect. 2.1). At the aggregate level, we estimated constant-boundary MA GDP and personal income as the product of the corresponding per capita values by the appropriate population numbers. It would not be adequate to make this kind of adjustment at the individual industry level. So the solution adopted was to estimate the effect of boundary changes on overall regional GDP as the ratio in 2005 of GDP according to the 2006 MA boundaries to GDP according to the 2001 MA boundaries. In the analysis, this "boundary change effect" was simply subtracted from the residual component of growth. No adjustment was made in year 2000 for boundary changes that have taken place in census year 2001, because the changes were minor.

The other challenge that had to be met is the effect of disclosure rules. The ISQ applies Statistics Canada's so-called "Duffett rules"[22] of data dissemination. These are designed to prevent the disclosure of information obtained under the authority of the Statistics Act that pertains to individual persons or organizations. An examination of the detailed tables[23] shows that, except for aggregates such as Manufacturing as a whole, almost every industry is undisclosed for at least one region. Consequently, if all industries that are undisclosed for some region were aggregated for all regions, analysis could be performed only at a highly aggregated level. In order to make full use of all the available information, four residual industries were created for each analytical region: Other Agriculture etc. (confidential data); Other Mining, Util. &Construc. (confidential data); Other Manufacturing (confidential data); and Other Services (confidential data). The gross value added of these residual industries was obtained by subtracting known data from the

[22] Named after Walter Duffett, who was Chief Statistician in the early 1970s when these rules were adopted.

[23] The detailed tables are available on demand from the first author.

aggregates. Now, as we already mentioned, the pattern of disclosure is not the same for all regions, which implies that the composition of the residual industries differs from one region to another. So, to calculate the structural component of growth of a residual industry, the growth rate of the corresponding aggregate at the provincial level was computed separately for each analytical region. Table 6.3 synthesizes the results of the shift-share analysis.[24]

In Table 6.3, the sum of the components in columns 4–6 is equal to the differential growth rate in column 3. In columns 7–9, the components from columns 4–6 are expressed in percentage shares of total differential growth (the sum of shares is 100 % or −100 %, depending on whether differential growth is positive or negative). First, we note that the effect of MA boundary changes (col. 5) has indeed been quite substantial for the Sherbrooke MA and the surrounding area. Otherwise, it has mostly had a negative effect on regions around metropolitan areas due to the transfer to the metropolitan area of territory with a relatively high density of economic activity. But we are mostly concerned with the structural component (col. 4). It is interesting to note that the structural component is negative for most regions except for the part of the Montreal MA which doesn't belong to the central Montreal AR, and the two administrative and political centers of Gatineau (part of the Ottawa-Gatineau MA, where the federal capital of Canada is located) and Québec (capital city of the Province of Quebec). The structural component is also positive for the Rest-of-the-North peripheral region; Table 6.4 below reveals that this is mostly associated to Mining etc., Utilities and Construction.

Table 6.4 gives the sectoral composition of structural growth. The sum of the sectoral components in columns 2–5 is equal to the structural component of regional GDP growth in column 1, which is identical to the fourth of Table 6.3. Each of the sectoral components is the sum of industry components computed at the finest level of detail possible (using the "Other" residual industries), given the confidentiality rules. By its very definition, the province-wide structural component is zero, because the weighted average growth rate of industries is equal to the overall growth rate of provincial GDP. But individual industries or groups of industries do not grow equally. It can be seen that, in Quebec like in all mature economies, the manufacturing sector has grown less than services. Accordingly, the contribution of manufacturing to the structural component of growth is negative everywhere, and larger (in absolute value) in regions where manufacturing is concentrated. Conversely, the service sector has grown faster than the overall economy, so its contribution is positive everywhere, and greater in metropolitan areas, where the service economy tends to concentrate. Interestingly, it can be seen from the detailed results (not displayed in this article) that the contribution of Services to the structural component of growth in the Gatineau and Quebec administrative MAs is not primarily associated with growth in Public Administration. Regarding Mining etc., Utilities and Construction, as far as can be

[24] The growth decomposition is for the 1997–2008 period, rather than 1997–2009, because the regional gross value added data for 2009 were not yet available by industry at the time of writing this article.

Table 6.3 Shift-share analysis of regional growth, 1997–2008

	Total growth = Provincial + differential			Components of differential growth			. . . in % of differential growth[a]		
Analytical region	Total growth (%)	Provincial growth (%)	Differential growth (%)	Structural component (%)	MA Boundary change effect (%)	Residual component (%)	Structural component (%)	MA Boundary change effect (%)	Residual component (%)
01 Montreal AR	52.0	62.3	−10.3	−0.3	0.0	−10.1	−2.6	0.0	−97.4
02 Laval AR	90.2	62.3	27.8	6.0	0.0	21.9	21.4	0.0	78.6
03 Rest of Montreal MA	87.3	62.3	25.5	1.8	3.2	20.5	7.1	12.4	80.5
04 Peri-metro. Region	65.1	62.3	2.8	−2.7	−4.2	9.7	−95.0	−149.7	344.7
05 Quebec MA	86.4	62.3	24.0	4.4	1.6	18.0	18.4	6.9	74.7
06 Around Quebec MA	39.5	62.3	−22.8	−4.3	−9.0	−9.5	−19.1	−39.5	−41.5
07 Gatineau MA	77.2	62.3	14.9	4.4	0.7	9.7	29.8	4.8	65.4
08 Around Gatineau MA	44.9	62.3	−17.5	−7.2	−1.8	−8.4	−41.5	−10.6	−47.9
09 Sherbrooke MA	86.5	62.3	24.2	−1.3	20.4	5.1	−5.3	84.3	21.0
10 Around Sherbrooke MA	1.5	62.3	−60.8	−13.7	−23.1	−24.0	−22.5	−37.9	−39.5
11 Trois-Revières MA	58.1	62.3	−4.2	−2.2	0.1	−2.1	−53.3	2.8	−49.4
12 Around Trois-Riv. MA	49.3	62.3	−12.5	−5.7	−0.1	−6.8	−45.2	−0.5	−54.2
13 Saguenay MA	60.2	62.3	−2.1	−3.5	0.4	1.0	−167.1	18.9	48.3
14 Around Saguenay MA	48.7	62.3	−13.6	−5.0	−0.7	−8.0	−36.5	−4.8	−58.7
15 Rest-of-the-North	58.3	62.3	−4.0	3.7	0.0	−7.7	91.2	0.0	−191.2
16 East	41.3	62.3	−20.5	−1.0	0.0	−19.5	−4.8	0.0	−95.2
Province of Quebec	62.3	62.3	0.0	0.0	0.0	0.0			

[a]The component shares are multiplied by −1 when differential growth is negative

Table 6.4 Sectoral detail of the structural component of differential regional growth, 1997–2008 (variation of gross value added, in thousands of current Canadian dollars)

Analytical region	Total structural component (%)	Agriculture, Forestry, Fishing and Hunting (%)	Mining etc., Utilities & Construc. (%)	Manufacturing (%)	Services-producing industries (%)
01 Montreal AR	−0.3	0.0	1.7	−8.8	6.9
02 Laval AR	6.0	−0.1	3.5	−5.0	7.6
03 Rest of Montreal MA	1.8	−0.2	3.6	−7.3	5.6
04 Peri-metro. region	−2.7	−0.9	3.2	−9.8	4.9
05 Quebec MA	4.4	−0.1	2.1	−3.5	6.1
06 Around Quebec MA	−7.2	−1.1	2.0	−11.4	3.3
07 Gatineau MA	4.4	−0.3	2.2	−2.8	5.3
08 Around Gatineau MA	−4.3	−4.4	3.5	−6.2	2.7
09 Sherbrooke MA	−1.3	−0.2	1.6	−7.5	4.7
10 Around Sherbrooke MA	−13.7	−1.1	2.4	−17.4	2.3
11 Trois-Rivières MA	−2.2	−0.2	5.1	−10.8	3.7
12 Around Trois-Riv. MA	−5.7	−1.1	2.5	−10.9	3.9
13 Saguenay MA	−3.5	−0.4	2.7	−11.1	5.4
14 Around Saguenay MA	−5.0	−2.2	2.2	−10.0	4.9
15 Rest-of-the-North	3.7	−1.1	8.6	−7.2	3.4
16 East	−1.0	−1.4	2.9	−6.3	3.9
Province of Quebec	0.0	−0.4	2.6	−8.1	5.9

seen from the non-disclosed data, its more rapid growth is mostly due to the construction boom that Quebec has experienced in the latter years of the period examined.

Returning to our initial question, we conclude that yes, is it possible to identify industries or groups of industries whose relatively high or low provincial growth rates may have influenced regional growth. However, Table 6.3 shows that the residual, unexplained, component of differential growth overweighs the structural component in most cases, even after subtracting the effect of MA boundary changes in 2006.

6.5 Summary and Conclusion

This article presents the method developed for the Institut de la statistique du Québec to estimate regional GDP in the Province of Quebec. It is a top-down method, which consists in allocating total labour income and net income of unincorporated business (NIUB) by industry among regions using allocators constructed from fiscal data on wages and salaries and NIUB obtained from the Quebec ministry of revenue (responsible for tax collection). For each industry, other components of value added (corporation profits, interest, capital consumption allowances, inventory valuation adjustment, and net indirect taxes on production) are then distributed in proportion to the sum of total labour income and NIUB. The key ingredients in the method are a compilation of fiscal data on incomes, and reliable home-to-work commuting tables by industry. We believe a similar approach could be used anywhere comparable data are available.

In order to examine the changing distribution of population, production and income in the Province of Quebec, we have reorganized the data to construct analytical regions according to the core-periphery view. This was achieved by summation and subtraction of data pertaining to the administrative regions and the metropolitan areas. A relatively clear picture emerged from our examination of the data, both through a comparison of the first- and last-year data, and through a graphical inspection of the evolution in-between.

Almost half of the population resides in the Montreal metropolitan area, and around 60 % in the combined Montreal metropolitan and peri-metropolitan area. Corresponding GDP and personal income shares are slightly higher. Population, GDP and income shares are quite stable throughout the period. Within the Montreal MA, however, there is a movement of deconcentration, as the Montreal administrative region, which constitutes the core of the metropolitan area, loses about 6 % of its share in production and income, while more or less maintaining its share of population. In other parts of the Province, only the Quebec and Gatineau MAs increase their shares of GDP and income. Both metropolitan areas include administrative and political centers: Quebec City is the capital of the Province, and Gatineau is part of the Ottawa-Gatineau MA where Canada's federal capital is located. Almost all other analytical regions show falling shares of population, production and income. In particular, peripheral regions' shares of population dropped by 12 %–13 % over the 1997–2009 period; their shares of GDP and personal income have fallen also, albeit less dramatically. And within the Quebec heartland, there is a redistribution from non-metropolitan to metropolitan areas.

How much of that spatial redistribution can be related to changes in the industry mix of the economy? In an attempt to throw some light in the issue, shift-share analysis was applied to gross value added by industry in order to decompose differential GDP growth into its two conventional components, "structural growth" and "residual growth". We found that it is possible to identify industries or groups of industries whose differential growth rates may have influenced regional growth. However, the residual, unexplained, component of differential growth outweighs the structural component in most cases, even after subtracting the effect of MA boundary changes. These results show, not too surprisingly, that the core-periphery logic underlying the changing geography of production cannot be reduced to a mere reflection of changes in the industry mix of the economy. Space matters.

Appendix 1: Definition of Analytical Regions

Montreal AR	Montreal AR
Laval AR	Laval AR
Rest-of-Montreal MA	Montreal MA − [Montreal AR + Laval AR]
Perimetropolitanarea	[Laurentides AR + Lanaudière AR + Montérégie AR] − Rest-of-Montreal MA
Québec MA	Québec MA
Non-metro around Québec MA	[Chaudière-Appalaches AR + Capitale-Nationale AR] − Québec MA
Gatineau MA	Part of the Ottawa-Gatineau MA located in the Province of Quebec
Non-metro around Gatineau MA	Outaouais AR − Gatineau MA
Sherbrooke MA	Sherbrooke MA
Non-metro around Sherbrooke MA	Estrie AR − Sherbrooke MA
Trois-Rivières MA	Trois-Rivières MA
Non-metro around Trois-Rivières MA	[Mauricie AR + Centre-du-Québec AR] − Trois-Rivières MA
Saguenay MA	Saguenay MA
Non-metro around Saguenay MA	Saguenay – Lac-Saint-Jean AR − Saguenay MA
Rest-of-the-North	Abitibi-Témiscamingue AR + Nord-du-Québec AR + Côte-Nord AR
East	Bas-Saint-Laurent AR + Gaspésie – Îles-de-la-Madeleine AR
Subtotals:	
Montreal and peri-metro area	Montreal MA + Perimetropolitan area
Montreal MA	Montreal AR + Laval AR + Rest-of-Montreal MA
Other metropolitan areas	Québec MA + Gatineau MA + Sherbrooke MA + Trois-Rivières MA + Saguenay MA
Montreal MA	Montreal AR + Laval AR + Rest-of-Montreal MA
Other non-metro	Perimetropolitan area + Non-metro away from Montreal
Non-metro away from Montreal	Non-metro around Québec MA + Non-metro around Gatineau MA + Non-metro around Sherbrooke MA + Non-metro around Trois-Rivières MA + Peripheral regions
Peripheral regions	Non-metro around Saguenay MA + Rest-of-the-North + East

References

Bureau of Economic Analysis (2008) Local area personal income and employment methodology (LAPI). U.S. Department of Commerce, Economics and Statistics Administration. www.bea.gov/regional/pdf/lapi2006/lapi2006.pdf. Accessed October 11, 2012

Delisle JP, avec la participation de Lelong C, Kirthichandra A (2000) Méthode de calcul des valeursajoutéesrégionales par branche et des PIB régionaux SEC 1995, années de base 1994–1996. InstitutNational de la Statistique et des ÉtudesÉtconomiques (INSEE), Paris

Golan A, Judge G, Miller D (1996) Maximum entropy econometrics: robust estimation with limited data. Wiley, New York

Günlük-Şenesen G, Bates JM (1988) Some experiments with methods of adjusting unbalanced data matrices. J R Stat Soc Ser A (Stat Soc) 151(3):473–490

Jaynes ET (1957) Information theory and statistical mechanics. Phys Rev 106:620–630, vol 108, pp 171–190

Junius T, Oostrehaven J (2003) The solution of updating or regionalizing a matrix with both positive and negative entries. Econ Syst Res 15(1):87–96

Kapur JN, Kesavan HK (1992) Entropy optimization principles with applications. Academic Press, San Diego

Lacey D (2000) UK regional gross domestic product (GDP): methodological guide. Economic Trends, No. 565, Office for National Statistics (ONS), London. http://www.ons.gov.uk/ons/index.html. Accessed October 11, 2012

Lemelin A (2010) A GRAS variant solving for minimum information loss. Econ Syst Res 21 (4):399–408

Lemelin A, Mainguy P (2005) Estimating regional gross domestic product (RGDP) 1997–2000, Institut de la statistique du Québec. http://www.stat.gouv.qc.ca/publications/cptes_econo/pdf2007/Estimation_PIRreg_an.pdf. Accessed October 11, 2012

Lemelin A, Mainguy P (2009a) Estimation du produitintérieur brut régional des 17 régionsadministratives du Québec, Cahier technique etméthodologique, Institut de la statistique du Québec. http://www.stat.gouv.qc.ca/publications/cptes_econo/pdf2009/estimation_PIB.pdf. Accessed October 11, 2012

Lemelin A, Mainguy P (2009b) Revue des méthodesd'estimation du produitintérieurbrutrégional, Cahiertechnique et méthodologique, Institut de la statistique du Québec. http://www.stat.gouv.qc.ca/clacon/Revue_des_methodes.pdf. Accessed October 11, 2012

OECD (2011) "Glossary of Main Terms", in National Accounts at a Glance 2010, OECD Publishing. http://dx.doi.org/10.1787/na_glance-2010-29-en

Panek SD, Baumgardner FT, Mathew JM (2007) Introducing new measures of the metropolitan economy. Prototype GDP-by-metropolitan-Area estimates for 2001–2005. Surv Curr Bus 87: 79–114

Polèse M, Shearmur R (2002) The periphery in the knowledge economy. The spatial dynamics of the Canadian economy and the future of non-metropolitan regions in Quebec and the Atlantic Provinces, Coll. Regions in the Knowledge Economy, Montréal, Institut National de la RechercheScientifique – Centre Urbanisation, Culture et Société. http://www.ucs.inrs.ca/ucs/publications/collections/collection-regions-et-economie-du-savoir. Accessed October 11, 2012

Rosenthal RE (2010) GAMS – a user's guide. GAMS Development Corporation, Washington, DC. http://www.gams.com/docs/document.htm. Accessed October 11, 2012

Shearmur R, in collaboration with Polèse M (2001) Economic development in Canadian peripheral regions, 1971 to 1996. A statistical overview, Institut National de la RechercheScientifique, Centre Urbanisation, Culture etSociété, and the Canadian Institute for Research on Regional Development. http://www.inrs-ucs.uquebec.ca/default.asp?p=res

Theil H (1967) Economics and information theory, vol 7, Studies in mathematical and managerial economics. Rand McNally, Chicago, 7

Chapter 7
An Application of the Disequilibrium Adjustment Framework to Small Area Forecasting and Impact Analysis

Jae Hong Kim and Geoffrey J.D. Hewings

7.1 Introduction

Regional disequilibrium adjustment frameworks, pioneered by Carlino and Mills (1987), have been widely employed for a broad range of regional and more disaggregated level research. In particular, the method has been more extensively used, after Boarnet (1994a) extended the original form of the adjustment model by introducing a spatial weight matrix into the equation system in order to explicitly consider the intrinsic spatial interdependence. So far, the applications include a variety of empirical analyses of growth dynamics, ranging from the examinations of the population-employment interaction (see e.g., Carlino and Mills 1987; Boarnet 1994b; Clark and Murphy 1996; Vias 1999) to the studies on spatial linkages (see e.g., Henry et al. 1997, 1999, 2001; Feser and Isserman 2005) and the investigations on development policy issues (see e.g., Bollinger and Ihlanfeldt. 1997; Edmiston 2004; Ke and Feser 2010).

This study identifies key advantages of the disequilibrium adjustment model, particularly its dynamic nature, and explores the possibility of a further expansion of the model applications, beyond its existing uses for empirical assessments. An attempt is made to exploit a spatial econometric version of the regional disequilibrium adjustment model (Boarnet 1994a) for small area socio-economic forecasting and impact analysis. Rather than using the adjustment model as it stands for a forecasting purpose, the study introduces an idea of combining it hierarchically with a regional econometric input–output model (REIM) that provides a long-term trajectory of economic growth that is not reflected by the adjustment model alone.

J.H. Kim (✉)
Department of Planning, Policy, and Design, University of California, Irvine, USA
e-mail: jaehk6@uci.edu

G.J.D. Hewings
REAL – Regional Economics Applications Laboratory, University of Illinois, Urbana, IL, USA

E. Fernández Vázquez and F. Rubiera Morollón (eds.), *Defining the Spatial Scale in Modern Regional Analysis*, Advances in Spatial Science,
DOI 10.1007/978-3-642-31994-5_7, © Springer-Verlag Berlin Heidelberg 2012

Further, it demonstrates an application of the combined framework for a small area population and employment forecasting under various scenarios.

The remainder of this chapter is structured as follows. Section 7.2 briefly explains the disequilibrium adjustment model and reviews existing model applications for various purposes. In Sect. 7.3, attention is paid to the dynamic nature of the adjustment model and its potential as a tool to support socio-economic forecasting and dynamic impact analysis. Section 7.4 presents a strategy for combining the adjustment model with the REIM to construct an integrated forecasting and dynamic impact analysis framework. Finally, a small area population and employment forecasting based upon the integrated framework is demonstrated in Sect. 7.5, followed by a conclusion where some challenges in this type of the application are discussed.

7.2 Regional Disequilibrium Adjustment Model

7.2.1 Logic and Formulation

In common with other applied analysis frameworks in regional studies, generally the disequilibrium adjustment model describes growth dynamics of certain geographic areas, with a focus on population and employment changes. The most critical feature of the model is that it characterizes the population and employment changes in the real world as an incremental adjustment process, rather than assuming that the observed patterns of population and employment distributions are in a state of spatial equilibrium. In other words, it recognizes not only the equilibrating forces arising from the rational behaviors of economic agents but also the difficulties of attaining equilibrium in reality. Even though this characterization (i.e., the denial of the assumption of the equilibrium status) somewhat complicates the model formulation and estimation, this way of understanding the process is appealing, given that there are many factors (e.g., irreversible investments, moving costs, etc.) that prevents households and businesses from relocating frequently in response to the evolving environments. In addition, the model provides a more general framework that can include the case of a perfect adjustment (i.e., the equilibrium state) depending on the data, the context, and the time period involved.

The fundamental concepts of this model had been expressed by the following two sets of the equations (i.e., Eqs. 7.1, 7.2, 7.3, and 7.4). The equations basically imply (a) that the equilibrium level of population and employment can be determined by household and business location choice factors and their influences on each other and (b) that actual population and employment changes in reality are the result of adjustment processes from a current state towards an equilibrium state that may only be partially attained.

$$P_{i,t}^{*} = f(H_{i,t-1}, E_{i,t}^{*}) \tag{7.1}$$

$$E^*_{i,t} = g(B_{i,t-1}, P^*_{i,t}) \tag{7.2}$$

$$\Delta P_{i,t} = P_{i.t} - P_{i.t-1} = \lambda_P \cdot (P^*_{i,t} - P_{i.t-1}) \tag{7.3}$$

$$\Delta E_{i,t} = E_{i.t} - E_{i.t-1} = \lambda_E \cdot (E^*_{i,t} - E_{i.t-1}) \tag{7.4}$$

where $P^*_{i,t}$ and $E^*_{i,t}$ represent the equilibrium level of population and employment in area i in year t, while $P_{i.t}$ and $E_{i.t}$ indicate actual population and employment (i.e., observations). $H_{i,t-1}$ and $B_{i,t-1}$ are exogenous variables, representing household and business location choice factors, respectively. λ_P and λ_E are adjustment rates of population and employment towards the equilibrium points (i.e., $P^*_{i,t}$ and $E^*_{i,t}$). The two adjustment rates, between 0 and 1 by definition, represent the essence of this model. They are also the parameters, among many others, to be estimated.

To extend this form of the framework, Boarnet (1994a) suggested that population and employment in a certain area can also have effects on adjacent zones. In other words, for instance, $P^*_{i,t}$ is not only influenced by $E^*_{i,t}$ but also by $E^*_{j,t}$, as workers may decide to live in zone i, because they are working in zone j close to i. Boarnet (1994a) expressed this point by replacing Eqs. 7.1 and 7.2 with the following formulation.

$$P^*_{i,t} = f(H_{i,t-1}, \bar{E}^*_{i,t}) \tag{7.5}$$

$$E^*_{i,t} = g(B_{i,t-1}, \bar{P}^*_{i,t}) \tag{7.6}$$

where $\bar{P}^*_{i,t}$ and $\bar{E}^*_{i,t}$ represent the equilibrium level of population and employment in the local labor market centered on zone i in year t.

Boarnet (1994a) further specified the population and employment in the larger labor shed area, centered on each zone, using a spatial weight matrix (W), as follows:

$$\bar{P}^*_{i,t} = (I + W) \cdot P^*_{i,t} \tag{7.7}$$

$$\bar{E}^*_{i,t} = (I + W) \cdot E^*_{i,t} \tag{7.8}$$

Based upon these basic settings, an estimable disequilibrium adjustment model can be derived. In detail, from Eqs. 7.3, 7.4, 7.7, and 7.8, the following set of equations can be developed:

$$\bar{P}^*_{i,t} = (I + W) \cdot P_{i.t-1} + \frac{1}{\lambda_P} \cdot (I + W) \cdot \Delta P_{i,t} \tag{7.9}$$

$$\bar{E}^*_{i,t} = (I + W) \cdot E_{i.t-1} + \frac{1}{\lambda_E} \cdot (I + W) \cdot \Delta E_{i,t} \tag{7.10}$$

In addition, from Eqs. 7.3, 7.4, 7.5, and 7.6, the observed population and employment can be written as follows, assuming a linear relationship in Eqs. 7.5 and 7.6.

$$\Delta P_{i,t} = H_{i,t-1} \cdot \beta_P \cdot \lambda_P + \lambda_P \cdot \theta_P \cdot \bar{E}^*_{i,t} - \lambda_P \cdot P_{i,t-1} + u_{i,t} \tag{7.11}$$

$$\Delta E_{i,t} = B_{i,t-1} \cdot \beta_E \cdot \lambda_E + \lambda_E \cdot \theta_E \cdot \bar{P}^*_{i,t} - \lambda_E \cdot E_{i,t-1} + v_{i,t} \tag{7.12}$$

where β_P, β_E, θ_P, and θ_E indicate the parameters for corresponding determinants of the level of population and employment; and both $u_{i,t}$ and $v_{i,t}$ are the independent and identically distributed random error terms.

If Eqs. 7.9 and 7.10 are plugged into Eqs. 7.11 and 7.12, a spatial econometric version of the disequilibrium adjustment model, which is a structural equation system, can be derived, as shown by Boarnet (1994a).

$$\begin{aligned}\Delta P_{i,t} = {} & H_{i,t-1} \cdot \beta_P \cdot \lambda_P + \lambda_P \cdot \theta_P \cdot (I + W) \cdot E_{i,t-1} \\ & + \frac{\lambda_P \cdot \theta_P}{\lambda_E} \cdot (I + W) \cdot \Delta E_{i,t} - \lambda_P \cdot P_{i,t-1} + u_{i,t}\end{aligned} \tag{7.13}$$

$$\begin{aligned}\Delta E_{i,t} = {} & B_{i,t-1} \cdot \beta_E \cdot \lambda_E + \lambda_E \cdot \theta_E \cdot (I + W) \cdot P_{i,t-1} \\ & + \frac{\lambda_E \cdot \theta_E}{\lambda_P} \cdot (I + W) \cdot \Delta P_{i,t} - \lambda_E \cdot E_{i,t-1} + v_{i,t}\end{aligned} \tag{7.14}$$

It needs to be noted that population and employment densities, as opposed to the absolute numbers, are often used as the dependent variables (e.g., Carlino and Mills 1987; Clark and Murphy 1996). In some cases, a log linear form of the model specification has also been adopted (e.g., Carruthers and Mulligan 2007, 2008). Moreover, a considerable number of studies, using the framework, include additional dependent variables (e.g., land area, housing, wage, employment by industry, etc.) in the simultaneous equation system, in addition to population and employment (e.g., Carruthers and Mulligan 2008; Vermeulen and Ommeren 2009). However, the fundamental logic (i.e., characterizing the changes as a disequilibrium adjustment process) and the simultaneous structural equation form are the common grounds of this methodology. Mulligan et al. (1999), Vias and Mulligan (1999), and Boarnet et al. (2005) have provided meaningful suggestions on how this model can be used in a more effective and appropriate manner.

7.2.2 Applications

Although the model requires a special treatment in estimation due to the simultaneity involved, it is useful for a variety of research purposes, thus employed for a wide range of empirical analyses. Among others, the model's explicit consideration of the population–employment interactions enables researchers test whether (a) jobs follow people, (b) people follow jobs, or (c) reciprocal causality exists. Here, of interest are the signs and significance of $(\lambda_P \cdot \theta_P)/\lambda_E$ and $(\lambda_E \cdot \theta_E)/\lambda_P$in Eqs. 7.13 and 7.14 that show the influences of population changes on employment and vice versa. Carlino and Mills (1987), in one of the pioneering studies that applied the adjustment model, examined this population–employment interaction issue. They analyzed county-level population and employment changes in the United States from 1970 to 1980, and found a positive reciprocal interrelationship between population and total employment. They also paid attention to the interaction between population and manufacturing employment. According to the results of their analysis, population shows a significant negative effect on manufacturing employment growth, while the relationship the other way around is positive. Boarnet's (1994b) study, using municipality-level data in a part of New Jersey, also examined the population–employment interaction. The estimation result (p. 93) suggests that jobs are likely to follow people rather than vice versa, consistent with the result of a well-known study by Steinnes (1977). Using the adjustment model, many other studies also have empirically investigated the issue in other contexts. For instance, Vias (1999) examined rural Rocky Mountain region and found a similar conclusion that jobs tend to follow people, whereas the effect of employment changes on population appears weak. Clark and Murphy (1996) conducted a county-level analysis with 1980s data in the United States. Their analysis suggests that the effects of population on employment change vary by sector.[1] Although population change exhibited positive estimates on the employment in all five sectors that they test (i.e., Manufacturing, Construction, Service, Trade, and FIRE: Finance, insurance and real estate), the only effects that were statistically significant were in the Construction and FIRE sectors.

In addition to the investigation of the population-employment interactions, the spatial econometric version of the disequilibrium adjustment model has been employed to analyze how the growths or declines in different areas are interrelated. Henry et al. (1997) study of the spread vs. backwash effects is a notable example of this kind. They analyzed how the growth in urban core and fringe areas affected the population and employment changes in surrounding hinterlands based upon the disequilibrium adjustment model, and found a mix of spillover and backwash effects from the case study of southern regions in the United States (covering parts of Georgia, North Carolina and South Carolina). Specifically, according to their analysis, rural hinterlands were more likely to grow faster, if the adjacent urban fringes were rapidly growing, with a slower pace of growth in the core. In such a context, employment in the hinterlands was also found to increase more. Later, the authors (Henry et al. 1999, 2001) analyzed the urban–rural linkage in France. By testing

[1] By conducting a large number of experiments, Hoogstra et al. (2011) finds that the outcomes of the empirical studies on this population–employment interactions can differ by many other factors, such as measurements and spatial weight matrix specification.

various specifications of the adjustment model, Henryet al. (2001) detected a strong spillover effect of urban growth on rural areas in terms of population, but not that much significant impact on employment. Feser and Isserman (2005) investigated the case of 48 states using more recent (i.e., 1990–2000) county-level data. Their empirical analysis also revealed mixed effects, namely positive spillovers from the growth in highly urbanized counties to rural areas, but a competition with suburban (i.e., mixed urban + rural) counties.

The disequilibrium adjustment model has also been widely adopted for a variety of policy and impact analyses. As it properly separates the influences arising from population–employment and spatial linkages from the effects of other factors of growth, the model can be used for a more precise assessment of population and employment changes in relation to many potential growth factors and policies of interest. For instance, Bollinger and Ihlanfeldt (1997) measured the impact of Atlanta's MARTA rail transit system on census tract level growth. Using employment by industry data with the adjustment model, their analysis examined not only whether population and total employment was affected by the transit stations but also how the industrial structure was influenced. They found no substantial effect of the transit system on the growth (i.e., population and total employment), while it appeared that the system "has altered the composition of employment in favor of the public sector . . . in those areas with high levels of commercial activity" (p. 202). Henry et al. (1997), in the study noted earlier, paid attention to rural development factors as well as the spatial linkages among urban, suburban, and rural areas. In this case, a variety of local amenity features (infrastructure, public service, housing, labor force, school quality, etc.) were considered using a set of representative variables (p. 486–487), in order to come up with policy recommendations for rural area development. Among others, the provision of key public service, housing, and school quality were found to be significant for population growth in rural areas. There are many other applications of the adjustment model for the policy and impact analysis purposes, particularly those aiming to understand the growth determinants in rural areas and to test the role of infrastructure and amenity. (see e.g., Duffy-Deno 1998; Deller et al. 2001; Edmiston 2004; Vermeulen and Ommeren 2009; Ke and Feser 2010).

7.3 Forecasting with a Disequilibrium Adjustment Model

As explained in the previous section, the regional disequilibrium adjustment framework has been widely used as a powerful analytic method for the empirical analyses of population–employment interactions, spatial interlinkages, the real effects of various potential growth factors or policy instruments, and so forth. In addition to its usefulness in such empirical assessments, the model's dynamic nature would present an opportunity to be employed as a tool for forecasting and dynamic impact analysis. The model's explicit description of population and employment changes as well as relatively flexible formation may be other merits that need to be recognized. Particularly, the model can well fit to socio-economic forecasting and analysis for small areas where population and employment often change dramatically due to dynamic relocation processes.

Once the model is properly estimated, it is possible to derive the predicted values of the dependent variables of Eqs. 7.13 and 7.14 (i.e., population and employment changes in each zone between t−1 and t) from the state in t−1. In addition, it is possible to estimate the impact of a particular shift in an explanatory variable on population and employment changes, using the estimated model.

Few studies have attempted to generate projections using this approach. Mills and Lubuele (Mills and Lubuele 1995) estimated a simultaneous three equation model (in which population, employment, and wage were dependent variables) with 1970, 1980, and 1990 data for U.S. metropolitan statistical areas (MSA), and projected the changes in the three variables for a following time span (i.e., 1990–2000) to see how the MSAs' growth pattern will be in future. In the case of Carruthers and Mulligan (2007), a different set of three variables, (1) acres of developed land per person, (2) acres of developed land per employee, and (3) acres of developed land, are modeled. Then, using the model estimated with 1982, 1987, and 1992 data for metropolitan counties in the United States, the future (i.e., 1992–1997) pattern of land absorption was projected.

Although technically the estimated model can be used for projection purposes, the adjustment model alone presents challenges when considered as a plausible long-term forecasting method. The framework is not designed to project a long-term growth trajectory, although it describes the dynamic process of population and employment changes. If the model is used as it stands to project the future, the captured adjustment processes will be iterated without the full consideration of the essential growth momentum (i.e., the system-wide effects of changes in an economy). This can be very problematic from a long-term forecasting perspective, while it may be still meaningful for short-term projections. For this reason, the previous studies may have generated the predicted values only for the next time span rather than extending the projection period, when they have used the estimated adjustment model for the future analysis.

This challenge to utilizing the model for a forecasting purpose, could be addressed by using the model together with another framework that can describe the fundamental growth forces more effectively. The next section presents a proposal to integrate the disequilibrium adjustment model hierarchically with a REIM to forecast municipality-level population and employment changes under a metropolitan-wide economic growth forces.

7.4 Integrating REIM and Disequilibrium Adjustment Model

As indicated above, the disequilibrium adjustment model can be better used for a forecasting and dynamic impact analysis, when it is combined with a growth model. Among others, the REIM may be an appropriate partner of the disequilibrium adjustment model, as it captures dynamic changes in the demands for regionally produced goods and services, one of the main driving forces of regional economic growth. The REIM can also benefit from the adjustment model that captures the internal dynamics of the economic system. As a macro-economic framework, generally REIM is dealing with spatially aggregated variables (i.e., region-wide values as opposed to the distribution within a region), thus as it stands has limited usefulness for more detailed spatial analysis. If the variables can be spatially disaggregated in an appropriate manner, being combined with the adjustment model, the REIM can become a method for a more complete forecasting and empirical analysis (Kim and Hewings 2012).

As discussed by West (1995), West and Jackson (1998), and Rey (2000), a REIM describes growth and structural changes of a regional economic system with explicit consideration of the dynamic regional industrial structure. The integration of econometrics, namely the structural time-series equation modeling, and regional input–output makes it possible to simulate the complex behavior of a regional economic system.[2] For instance, the Chicago REIM (Israilevich et al. 1997; Hewings et al. 1998), which will be combined with a disequilibrium adjustment model, depicts the growth and transformation of the Chicago economy under the influence of national economic changes. Please refer to Israilevich et al. (1997) for a detailed explanation of the model.

The simplest way to feed the adjustment model may be to determine the long-term regional growth forecasts from the REIM and to use the numbers in adjustment modeling. However, a consistency issue can arise, if this strategy of loose coupling is adopted. To be consistent, the sum of the population and employment in individual zones from the adjustment model need to be fixed at the predetermined numbers from the REIM. Although the consistency can be ensured with a strict top-down approach, the integration in this manner is not desirable, as it does not attain the full potential of the integrated framework. As discussed in Kim and Hewings (2011, p. 691) "Because the integration [approach] neglects the effect of spatial structure and other subregional conditions on the performance of the regional economy or the regional demographic changes, the forecasting or simulation outcomes may be challenged in the sense that they may under- or overstate the more probable outcomes. Moreover, the framework based on the top-down integration inevitably has limited usefulness in the sense that the macroeconomic effects of different land-use policies or other actions at the subregional level cannot be appropriately assessed due to the lack of the bottom-up or feedback linkages."

A better integration approach is to derive the potential regional growth momentum, as opposed to the determined region-wide population and employment changes, from the REIM and then incorporate the potential growth variable into the disequilibrium adjustment model that describes the changes of the subzones within

[2] Given the methodological advantages, REIMs have long been widely used for regional socio-economic forecasting and various types of advanced policy and/or impact analyses (Kim and Hewings 2011).

the region. Then, the adjustment model can project an individual subarea's future with consideration of long term regional economic growth forces, but not strictly bounded by a certain number. This strategy of the integration can allow the adjustment model to modify the region-wide values with the consideration of a variety of subregional conditions, including resource constraints, transportation networks, the efficiency of the spatial structure of the region, etc., whereas under the top-down approach it will merely act as an allocation tool, assuming that the predetermined changes from the REIM will be realized regardless of the intraregional conditions.

This approach to a better integration can be implemented using the concept of expected growth that exists in some types of REIM. The expected level of growth, available in the Chicago REIM, can be plugged into the adjustment model to address the lack of the consideration of the fundamental growth momentum. Then, the disequilibrium adjustment model will determine whether or not each subzone will achieve the expected level of growth, that is derived considering the long-term growth momentum of the regional economy from REIM (see Kim and Hewings 2012). In other words, the adjustment model part can act as a module of an integrated framework that effectively describes the real changes in small areas with consideration of long-term growth forces as well as many location choice factors, population-employment interactions, spatial linkages, etc.

7.5 Forecasting and Impact Analysis with an Applied Model

The idea to develop an integrated framework, where REIM and a disequilibrium adjustment model are hierarchically combined, had been first implemented for the Chicago metropolitan area (Kim and Hewings 2012). The study area for the model consists of 7 counties and 296 municipalities in Illinois, USA (Fig. 7.1). The individual municipalities and seven unincorporated areas of the counties are the units of disequilibrium adjustment modeling. In other words, there are total 303 (=296 + 7) small areas, included in the modeling.

The integrated framework, presented in Kim and Hewings (2012), can be used for small area (i.e., municipality level) population and employment forecasting. Furthermore, the framework can support dynamic impact analysis, showing how the growth trajectories will differ under various conditions of interest. Admittedly, population and employment projections under a set of scenarios are essential for a broad range of development policy making and regional or community level planning practices.

Using the applied model, this study projects small area population and employment changes under three different scenarios, that represent a range of future economic performance of the U.S. economy – (1) High growth, (2) Baseline, and (3) low growth national economic. For the baseline scenario, national economic forecasts, produced by IHS Global Insight, are adopted; and for two other scenarios, ±0.3 % of variation in growth rates is considered. The three scenarios have 2.4 %, 2.7 %, and 3.0 % of the compound annual GDP (Gross domestic products) growth

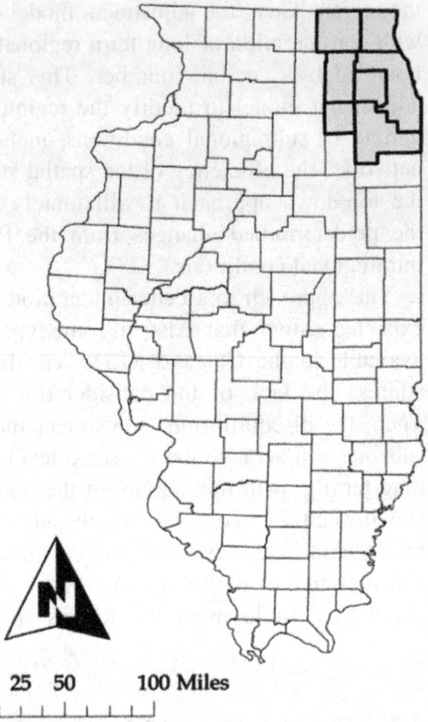

Fig. 7.1 Study region and the State of Illinois

rate for the following 30 years, respectively. Since a wide range of national economic indicators (GDP, investments, unemployment rate, etc.) are used in the Chicago REIM as key explanatory variables, the regional economic growth projection depends on the performance of national economy. Further, different trajectories of potential regional economic growth, projected by the Chicago REIM under different scenarios, will result in the shifts in small area population and employment.

Figures 7.2 and 7.3 show how total regional population and employment (i.e., the aggregated values of all 303 units of analysis) will change in future under the three different scenarios and reveal the net impact of the changes in national economic conditions. Under the growth scenarios with varying growth rates in national macroeconomic variables, it appears that municipalities' employment growth trajectories can be influenced significantly, while population growth shifts are somewhat modest. More specifically, ±0.3 % of variation in national economic growth rates causes ±0.1 % changes in population but changes ±0.4 % in employment growth rate in the study area, according to the projections generated by the integrated framework. As the national economy grows faster, first the Chicago region's economy can enjoy increasing demands for exporting goods and services. This stimulates an expansion of the regional production and further creates the ripple effects through inter-industrial linkages. A greater amount of production expansion certainly induces larger employment growth; and increasing job opportunities promotes population growth but not as much as the rate of employment increase.

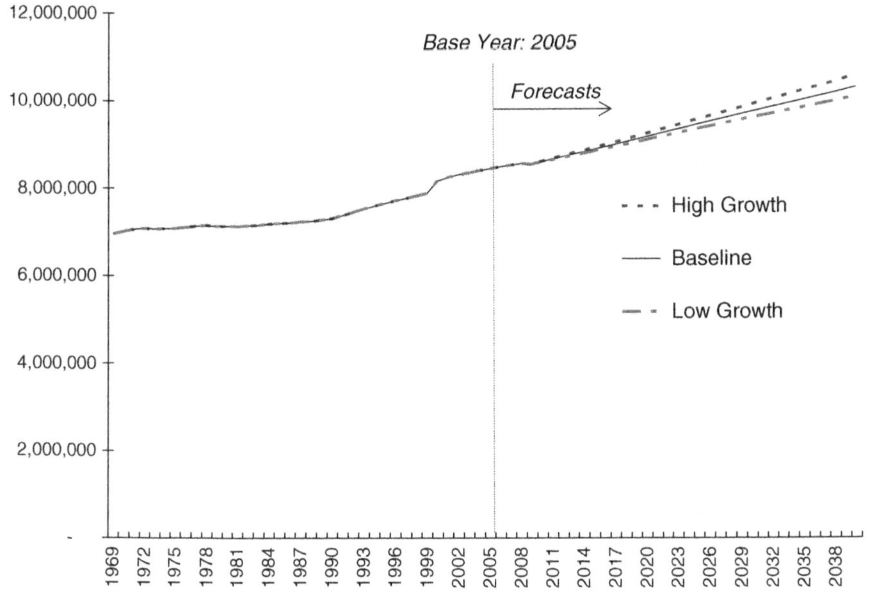

Fig. 7.2 Aggregated population growth under three scenarios

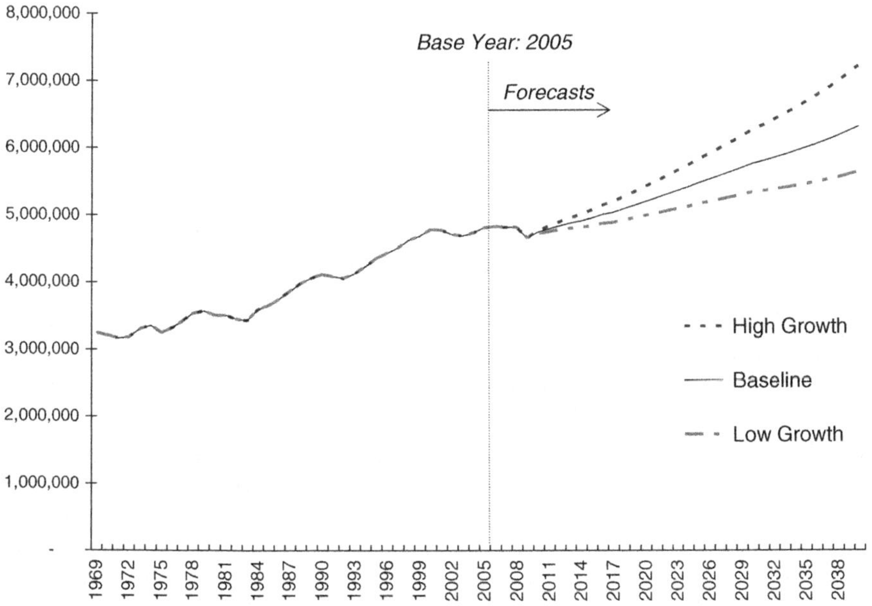

Fig. 7.3 Aggregated employment growth under three scenarios

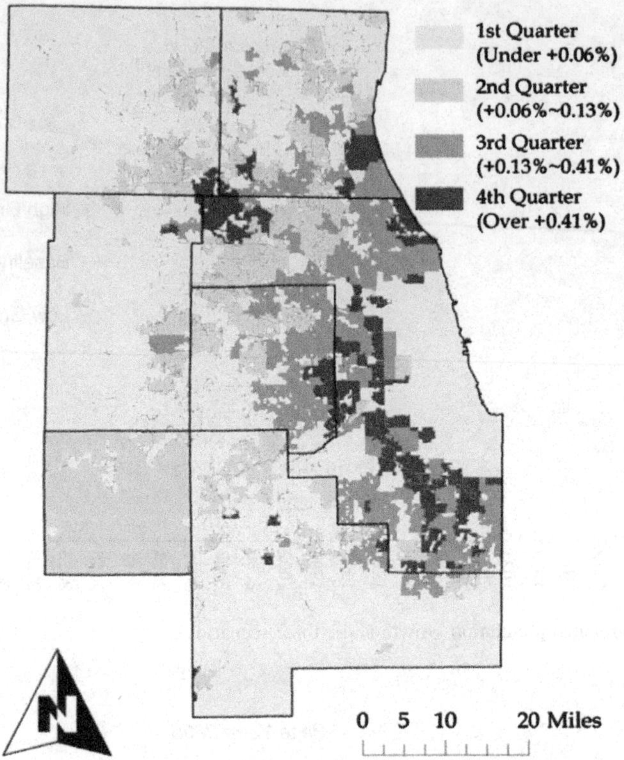

Fig. 7.4 Differences in the forecasted annual population change rates between high and low growth scenarios

Figures 7.4 and 7.5 present how individual municipalities will grow differently under different national economic situations by demonstrating the gaps in compound annual population and employment change rates between high and low growth scenarios. Although most municipalities exhibit faster growth rates under higher national economic growth conditions, the effects are not the same across space. Particularly, a large degree of variation is found in the changes in municipality-level population growth rates, according to this simulation.

A close look at the simulation outcomes reveals that the places adjacent to the City of Chicago, that have experienced decline in recent years, are likely to receive relatively larger benefits in population increase than suburban or exurban communities. This is at least partly attributable to a significant number of employment increases in the City of Chicago under the high growth scenario. In other words, the spillover effects of the Chicago's enhanced performance in employment tend to benefit the municipalities whose main comparative advantage is the proximity to Chicago, a dominantly large job center in this metropolitan region.

In contrast to such places, suburban and/or exurban communities attract development based on different comparative advantages such as abundant developable

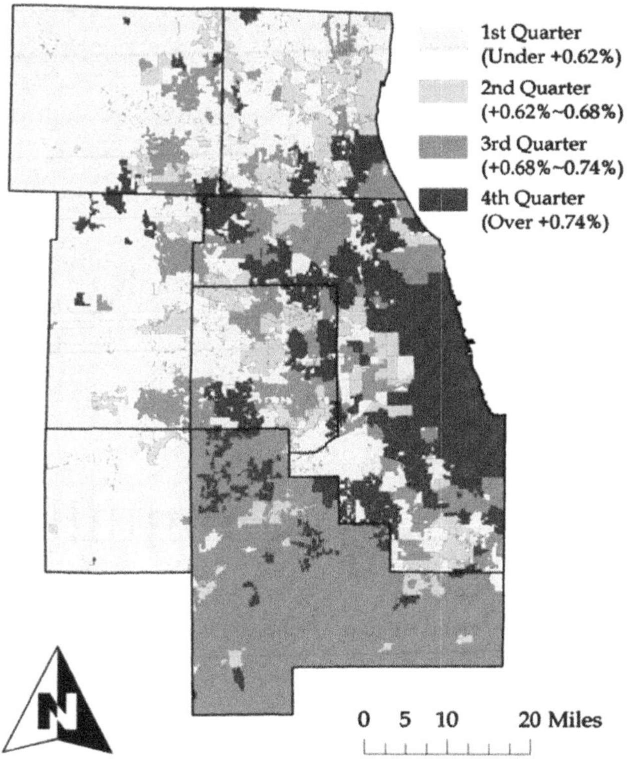

Fig. 7.5 Differences in the forecasted annual employment change rates between high and low growth scenarios

land within their jurisdictions and incorporable hinterlands, factors that are found significant in the estimation of the population change equation of the disequilibrium adjustment model. As the potential regional growth rate shifts up in respond to favorable national economic conditions, first their growth can be accelerated. However, as developable land and hinterland areas are depleted more rapidly (i.e., as the areas are built up more rapidly), they will tend to lose the initial driving forces for growth (i.e., diminishing comparative advantages). As a result, in a longer term, the benefits that these communities receive are not as large as expected, unlike the municipalities next to the central city.

This point is clearly illustrated in Figs. 7.6 and 7.7, showing the population forecasts under high and low growth scenarios for two different cities. Whereas the Village of Elmwood Park (Fig. 7.6), located right next to the City of Chicago, consistently obtains the benefits from the high growth scenario conditions, the Village of Monee (Fig. 7.7), a suburban town in Will County, may not be able to enjoy the benefits in the longer run. In fact, according to the forecasts generated by the integrated framework, Year 2040 population in this village is greater under the low growth scenario than high growth scenario. The rapid economic growth merely

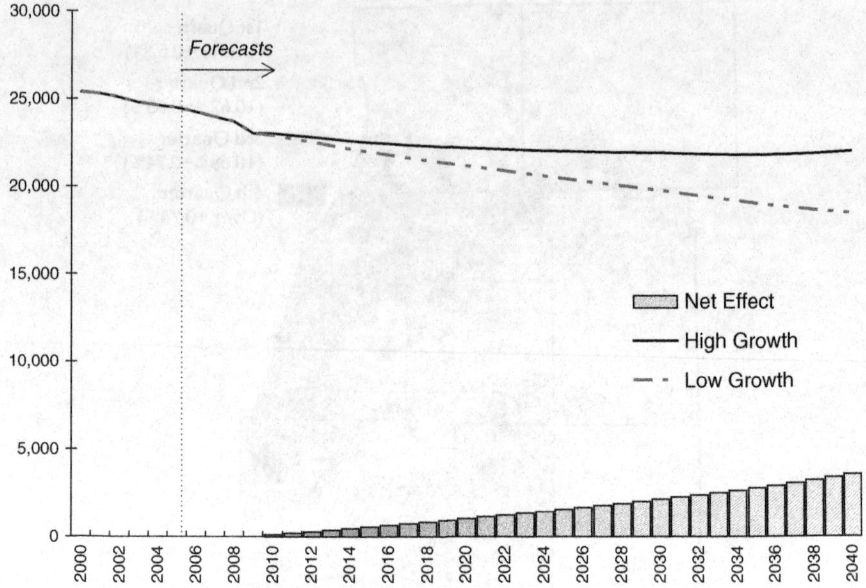

Fig. 7.6 Population forecasts for the Village of Elmwood Park

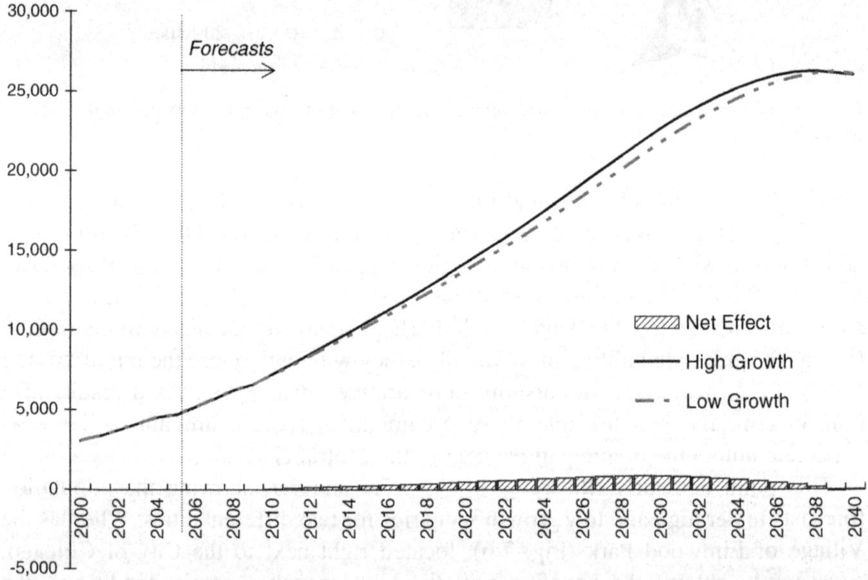

Fig. 7.7 Population forecasts for the Village of Monee

shifts the utilization of the development opportunities ahead rather than promoting sustainable growth.

The advantages of using the disequilibrium adjustment model, combined with a REIM, are well highlighted in this application. The growth trajectories of small areas seem to be effectively described and simulated with the consideration of various critical determinants, ranging from the potential regional growth forces to population–employment interactions, the interlinkages among the small areas, and other spatially explicit conditions (e.g., developable land stock).

7.6 Summary and Discussion

This study presents an application of a disequilibrium adjustment model for small area socio-economic forecasting and impact analysis, when it is part of an integrated multi-level framework that describes both long-term regional economic growth and dynamic internal changes. The integration of the multi-level variables in a single framework expands the capabilities of the adjustment model and provides a possibility of a more systematic examination small area growth dynamics.

Some challenges, however, do exist. Data availability for the small area is one obstacle for most empirical disaggregated level analysis. Model calibration often has to be made with a short period of observations, although long-range data series are required for a more robust long-term forecasting. Some important determinants of small area growth, particularly government policy variables cannot be considered, simply because such information over past years is not available. A good example would be the potential for the small area (communities) to adjust their land use policies in anticipation of attracting more growth. However, the problem is not simple in that the behavior of surrounding communities would also have to be considered generating a dynamic spatial game theoretic framework in which communities try to maximize the potential benefits of employment growth in neighboring locations.

In addition to the data availability problem, the linear fashion of the disequilibrium adjustment model adopted here may not be ideal for describing the intrinsic non-linearity of a dynamic metropolitan system. Another notable challenge is model stability, which is critical in forecasting. As all adjustment model equations (i.e., more than 600 equations in the case of the presented application) are interlinked with each other, it is much more difficult to generate reasonable future growth trajectories than the case of modeling individual variables separately. This is especially challenging, when the areas considered have a large degree of variation in terms of the size.

Finally, there is the problem of the use of a traditional spatial weight matrix, that is fixed and assumed to represent the spatial interdependence correctly.[3] Given that

[3] Cuaresma and Feldkircher (2010) has provided significant evidence that the choice of weight matrix is important, challenging the received assumption to the contrary.

the pattern of spatial interaction may evolve over time, it would not be desirable to use the fixed weight matrix for a long forecast period. A significant decline in transportation cost and changes in the transportation network (including in Chicago the construction of suburb-to-suburb passenger rail lines) will change the connections between the areas. With increases in two-career households, residential location decision-making in metropolitan areas is now more complicated creating further challenges for modeling population-employment dynamics. Thus, attention needs to be paid to how these issues can be addressed by reconciling the conflicts between the weight matrix requirements in econometric estimation and an increasing need for describing the evolution of spatial interdependence in long-term forecasting or simulation. Little is known about how we can better consider and handle this temporal variation of the spatial relation, although recently many studies have attempted to carefully choose a spatial weight matrix among many possible ways of constructing the matrix (e.g., contiguity-based, distance-based, flow-data-based, etc.) in empirical analysis (see e.g., Boarnet et al. 2005; Hoogstra et al. 2011). The use of spatio temporal weight matrices (see e.g., Sun et al. 2005; Maddison 2009) and few other attempts (e.g., Cheng et al. 2012) recently emerged in the literature could be a meaningful step towards a more robust consideration of the dynamic spatial dependence.

References

Boarnet MG (1994a) An empirical model of intrametropolitan population and employment growth. Pap Reg Sci 73:135–152

Boarnet MG (1994b) The monocentric model and employment location. J Urban Econ 36:79–97

Boarnet MG, Chalermpong S, Geho E (2005) Specification issues in models of population and employment growth. Pap Reg Sci 84:21–46

Bollinger C, Ihlanfeldt KR (1997) The impact of rapid rail transit on economic development: the case of Atlanta's MARTA. J Urban Econ 42:179–204

Carlino GA, Mills ES (1987) The determinants of county growth. J Reg Sci 27:39–54

Carruthers JI, Mulligan GF (2007) Land absorption in U.S. metropolitan areas: estimates and projections from regional adjustment models. Geogr Anal 39:78–104

Carruthers JI, Mulligan GF (2008) A locational analysis of growth and change in American metropolitan areas. Pap Reg Sci 87:155–171

Cheng T, Haworth J, Wang J (2012) Spatio-temporal autocorrelation of road network data. J Geogr Syst 14:389–413

Clark DE, Murphy CP (1996) Countywide employment and population growth: an analysis of the 1980s. J Reg Sci 36:235–256

Cuaresma JC, Feldkircher M (2010) Spatial filtering, model uncertainty and the speed of income convergence in Europe. Working Paper 160, Öesterreichische National Bank, Vienna

Deller SC, Tsai TH, Marcouiller DW, English BK (2001) The role of amenities and quality of life in rural economic growth. Am J Agric Econ 83:352–365

Duffy-Deno KT (1998) The effect of federal wilderness on county growth in the intermountain western United States. J Reg Sci 38:109–136

Edmiston KD (2004) The net effects of large plant locations and expansions on county employment. J Reg Sci 44:289–319

Feser E, Isserman A (2005) Urban spillovers and rural prosperity. http://www.urban.uiuc.edu/faculty/feser/Pubs/Urban_spillovers_and_rural_prosperity.pdf Accessed date: October 15, 2008
Henry MS, Barkley DL, Bao S (1997) The hinterland's stake in metropolitan growth: evidence from selected southern regions. J Reg Sci 37:479–501
Henry MS, Schmitt B, Kristense K, Barkley DL, Bao S (1999) Extending Carlino-Mills models to examine urban size and growth impacts on proximate rural areas. Growth Change 30:526–548
Henry MS, Schmitt B, Piguet V (2001) Spatial econometric models for simultaneous systems: application to rural community growth in France. Int Reg Sci Rev 24:171–193
Hewings GJD, Sonis M, Guo J, Israilevich PR, Schindler GR (1998) The hollowing out process in the Chicago economy, 1975–2015. Geogr Anal 30:217–233
Hoogstra GJ, van Dijk J, Florax RJGM (2011) Determinants of variation in population – employment interaction findings: a quasi-experimental meta-analysis. Geogr Anal 43:14–37
Israilevich PR, Hewings GJD, Sonis M, Schindler GR (1997) Forecasting structural change with regional econometric input–output model. J Reg Sci 37:565–590
Ke S, Feser E (2010) Count on the growth pole strategy for regional economic growth? Spread-backwash effects in greater central China. Reg Stud 44:1131–1147
Kim JH, Hewings GJD (2011) Framing urban systems and planning concerns as a multi-level problem: a review of the integrated urban system models with an emphasis on their hierarchical structures. In: Brooks N, Donaghy K, Knaap G (eds) Oxford Handbook of Urban Economics and Planning. Oxford University Press, New York, pp 674–700
Kim JH, Hewings GJD (2012) Integrating the fragmented regional and subregional socio-economic forecasting and analysis: a spatial regional econometric input–output framework. Ann Reg Sci 49:485–513
Maddison D (2009) A spatio-temporal model of farmland values. J Agric Econ 60:171–189
Mills ES, Lubuele LS (1995) Projecting growth of metropolitan areas. J Urban Econ 37:344–360
Mulligan GF, Vias AC, Glavac SM (1999) Initial diagnostics of a regional adjustment model. Environ Plann A 31:855–876
Rey SJ (2000) Integrated regional econometric + input–output modeling: issues and opportunities. Pap Reg Sci 79:271–292
Steinnes DN (1977) Causality and intra urban location. J Urban Econ 4:69–79
Sun H, Tu Y, Yu SM (2005) A spatio-temporal autoregressive model for multi-unit residential market analysis. J Real Estate Financ Econ 31:155–187
Vermeulen W, Ommeren J (2009) Does land use planning shape regional economies? A simultaneous analysis of housing supply, internal migration and local employment growth in the Netherlands. J Hous Econ 18:294–310
Vias AC (1999) Jobs follow people in the rural Rocky Mountain West. Rural Develop Perspect 14:14–23
Vias AC, Mulligan GF (1999) Integrating economic base theory with regional adjustment models: the nonmetropolitan Rocky Mountain West. Growth Change 30:507–525
West GR (1995) Comparison of input–output, input–output + econometric, and computable general equilibrium impact models at the regional level. Econ Syst Res 7:209–227
West GR, Jackson RW (1998) Input–output + econometric and econometric + input–output: model differences or different models? J Reg Anal Pol 28:33–48

Chapter 8
An Information Theoretic Approach to Ecological Inference in Presence of Spatial Dependence

Rosa Bernardini-Papalia

8.1 Introduction

This chapter introduces an Information Theory (IT)-based method for modeling economic aggregates and for obtaining estimates for small area (sub-group) or subpopulations when no sample units or limited data are available. The proposed approach offers a tractable framework for modeling the underlying variation in small area indicators, in particular when data set contains outliers and in presence of collinearity among regressors since the maximum entropy estimates are robust with respect to the outliers and also less sensitive to a high condition number of the design matrix. A basic ecological inference problem which allows for spatial heterogeneity and dependence is presented with the aim of estimating small area/ sub-group indicators by combining all available information at both macro and micro data level.

The latent small area indicators may be treated as random coefficients or modeled as a parametric functional relationship in the unit level model in which the observed aggregate is regressed on the explanatory variables both at the group and sub-group level.

By taking as a point of departure the approach presented in Johnston and Pattie (2000), Judge et al. (2004), Peeters and Chasco (2006) and Bernardini-Papalia (2010a, b), the basic idea is to introduce an estimator based on an entropy measure of information which provides an effective and flexible procedure for reconciling

This chapter has been completed within the BLU-ETS project "Blue-Enterprises and Trade Statistics", a small or medium-scale focused research project funded by the Seventh Framework Programme of the European Commission, FP7-COOPERATION-SSH (Cooperation Work Programme: Socio-Economic Sciences and the Humanities).

R. Bernardini-Papalia (✉)
Department of Statistical Sciences, University of Bologna, Italy, Via Belle Arti, 41, Bologna 40126, Italy
e-mail: rossella.bernardini@unibo.it

E. Fernández Vázquez and F. Rubiera Morollón (eds.), *Defining the Spatial Scale in Modern Regional Analysis*, Advances in Spatial Science,
DOI 10.1007/978-3-642-31994-5_8, © Springer-Verlag Berlin Heidelberg 2012

micro and macro data. The maximum entropy (ME) procedures (Levin 1980; Shannon 1948; Golan et al. 1994, 1996; Golan 2008) give the possibility to take into account out-of-sample information which can be introduced as additional constraints in the optimization program or by specifying particular priors for parameters and errors. An unique optimum solution can be achieved also if there are more parameters to be estimated than available moment conditions and the problem is ill-posed. If there exists additional non-sample information from theory and/or empirical evidence, over that contained in the consistency and adding-up constraints, for the unknown probabilities, it may be introduced in the form of known probabilities, by means of the cross-entropy formalism (Kullback 1959).

The chapter is structured as follows. In Sect. 8.2 an introduction to the traditional ecological inference (EI) problems is presented. Alternative approaches to ecological modeling that account for spatial heterogeneity and spatial dependence problems, are also introduced. Section 8.3 provides the formulation of the proposed information theoretic approaches incorporating both spatial heterogeneity and dependence. In Sect. 8.4, the IT-based disaggregation procedure is applied to Italian data. Finally, the last section provides concluding remarks and outlines some direction for further research.

8.2 Ecological Inference and Dependence Across Space

The traditional approach to ecological inference is based on the homogeneity across space hypothesis which assumes constancy of parameters across the disaggregate spatial units. This assumption is rarely tenable, since the aggregation process usually generates macro-level observations across which the parameters describing individuals may vary (Cho 2001). It is recognized that observations at an aggregate level of analysis do not necessarily provide useful information about lower levels of analysis, particularly when spatial heterogeneity is present. Moreover, the objective of recovering disaggregate information from aggregate data may produce "ill-posed" or "undetermined" inverse problems given that there are more unknowns than data points. In EI it is also important to deal with the "modifiable area unit problem" which refers to (1) the scale effect or aggregation effects, and (2) the grouping effect or zoning effect. In the first case the resulting aggregation bias may produce different results when data (or individuals) are grouped into increasingly larger areal units. In the second case, the resulting specification bias is connected to the variability in results due to alternative formulations of the areal units leading to differences in unit shape at the same or similar scales and arises when there is a non linear relationship that is not properly accounted for in the specification of the aggregated model. Many different possible relationships at the individual (or subgroup) level can generate the same observations at the aggregate (or group) level (King 1997; King et al. 2004). In the absence of individual (or subgroup) level measurement (in the form of survey data), such information need to be inferred. Estimates of the disaggregated values for the variable of interest can be inferred from aggregate data by using appropriate statistical techniques. However, in many situations, given that micro-data of interest are not available, the accuracy of any predicted value cannot be verified.

Moreover, in presence of spatial structures, (1) absolute location effects (that refer to the impact—for each unit—of being located at a particular point in space), and (2) relative location effects (that consider relevant the position of an unit relative to other units, Spatial dependence), have to be considered.

The absolute location effects can be introduced by assuming: (1) slope heterogeneity across spatial units, implying that parameters are not homogeneous over space but vary over different geographical locations; (2) the presence of cross-sectional correlation due to the presence of some common immeasurable or omitted factors.

The relative location effects are traditionally introduced by incorporating: a spatial autoregressive process in the error term, and/or a spatially lagged dependent variable. A Spatial Error Model specification assumes that the spatial autocorrelation is modeled by a spatial autoregressive process in the error terms. It follows that: spatial effects are assumed to be identical within each unit, but all the units are still interacting spatially through a spatial weight matrix. The presence of spatial dependence is then associated with random shocks (due to the joint effect of misspecification, omitted variables, and spatial autocorrelation). In alternative, a Spatial Autoregressive Model specification, (Spatial Lag Model) assumes that all spatial dependence effects are captured by the lagged term. The spatial autocorrelation is then modeled by including a spatially lagged dependent variable. Global and local measures of spatial autocorrelation are computed to determine whether the data exhibit spatial dependence and a series of test statistics based on the Lagrange Multiplier (LM) or Rao Score (RS) principle are used to determine whether the variables in the model sufficiently capture the spatial dependence in the data. If the variables do not fully model the dependence, the diagnostics indicate whether the researcher should estimate a model with a spatially lagged dependent variable, a spatially lagged error term, or both. The LM/RS principle can also be extended to more complex spatial alternatives, such as higher order processes, spatial error components and direct representation models. Paralleling and complementing the theoretical motivation may represent a useful guide for modelling the spatial dependence.

The objectives of the chapter are (1) to formulate an informational-theoretical approach to estimate small area/sub group variables or indicators in the presence of spatial structure and limited/incomplete information; (2) to provide an empirical application to real data.

As a first task, a functional relationship between the variable to be disaggregated and a set of variables/indicators at area level is specified by combining different macro and micro data sources. The model at the aggregate level is then estimated and the sub-group level variables/indicators are obtained by employing these parameter estimates. Different model specifications extended to include spatial effects are also introduced with the aim of testing the hypothesis of: (1) parameters homogeneity/heterogeneity; (2) uniform/varying spatial dependence.

We start by defining the aggregate indicator for group/region i, yi, as a weighted geometric mean of the latent small area or sub group indicator y_{ij} in group/region i: $\mathbf{y}_i = \prod_{j=1}^{J_i} (\mathbf{y}_{ij})^{\boldsymbol{\theta}_{ij}}$, that is:

$$\ln \mathbf{y}_i = \sum_{j=1}^{J_i} (\ln \mathbf{y}_{ij})\theta_{ij} \tag{8.1}$$

where $\mathbf{y}_{ij}$ is the indicator of the jth small area (sub group/region) in group/region i, $\boldsymbol{\theta}_{ij}$ is the weight of small area (sub group) j in i, with $\sum_{j=1} \boldsymbol{\theta}_{ij} = 1$, and where $i = 1,\ldots,N$ denotes the groups/regions and $j = 1,\ldots,Ji$ denotes the number of small areas (sub groups/regions) in i.

The latent sub group values are specified in a multiplicative form, which is consistent with a logarithmic type functional form.

The small area/sub-regional indicators are not observed, but the $\mathbf{y}_i$'s and $\boldsymbol{\theta}_{ij}$'s are. In addition, by introducing an observed vector of explanatory variables for group/region i, xi, an observed vector of explanatory variables for small area (sub group/region) j in group/region i, zij, the latent small area/sub-group indicators are expressed in a multiplicative form as follows:

$$\mathbf{y}_{ij} = \boldsymbol{\alpha}_{ij} \prod_{k=1}^{K} \mathbf{Z}_{ij,k}^{\boldsymbol{\beta}_{ij,k}} \prod_{h=1}^{H} \mathrm{X}_{\mathrm{i,h}}^{\boldsymbol{\gamma}_{ij,h}} e^{\boldsymbol{\varepsilon}_{ij}} \tag{8.2}$$

where zij,k (k = 1, K) are the covariates observed at the level of small area/sub group j within the group/region i, xi,h (h = 1,...H) are the covariates observed only at the level of group/region i, $\boldsymbol{\alpha}_{ij}$ are unobserved fixed effects, and $\boldsymbol{\varepsilon}_{ij}$ are error terms.

By substituting Eq. 8.2 into Eq. 8.1, we can obtain the following model:

$$\ln \mathbf{y}_i = \sum_{j=1}^{J_i} \left(\ln \boldsymbol{\alpha}_{ij} + \sum_{k=1}^{K} \boldsymbol{\beta}_{ij,k} \ln \mathbf{z}_{ij,k} + \sum_{h=1}^{H} \boldsymbol{\gamma}_{ij,h} \ln \mathbf{x}_{i,h} + \boldsymbol{\varepsilon}_{ij} \right) \boldsymbol{\theta}_{ij}$$

or

$$\ln \mathbf{y}_i = \sum_{j=1}^{J_i} \left(\ln \boldsymbol{\alpha}_{ij} + \sum_{k=1}^{K} \boldsymbol{\beta}_{ij,k} \ln \mathbf{z}_{ij,k} + \sum_{h=1}^{H} \boldsymbol{\gamma}_{ij,h} \ln \mathbf{x}_{i,h} \right) \boldsymbol{\theta}_{ij} + \mathbf{u}_i \tag{8.3}$$

where $\mathbf{u}_i = \sum_{j=1}^{J_i} \boldsymbol{\varepsilon}_{ij}\boldsymbol{\theta}_{ij}$ is a "composite" error term, which is heteroskedastic.

This model implies some kind of weighted regression, capturing "distributional effects" by using data on weights for each small area/sub group. It is important to point out that we assume: (1) unit specific coefficients for the small areas/sub groups (parameter heterogeneity); (2) a parametric specification of the unobserved spatial effects (spatial heterogeneity) through ε_{ij} 's, which can be positive or negative.

Using the estimated coefficients in Eq. 8.3 we can obtain estimates of the unobserved or latent small area/sub group indicators as follows:

$$\hat{\mathbf{y}}_{ij} = \hat{\boldsymbol{\alpha}}_{ij} \prod_{k=1}^{K} \mathbf{z}_{ij,k}^{\hat{\boldsymbol{\beta}}_{ij,k}} \prod_{h=1}^{H} \mathbf{x}_{i,h}^{\hat{\gamma}_{ij,h}} e^{\hat{\boldsymbol{\varepsilon}}_{ij}} \tag{8.4}$$

As proxies for the ignorance of the sources of spatial dependence, statistically significant parameters on dummy variables for geographic areas merely indicate that behaviours differ for units in these particular areas in contrast to the reference category (Anselin 1988). Such an approach cannot indicate whether the spatial dependence is consistent with diffusion or with the spatial clustering of the behaviour's sources. Spatial diffusion occurs because units' behaviour is directly influenced by the behaviour of "neighbouring units." This diffusion effect corresponds to a positive and significant parameter on a spatially lagged dependent variable capturing the direct influence between neighbours. In the diffusion case, neighbors influence the behavior of their neighbors and vice versa.

If one is unable to fully model the sources of spatial dependence in the data generating process (DGP), the spatial dependence in the error terms between neighboring locations is assumed. This spatial error dependence can be modeled via a spatially lagged error term. It is also possible to hypothesize that spatial dependence is produced both by the diffusion and by the independent adoption of behaviors by neighbors. This joint spatial dependence can be modeled by incorporating both a spatially lagged dependent variable and a spatial error term, with proper identifying restrictions imposed.

When the spatial autocorrelation is modeled by a Spatial Lag Model, Spatial Autoregressive Model (SAR Model), the previous model Eq. 8.3 can be generalized by introducing a spatial-lag term into the model. The resulting latent small area/sub group indicators are specified in a multiplicative form as follows:

$$\ln \mathbf{y}_i = \sum_{j=1}^{J_i} \left(\boldsymbol{\alpha}_{ij} + \sum_{K=1}^{K} \boldsymbol{\beta}_{ij,k} \ln \mathbf{z}_{ij,k} + \sum_{h=1}^{H} \boldsymbol{\gamma}_{ij,h} \ln \mathbf{x}_{i,h} + \rho \ln \mathbf{w}\mathbf{y}_j + \boldsymbol{\varepsilon}_{ij} \right) \boldsymbol{\theta}_{ij} \tag{8.5}$$

where ρ is a spatial lag coefficient (the parameter associated to the spatially lagged dependent variable at local level, ln **wy**), **w** is a proximity matrix of order N.

The definition of neighbors for each observation via a spatial weights matrix is a critical decision in modeling spatial autocorrelation. In empirical applications, it is common practice to derive spatial weights from the location and spatial arrangements of observation by means of a geographic information market. In this case, units are defined 'neighbors' when they are within a given distance of each other, i.e. $w_{\mathrm{ij}} = 1$ for $d\ ij \leq \delta$ and $i \neq j$, where dij is the distance function chosen, and δ is the critical cut-off value.

More specifically, a spatial weights matrix w* is defined as follow:

$$w^*_{ij} = \begin{cases} 0 & \text{if} \quad i = j \\ 1 & \text{if} \quad d_{ij} \leq \delta,\ i \neq j \\ 0 & \text{if} \quad d_{ij} \leq \delta,\ i \neq j \end{cases} \tag{8.6}$$

and the elements of the row-standardized spatial weights matrix w (with elements of a row sum to one) result:

$$w_{ij} = \frac{w^*_{ij}}{\sum_{j=1}^{N} w^*_{ij}}, \qquad i,j = 1, .., N. \tag{8.7}$$

The SAR model assumes that all spatial dependence effects are captured by the lagged term by showing how the performance of the dependent variable impacts all the other (neighbor) groups/regions through the spatial transformation.

In alternative, by assuming a spatial dependence is the error structure (in terms of a first order spatial autoregressive process), the resulting Spatial Error Model (SEM Model) specification relative to model Eq. 8.3 is derived as follows:

$$\ln \mathbf{y}_j = \sum_{j=1}^{J_i} \left(\ln \boldsymbol{\alpha}_{ij} + \sum_{k=1}^{K} \boldsymbol{\beta}_{ij,k} \ln \mathbf{z}_{ij,k} + \sum_{h=1}^{H} \boldsymbol{\gamma}_{ij,h} \ln \mathbf{x}_{i,h} + (\lambda \mathbf{w} \boldsymbol{\varepsilon}_{ij} + \boldsymbol{\tau}_{ij}) \right) \boldsymbol{\theta}_{ij} \tag{8.8}$$

where λ is a spatial autoregressive coefficient, $\mathbf{w}$ is a proximity matrix of order N, as previously defined, and $\boldsymbol{\tau}_{ij}$ are the usual stochastic error terms.

The Spatial Error Model leaves unchanged the systematic component and assumes spatially autocorrelated errors. In this respect, it is observed how a random shock in a small area/sub group affects performances in that small area/sub group and additionally impacts all the other small areas/sub groups through the spatial transformation. This model specification measures the joint effect of misspecification, omitted variables, and spatial autocorrelation.

8.3 The Information Theoretic Formulation

The application of Maximum Entropy methods and Information Theoretic techniques has been explored within the context of ecological. The first use of entropy-maximizing models concerned the application of gravity models and transportation flows. Recently, applications of Information Theoretic methods have focused on the analysis of spatial patterns of voting at the individual level (King et al. 2004).

However, the present study extends the IT approach to the case of Ecological Inference incorporating Spatial Dependence. Past studies have given little weight to the role of spatial effects in ecological inference analysis, and so this present study is going to introduce a basic framework for EI in the presence of spatial heterogeneity and dependence. It also deals with the specification of models that explicitly control for spatial effects, interpretation and IT-based formulation.

An Information Theoretic technique (Golan and Gzyl 2006; Peeters and Chasco 2006; Bernardini-Papalia 2010a, b) is suggested as an adequate solution in the present context since it provides an effective and flexible procedure for reconciling micro and macro data and for addressing problems related to spatial structures.

Implementation of these methods requires that the parameters and errors of the model in Eqs. 8.5 and 8.8 are specified as linear combinations of some predetermined and discrete support values and unknown probabilities (weights). Thus, all coefficients $\boldsymbol{\alpha}_{ij}$, $\boldsymbol{\beta}_{ij}$, $\boldsymbol{\gamma}_{ij}$, ρ, λ and unknown errors ε_{ij}, $\boldsymbol{\tau}_{ij}$ in Eqs. 8.5 and 8.8, are reparameterized and expressed in terms of proper probabilities. For each parameter, a set of M support points (with $2 \leq \mathrm{M} < \infty$) has been chosen: $\mathbf{s}_\alpha = \left(\mathbf{s}_1^\alpha,..,\mathbf{s}_M^\alpha\right)'$, $\mathbf{s}_\beta = \left(\mathbf{s}_1^\beta,..,\mathbf{s}_M^\beta\right)'$, $\mathbf{s}_\gamma = \left(\mathbf{s}_1^\gamma,..,\mathbf{s}_M^\gamma\right)'$, $\mathbf{s}_\rho = \left(\mathbf{s}_1^\rho,..,\mathbf{s}_M^\rho\right)'$, $\mathbf{s}_\lambda = \left(\mathbf{s}_1^\lambda,..,\mathbf{s}_M^\lambda\right)'$, and the corresponding unknown probabilities defined on these support spaces $\mathbf{p}_{\alpha,ij} = \left(\mathbf{p}_{ij,1}^\alpha,..,\mathbf{p}_{ij,M}^\alpha\right)'$, $\mathbf{p}_{\beta,ij} = \left(\mathbf{p}_{ij,1}^\beta,..,\mathbf{p}_{ij,M}^\beta\right)'$ $\mathbf{p}_{\gamma,ij} = \left(\mathbf{p}_{ij,1}^\gamma,..,\mathbf{p}_{ij,M}^\gamma\right)'$, $\mathbf{p}_{\rho,ij} = \left(\mathbf{p}_{ij,1}^\rho,..,\mathbf{p}_{ij,M}^\rho\right)$, $\mathbf{p}_{\lambda,ij} = \left(\mathbf{p}_{ij,1}^\lambda,..,\mathbf{p}_{ij,M}^\lambda\right)$. Similarly, the errors $\varepsilon_{ij}, \tau_{ij}$, are treated as unknowns, and a set of R support points $\mathbf{s}_\varepsilon = \left(\mathbf{s}_1^\varepsilon,..,\mathbf{s}_R^\varepsilon\right)'$, $\mathbf{s}_\tau = \left(\mathbf{s}_1^\tau,..,\mathbf{s}_R^\tau\right)'$, has been chosen, with $2 \leq \mathrm{j} < \infty$ with reference to the unknown probabilities $\mathbf{p}_{\varepsilon,ij} = \left(\mathbf{p}_{ij,1}^\varepsilon,..,\mathbf{p}_{ij,R}^\varepsilon\right)'$, $\mathbf{p}_{\tau,ij} = \left(\mathbf{p}_{ij,1}^\tau,..,\mathbf{p}_{ij,R}^\tau\right)'$.

For the sake of simplicity, the above support spaces are constructed as discrete, bounded entities. It is possible to construct unbounded and continuous supports within the same framework (Golan 2008).

The support points are chosen on the basis of a priori information as discussed in Golan and Gzyl (2006). However, such knowledge is not always available, and symmetric parameter supports around zero are generally used in the presence of scarce prior information about each parameter. With regard to errors, in most cases where the underlying distribution is unknown, one conservative way of choosing the error supports s_ε, s_τ, is to employ the "three-sigma rule" established by Pukelsheim.

Under the GCE framework, the full distribution of each parameter and of each error (within their support spaces) is simultaneously estimated under minimal distributional assumptions. More specifically, the parameters $\boldsymbol{\alpha}_{ij}, \boldsymbol{\beta}_{ij}, \boldsymbol{\gamma}_{ij}, \rho, \lambda$ and errors $\varepsilon_{ij}, \tau_{ij}$ are reparameterized as:

$$\begin{aligned} &\boldsymbol{\alpha}_{ij} = \mathbf{s}_\alpha' \mathbf{p}_{\alpha,ij}, \quad \boldsymbol{\beta}_{ij} = \mathbf{s}_\beta' \mathbf{p}_{\beta,ij}, \quad \boldsymbol{\gamma}_{ij} = \mathbf{s}_\gamma' \mathbf{p}_{\gamma,ij}, \quad \rho = \mathbf{s}_\rho' \mathbf{p}_\rho, \quad \lambda = \mathbf{s}_\lambda' \mathbf{p}_\lambda \\ &\boldsymbol{\varepsilon}_{ij} = \mathbf{s}_\varepsilon' \mathbf{p}_{\varepsilon,ij}, \quad \tau_{ij} = \mathbf{s}_\tau' \mathbf{p}_{\tau ij} \end{aligned} \tag{8.9}$$

with support vectors for parameters $\boldsymbol{\alpha}_{ij}, \boldsymbol{\beta}_{ij}, \boldsymbol{\gamma}_{ij}, \rho, \lambda$ and errors $\varepsilon_{ij},\ \tau_{ij}$ given by:

$$\mathbf{s}_\alpha = \left(\mathbf{s}_1^\alpha, .., \mathbf{s}_M^\alpha\right)', \quad \mathbf{s}_\beta = \left(\mathbf{s}_1^\beta, .., \mathbf{s}_M^\beta\right)', \quad \mathbf{s}_\gamma = \left(\mathbf{s}_1^\gamma, .., \mathbf{s}_M^\gamma\right)',$$
$$\mathbf{s}_\rho = \left(\mathbf{s}_1^\rho, .., \mathbf{s}_M^\rho\right)', \quad \mathbf{s}_\lambda = \left(\mathbf{s}_1^\lambda, .., \mathbf{s}_M^\lambda\right)'$$
$$\mathbf{s}_\varepsilon = \left(\mathbf{s}_1^\varepsilon, .., \mathbf{s}_R^\varepsilon\right)', \quad \mathbf{s}_\tau = \left(\mathbf{s}_1^\tau, .., \mathbf{s}_R^\tau\right) \tag{8.10}$$

and corresponding unknown probabilities given by:

$$\mathbf{p}_{\alpha,ij} = \left(\mathbf{p}_{ij,l}^\alpha, .., \mathbf{p}_{ij,M}^\alpha\right)', \quad \mathbf{p}_{\beta,ij} = \left(\mathbf{p}_{ij,l}^\beta, .., \mathbf{p}_{il,M}^\beta\right)', \quad \mathbf{p}_{\gamma,ij} = \left(\mathbf{p}_{ij,l}^\gamma, .., \mathbf{p}_{ij,M}^\gamma\right)',$$
$$\mathbf{p}_{\rho,ij} = \left(\mathbf{p}_{ij,l}^\rho, .., \mathbf{p}_{ij,M}^\rho\right)\mathbf{p}_{\lambda,ij} = \left(\mathbf{p}_{ij,l}^\lambda, .., \mathbf{p}_{ij,M}^\lambda\right), \quad \mathbf{p}_{\varepsilon,ij} = \left(\mathbf{p}_{ij,l}^\varepsilon, .., \mathbf{p}_{ij,R}^\varepsilon\right)',$$
$$\mathbf{p}_{\tau,ij} = \left(\mathbf{p}_{ij,l}^\tau, .., \mathbf{p}_{ij,R}^\tau\right) \tag{8.11}$$

with M, R ≥ 2.

In addition, prior information reflecting subjective information or any other sample and pre-sample information is introduced by specifying the priors for all parameters and errors: $\tilde{\mathbf{p}}_{\alpha,ij}, \tilde{\mathbf{p}}_{\beta,ij}, \tilde{\mathbf{p}}_{\gamma,ij}, \tilde{\mathbf{p}}_{\rho,ij}, \tilde{\mathbf{p}}_{\varepsilon,ij}, \tilde{\mathbf{p}}_{\tau,ij}$. These priors may come from prior data, theory, and/or other experiments.

The GCE optimization problem for the ecological spatial model corresponding to Eq. 8.5 can be reformulated by minimizing the following objective function H(.) as follows:

$$\begin{aligned} H = &\sum_i \sum_j (\mathbf{p}_{\alpha,ij})' \ \ln\left(\frac{\mathbf{p}_{\alpha,ij}}{\tilde{\mathbf{p}}_{\alpha,ij}}\right) + \sum_i \sum_j (\mathbf{p}_{\beta,ij})' \ \ln\left(\frac{\mathbf{p}_{\beta,ij}}{\tilde{\mathbf{p}}_{\beta,ij}}\right) \\ &+ \sum_i \sum_j (\mathbf{p}_{\gamma,ij})' \ \ln\left(\frac{\mathbf{p}_{\gamma,ij}}{\tilde{\mathbf{p}}_{\gamma,ij}}\right) + \sum_i \sum_j (\mathbf{p}_{\rho,ij})' \ \ln\left(\frac{\mathbf{p}_{\rho,ij}}{\tilde{\mathbf{p}}_{\rho,ij}}\right) \\ &+ \sum_i \sum_j (\mathbf{p}_{\varepsilon,ij})' \ \ln\left(\frac{\mathbf{p}_{\varepsilon,ij}}{\tilde{\mathbf{p}}_{\varepsilon,ij}}\right) \end{aligned} \tag{8.12}$$

subject to:

1. Data consistency conditions:

$$\begin{aligned} \text{In } \mathbf{y} = \sum_{j=1}^{J_i} \Bigg(&\mathbf{s}_\alpha{}'\mathbf{p}_{\alpha,ij} + \sum_{K=1}^{K}\left(\mathbf{s}_\beta{}'\mathbf{p}_{\beta,ij}\right) \text{In } \mathbf{z}_{ij,k} + \sum_{h=1}^{H}\left(\mathbf{s}_\gamma{}'\mathbf{p}_{\gamma,ij}\right) \text{In } \mathbf{X}_{i,h} \\ &+ \left(\mathbf{s}_\rho{}'\mathbf{p}_{\rho,ij}\right) \text{In } \mathbf{wy}_i + \left(\mathbf{s}_\varepsilon{}'\mathbf{p}_{\varepsilon,ij}\right)\Bigg)\boldsymbol{\theta}_{ij} \end{aligned} \tag{8.13}$$

2. Adding-up constraints for probabilities.

$$\sum \mathbf{p}_{\alpha,ij} = \sum \mathbf{p}_{\beta,ij} = \sum \mathbf{p}_{\gamma,ij} = \sum \mathbf{p}_{\rho,ij} = \sum \mathbf{p}_{\varepsilon,ij} = 1 \quad \forall i,j$$

$$\sum \hat{\mathbf{p}}_{\alpha,ij} = \sum \hat{\mathbf{p}}_{\beta,ij} = \sum \hat{\mathbf{p}}_{\gamma,ij} = \sum \hat{\mathbf{p}}_{\rho,ij} = \sum \hat{\mathbf{p}}_{\varepsilon,ij} = 1 \quad \forall i,j$$

Analogously, the GCE optimization problem for the ecological spatial model corresponding to Eq. 8.8 can be reformulated by minimizing the following objective function H(.) as follows:

$$\begin{aligned} H = & \sum_i \sum_j (\mathbf{p}_{\alpha,ij})' \ \ln \left(\frac{\mathbf{p}_{\alpha,ij}}{\tilde{\mathbf{p}}_{\alpha,ij}} \right) + \sum_i \sum_j (\mathbf{p}_{\beta,ij})' \ \ln \left(\frac{\mathbf{p}_{\beta,ij}}{\tilde{\mathbf{p}}_{\beta,ij}} \right) \\ & + \sum_i \sum_j (\mathbf{p}_{\gamma,ij})' \ \ln \left(\frac{\mathbf{p}_{\gamma,ij}}{\tilde{\mathbf{p}}_{\gamma,ij}} \right) + \sum_i \sum_j (\mathbf{p}_{\lambda,ij})' \ \ln \left(\frac{\mathbf{p}_{\lambda,ij}}{\tilde{\mathbf{p}}_{\lambda,ij}} \right) \\ & + \sum_i \sum_j (\mathbf{p}_{\varepsilon,ij})' \ \ln \left(\frac{\mathbf{p}_{\varepsilon,ij}}{\tilde{\mathbf{p}}_{\varepsilon,ij}} \right) + \sum_i \sum_j (\mathbf{p}_{\tau,ij})' \ \ln \left(\frac{\mathbf{p}_{\tau,ij}}{\tilde{\mathbf{p}}_{\tau,ij}} \right) \end{aligned} \tag{8.14}$$

subject to:

1. Data consistency conditions:

$$\begin{aligned} \ln \mathbf{y} = & \sum_{j=1}^{J_i} \left(\mathbf{s}_\alpha{}' \mathbf{p}_{\alpha,ij} + \sum_{K=1}^{K} \left(\mathbf{s}_\beta{}' \mathbf{p}_{\beta,ij} \right) \ln \mathbf{z}_{ij,k} + \sum_{h=1}^{H} \left(\mathbf{s}_\gamma{}' \mathbf{p}_{\gamma,ij} \right) \ln \mathbf{X}_{i,h} + \right. \\ & \left. \left(\mathbf{s}_\lambda{}' \mathbf{p}_{\lambda,ij} \right) w \left(\mathbf{s}_\varepsilon{}' \mathbf{p}_{\varepsilon,ij} \right) + \left(\mathbf{s}_\tau{}' \mathbf{p}_{\tau,ij} \right) \right) \boldsymbol{\theta}_{ij} \end{aligned} \tag{8.15}$$

2. Adding-up constraints for probabilities:

$$\sum \mathbf{p}_{\alpha,ij} = \sum \mathbf{p}_{\beta,ij} = \sum \mathbf{p}_{\gamma,ij} = \sum \mathbf{p}_{\lambda,ij} = \sum \mathbf{p}_{\tau,ij} = 1 \quad \forall i,j$$
$$\sum \hat{\mathbf{p}}_{\alpha,ij} = \sum \hat{\mathbf{p}}_{\beta,ij} = \sum \hat{\mathbf{p}}_{\gamma,ij} = \sum \hat{\mathbf{p}}_{\lambda,ij} = \sum \hat{\mathbf{p}}_{\tau,ij} = 1 \quad \forall i,j$$

The optimal solutions depend on the prior information, the data and a normalization factor. If the priors are specified such that each choice is equally likely to be selected (uniform distributions), then the GCE solution reduces to the GME one. As with the GME estimator, numerical optimization techniques should be used to obtain the GCE solution.

In order to determine whether additional information in the data, expressed in the form of constraints, produce a departure from the condition of total uncertainty and a consequent reduction of uncertainty related to the phenomenon, the standard normalized entropy measure can be used (Golan et al. 1996).

Note that one can simultaneously consider the choice of the model, that is the functional relationship linking the variable to be disaggregated and a set of variables/indicators at area level, and the choice associated with the macro and micro data sources.

8.4 An Empirical Application

We present the application of the GME formulation introduced in Sect. 8.2 to the case of an Italian data set. The GME-based formulation is used to disaggregate the value-added of Umbria's local labour markets (LLM) in nine macro-sectors of manufacturing industry, in the year 2001. Nine manufacturing sectors are dealt with: (1) Food, beverages and tobacco; (2) Textiles and clothing; (3) Wood products; (4) Paper, printing and publishing; (5) Coke and refined petroleum products, chemicals; (6) Non-metallic mineral products; (7) Basic metals, fabricated metal products; (8) Machinery, computing, precision medical instruments, transport; (9) Rubber, plastic and other manufacturing sectors.

The case study is particularly suitable to represent the usefulness of our approach to study the local labour markets. The Umbria region assumes the character of the region-not region, that is, a political-administrative unit dominated by centripetal and centrifugal forces which thus tend to enhance linkages and integration with neighboring regions (Fig. 8.1). The different areas are characterized by specific features: (1) the rural high Valnerina area (Norcia and Cascia) projected to enhance the economic potential of cultural and environmental specificities; (2) Città di Castello and Umbertide characterized by an territorial organization of district type, (3) the area of Tevere's valley, re-organized into several spatial components (the rural Todi, the area relative to Perugia, Deruta, and an area of small and medium enterprises with a significant systemic organizational structure, Marsciano) and (4) the territories of the Lake Trasimeno, Orvieto, those of the Valle Umbra (Assisi, Foligno), and so on (the Terni, in the Gubbio area Gualdese), each with its own characteristics and distinct growth path characterized by distinctive specifities.

The basic formulation assumes that: (1) the GME estimates of the value-added of Umbria's LLM, disaggregated by sector, are consistent with the total value-added observed at the regional level; (2) the value-added of the LLM by sector are measured with error.

By introducing the baseline statistical model, $y_i = \sum_{j=1}^{J_i} (\alpha_{ij} + \sum_{K=1}^{K} \beta_{ij,k} z_{ij,k} + \sum_{h=1}^{H} \gamma_{ij,h} x_{i,h} + \rho w y_j + \varepsilon_{ij})\theta_{ij}$ we estimate the total value-added for each sector at the level of Umbria's LLM, by employing all available information, that is: sub-area (LLM) level information about K explanatory variables Zj, that refer to: total value-added of manufacturing's local labour markets, employment rate, ER; Job placement rate, JPR, but also refer to measures of spatial externalities. The sectors' shares of the total number of manufacturing firms here is used for θ_{ij}. From the macro perspective, the total value-added for each sector within Umbria is a known quantity, and is

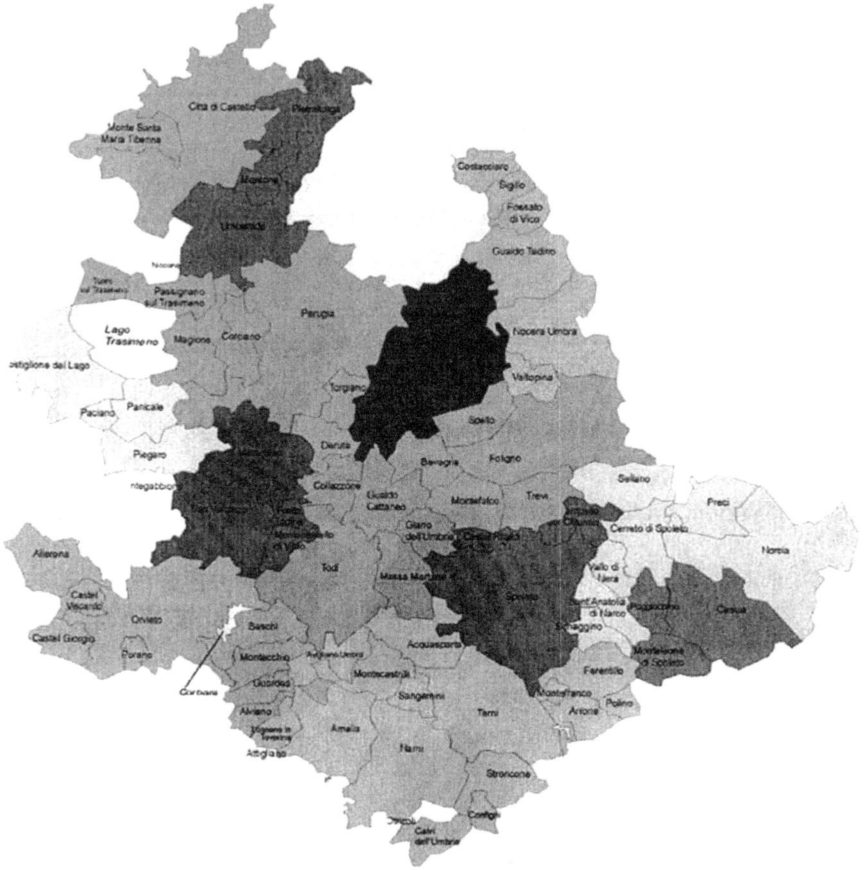

Fig. 8.1 Labour local markets of Umbria

regarded as a fixed regional total. The latent sub group values are specified in a multiplicative form, which is consistent with a Cobb-Douglas type of production function.

Spatial dependence of the LLM's value added is confirmed; specifically, Moran's I and Geary's C tests cannot accept the null hypothesis of global spatial independence (0.0593; p-value: 0.061 for the former; 0.0493; p-value: 0. 0003 for the latter). In our analysis, the weight matrix is computed by means of the distance of each LLM from Perugia, where the critical cut-off value is given by the first quartile of the distance's distribution as well as by means of weights based on contiguity measures of LLM. Results produced by different weight matrices are robust for all model specifications. Alternative specifications, also related to spatial LAG model and spatial Error model have been the objective of a preliminary analysis.

Table 8.1 Comparison of alternative model specifications in terms of normalized entropy measures

	Model 1	Model 2	Model 3	Model 4	Model 5	Model 6	Model 7	Model 8
Explanatory variables								
LLM_Employment rate: ER	X	X	X			X	X	X
LLM_Job Placement rate: JPR	X			X	X	X		
LLM_Value added: VA	X	X	X	X	X	X	X	X
LLM_Spatial-lag Value added: WVA	X		X		X		X	X
Spatial fixed effects		X	X	X	X	X	X	
Weight of sub group indicator: θ_{ij}	Sectors' shares of the total number of manufacturing firms							
Normalized entropy measure	0.5401	0.5235	0.5401	0.5243	0.5403	0.5402	0.5543	0.5231

The ME principle is used to yield the most uninformed distribution in keeping with the observed sample data, with minimal assumptions made regarding the underlying distribution generating the data. We choose symmetric parameter supports around zero, given that we have very little prior information about each parameter, and M = 5 support points for each parameter, since estimation is not improved by choosing more than about five support points. We choose j = 3 support points for each error, and we specify error supports according to Pukelsheim's "Three Sigma Rule". The estimation procedure is implemented using the GAMS software and a nonlinear solver, CONOPT2.

Using the measure of normalized entropy (NE) (Golan et al. 1996) relative to different scenarios, alternative formulations are compared with the aim of choosing the model specification that, conditional to the information available, incomplete and limited, contributes in reduction of uncertainty concerning the phenomenon of interest. The NE of the Model 8 is the smallest one (Table 8.1), indicating that it has the lowest uncertainty of all models considered. These results show the sensitivity of variable selection relative to the data generation process.

Results of the selected model (see Table 8.2) seem to be relatively robust with respect to the parameter supports: the GME parameter estimates do not vary a great deal as parameter supports are modified. The choice of support vectors for the parameters, within the intervals (−100,100) and (−20,20), has a negligible effect on the coefficients. The asymptotic standard errors are calculated using the method proposed by Golan et al. 1996.

The distribution of the value-added of Umbria's 16 LLM, disaggregated by sector for 2001, seems to be quite heterogeneous. Our analysis validates the hypothesis of spatial heterogeneity across the LLMs, as well as the contribution of the indicator chosen as weight for the small area latent indicators that is the share of the total number of firms operating in each sector i and located in local labour market j.

Table 8.2 Estimates of the value added of Umbria's LLM disaggregated by manufacturing sector for the year 2001

Manufacturing sectors									
Local labour markets (Umbria region)	Food, beverages tobacco	Textiles and clothing	Wood products	Paper, printing publishing	Coke, chemicals	Non-metallic mineral products	Basic metals, metal products	Machinery, computing, transport	Rubber, plastic, other manufacturing sectors
ASSISI	32.95	68.65	0.00	14.29	0.00	30.34	36.28	33.60	31.29
CASCIA	1.43	0.00	0.00	0.00	0.00	0.38	1.05	0.53	1.08
CASTIGLIONE DEL LAGO	13.59	14.31	0.41	2.27	0.00	4.13	18.23	18.41	10.43
CITTA' DI CASTELLO	10.71	21.68	0.84	44.14	0.00	8.45	21.62	25.82	32.55
FOLIGNO	53.06	33.45	0.00	20.31	24.33	20.53	45.82	49.19	28.16
GUALDO TADINO	15.00	7.94	6.55	7.23	12.99	57.01	16.82	29.36	9.50
GUBBIO	26.27	20.24	0.61	6.70	12.05	30.50	18.43	12.89	13.21
MARSCIANO	9.52	16.38	2.08	4.59	8.25	8.35	14.57	9.81	11.56
NORCIA	10.02	0.90	0.00	0.27	0.00	1.32	4.14	1.16	3.09
PERUGIA	69.02	148.96	6.78	56.15	67.25	136.21	95.00	111.98	69.65
SPOLETO	31.35	21.62	0.65	5.37	12.86	13.03	22.71	19.89	15.67
TODI	28.08	22.99	0.00	7.61	13.68	9.24	28.99	22.78	16.67
UMBERTIDE	10.77	27.25	0.00	4.38	0.00	7.08	30.88	9.98	10.23
FABRO	3.87	1.97	0.00	0.80	0.00	1.94	2.37	1.37	2.54
ORVIETO	16.46	10.38	0.38	3.15	7.55	19.11	9.78	9.88	10.12
TERNI	151.99	79.18	0.00	51.52	74.04	49.99	165.62	176.14	94.74
Regional VA	484	496	18	229	233	398	532	533	360

8.5 Conclusions

In this chapter we have tackled the problem of providing reliable estimates of a target variable in a set of small geographical areas, by exploring spatially relationships at the disaggregate level. Controlling for spatial effects means introducing models whereby the assumption is that values in adjacent geographic locations are linked to each other by means of some form of underlying spatial relationship. Given researchers' uncertainty about spatial data sampling processes and error-correlation structures, it seems reasonable to explore more flexible estimation and inference frameworks that reduce the assumptions about some or all of these features while, at the same time, allowing them to incorporate knowledge about the spatial structure in a sample.

In certain cases, in order to account for spatial dependency we need to grasp the spatial variations in the regression coefficients, since empirical predictions based on global parameters may be biased, and thus misrepresent local behavior. This is particularly problematic in the case of regional analysis, where locally representative regression coefficients are required for micro-level policy decisions to be taken.

We have discussed the importance of taking into account individually- and spatially-correlated small area level variations, and we have recommended the use of Information Theoretic-based methods for the estimation of variables within the small groups of interest.

The proposed ME-based methods of disaggregation are capable of yielding disaggregate data consistent with prior information, resulting from different sources of data in the absence of high quality and detailed data as well as in the presence of problems of collinearity and endogeneity, without imposing strong distributional assumptions. Within this framework, we have shown how partial information at the disaggregated level can be combined with aggregated data to provide estimates of latent variables or indicators which are of interest at the small area/sub group level.

Two interesting points emerge here. Firstly, the ME-based formulation has the advantage of being consistent with the underlying spatial dependence in the data-generating process, and eventually with the restrictions implied by certain non-sample information, or by previous empirical experience. Compared to traditional estimation methods, this approach is characterized by its robustness to ill-conditioned designs, and by its ability to fit over-parametrized models such as those pertaining to data disaggregation problems and small area estimation. It is also particularly effective to deal with problems of skewed distributions and outliers and also represents a good choice in presence of collinearity and endogeneity problems.

Secondly, within a ME-based framework, the informative contribution in reduction of uncertainty of the phenomenon under study, made by each restriction and by each variable included in the basic problem formulation can be verified simultaneously.

The GME formulation has been employed in relation to an Italian data set in order to compute the value-added of Umbria's local labour markets in 2001 for nine manufacturing sectors which are consistent with the total regional value-added per sector, and by formulating a suitable set of constraints for the optimization problem in the presence of errors in the aggregates at sub-area level.

The results show that this approach provides a flexible, powerful data-disaggregation method, since it enables us to: (1) consider prior knowledge introduced by adding linear and nonlinear inequality constraints, errors in equations, and error in variables; (2) allow for the efficient use of information from a variety of sources; (3) reconcile data at different levels of aggregation within a coherent framework.

Further work should be done in order to explore IT methods by considering (1) small area parameters which are a non linear functions of the small area total variable (small rates and proportions) in presence of spatial structures; and (2) temporal dependence. Possible extensions of the proposed procedure include estimation using composite IT methods incorporating both GME and GCE estimators (Bernardini-Papalia 2008) that can be used when some of the small areas have no sample units.

References

Anselin L (1988) Spatial econometrics: methods and models. Kluwer, Boston

Bernardini-Papalia R (2008) A composite generalized cross entropy formulation in small samples estimation. Econ Rev 27(4–6):596–609

Bernardini-Papalia R (2010a) Incorporating spatial structures in ecological inference: an information theoretic approach. Entropy 12(10):2171–2185

Bernardini-Papalia R (2010b) Data disaggregation procedures within a maximum entropy framework. J Appl Stat 37(11):1947–1959

Cho WKT (2001) Latent groups and cross-level inferences. Elect Stud 20:243–263

Golan A (2008) Information and entropy econometrics: a review and synthesis. Foundations and trends® in econometrics, vol 2, No 1–2. Now Publishers, Hanover, pp 1–145

Golan A, Gzyl H (2006) Ill conditioned linear estimation problems: an information theoretic approach with priors. Working paper

Golan A, Judge G, Robinson S (1994) Recovering information from incomplete or partial multisectoral economic data. Rev Econ Stat 76(3):541–549

Golan A, Judge G, Miller D (1996) Maximum entropy econometrics: robust estimation with limited data. Wiley, New York

Johnston R, Pattie C (2000) Ecological inference and entropy-maximizing: an alternative estimation procedure for split-ticket voting. Pol Anal 8:333–345

Judge G, Miller D, Cho WKT (2004) An information theoretic approach to ecological inference. In: King G, Rosen O, Tanner MA (eds) Ecological inference: new methodological strategies. Cambridge University Press, Cambridge, pp 162–187

King G (1997) A solution to the ecological inference problem: reconstructing individual behavior from aggregate data. Princeton University Press, Princeton

King G, Rosen O, Tanner MA (2004) Ecological inference: new methodological strategies. Cambridge University Press, Cambridge, pp 1–422

Kullback J (1959) Information theory and statistics. Wiley, New York

Levine RD (1980) An information theoretical approach to inversion problems. J Phys A 13:91–108

Peeters L, Chasco C (2006) Ecological inference and spatial heterogeneity: an entropy-based distributionally weighted regression approach. Pap Reg Sci 85(2):257–276

Shannon CE (1948) A mathematical theory of communication. Bell Syst Tech J 27:379–423

Chapter 9
Short-Run Regional Forecasts: Spatial Models through Varying Cross-Sectional and Temporal Dimensions

Matías Mayor and Roberto Patuelli

9.1 Introduction

Forecasting economic values in administrative units provides very important information for political, institutional and economic agents for their respective planning processes. A crucial stage is the choice of the econometric method to obtain these future values taking into account the diversity and complexity of the real economy. Two aspects may be considered when choosing an econometric specification. Firstly, disparities in economic development and welfare within countries (i.e., at the regional level) are often bigger than between countries (Elhorst 1995; Taylor and Bradley 1997; Ertur and Le Gallo 2003; Patuelli 2007; see, for example, the cases of Germany and Spain), and they often show typical geographical/spatial structures. Secondly, with regard to regional unemployment disparities, policy makers need, in order to correctly target their actions and policies, to understand two aspects of such disparities: (a) the determinants of 'equilibrium' unemployment and its variation; and, (b) the region-specific and the cross-regional dynamics of unemployment. On the one hand, the need for an explicit consideration of the existence of spatial interdependence in econometric models, which is consistent with regional science theories asserting the importance of spatial linkages in local economic processes, led to what is nowadays quite a large literature of empirical papers. On the other hand, the temporal perspective of the problem has attracted less attention in spatial models, but should be considered jointly.

The spatial perspective has achieved an increasing relevance within the field of labour market studies. Some recent contributions have taken into account the

M. Mayor (✉)
School of Economics and Business, Department of Applied Economics, Regiolab, University of Oviedo, Avda del Cristo s/n, Oviedo 33006, Spain
e-mail: mmayorf@uniovi.es

R. Patuelli
The Rimini Centre for Economic Analysis (RCEA), University of Bologna, Bologna, Italy

E. Fernández Vázquez and F. Rubiera Morollón (eds.), *Defining the Spatial Scale in Modern Regional Analysis*, Advances in Spatial Science,
DOI 10.1007/978-3-642-31994-5_9, © Springer-Verlag Berlin Heidelberg 2012

spatial dimension of regional labour markets and pointed out the high degree of interdependence of local labour markets (Molho 1995; López-Bazo et al. 2002; Overman and Puga 2002). Furthermore, Patacchini and Zenou (2007) analyse the reasons for spatial dependence in local unemployment rates. This spatial autocorrelation is mainly due to the fact that unemployed individuals may seek and find work in different areas, so that spatial interactions result from their mobility. When the data is collected at the administrative level, as it is often the case, spatial autocorrelation is likely to be a relevant issue.

The contribution to forecasting power of the inclusion of dependence across spatial units has been analysed in several papers. For example, Giacomini and Granger (2004, p. 7), on the one hand, stress that 'ignoring spatial autocorrelation, even when it is weak, leads to highly inaccurate forecasts'. On the other hand, Hernández-Murillo and Owyang (2006) find reductions in the out-of-sample mean squared error (MSE) when employment forecasts are obtained using disaggregated data in a space-time autoregressive model without contemporaneous influence from a region's neighbours[1] incorporating spatial interaction. Different econometric techniques have been proposed in the literature. Using static spatial panel data models, Baltagi and Li (2004, 2006), as well as Longhi and Nijkamp (2007), forecast employment in West German regions, while Fingleton (2009) predicts the average wage rate across all occupations in local administrative units in Great Britain. A dynamic spatial panel data model is used by Kholodilin et al. (2008) conclude that accounting for spatial effects improves forecast performance, and this improvement is more important when the forecasting horizon is longer. Schanne et al. (2010) reach a similar conclusion comparing a univariate spatial GVAR model with univariate time series methods.

In this chapter, we compare different methods to obtain short-run unemployment forecasts in (small) administrative units, and observe their performance between different countries. Our interest here is to exploit the strong heterogeneity in the size (e.g., in terms of population or area) of NUTS regions at the same level of aggregation across countries to investigate the variation in the performance of different spatial econometric methods. We analyse the forecasting performance of two competing econometric methods: a spatial vector autoregressive (SVAR) model (Beenstock and Felsenstein 2007; Kuethe and Pede 2011) and a dynamic heterogeneous-coefficients panel data model based on an eigenvector-decomposition spatial filtering (SF) procedure (Griffith 2000, 2003). The two models chosen belong to two separate traditions: VAR models represent the mainstream (time-series) forecasting tradition, while the SF-enhanced dynamic panel model attempts to merge the panel data modelling tradition to the spatial statistics one, within a semi-parametric framework.

The remainder of the chapter is structured as follows. In Sect. 9.2, SVAR models and SF procedures are described, pointing out their advantages and disadvantages. In Sect. 9.3, we discuss the spatial configuration of regional unemployment data in

[1] These authors assume that these effects are expanded through the time dimension.

Spain and Switzerland, and subsequently present our forecasting experiment strategy. Our forecasting results and related multidimensional measures of accuracy are included in Sect. 9.4. Section 9.5 concludes the chapter.

9.2 Modelling Spatio-Temporal Data: Spatial VAR Models and Spatial Filtering

A VAR model (Sims 1980) can be written as a set of symmetric equations in which each (dependent) variable is described by a set of its own lags and the lags of other variables in the system. VAR models are considered as the most popular method to study the linkages among several variables with high flexibility since these types of models are not based on any theoretical structure. Restrictions are imposed to a large extent by statistical tools rather than by prior beliefs supported by uncertain theoretical models. However, this flexibility is only certain in the temporal dimension.[2] A standard VAR model does not assume the existence of spatial spillovers: for example, a shock in one region only influence the economic behaviour of this administrative area.

There are a few proposals in the literature on how to introduce spatial relationships in a VAR framework, thus relaxing this limitation. This is due to the fact that the number of parameters which are needed to collect such neighbouring relations increase quadratically with the number of spatial units. Therefore, some of the existing proposals use spatial contiguity information to limit the number of parameters. In particular, Pan and LeSage (1995) propose to use spatial contiguity information as an alternative prior in a Bayesian VAR model. Following the same idea, Di Giacinto (2003) defines parameter constraints in a structural VAR model based on neighbouring structure, allowing the identification and estimation of the spatial VAR model. A further option is developed by Schanne et al. (2010), based on the Global VAR (GVAR) model proposed by Pesaran et al. (2004), where geographical information is used to include spatial connections between regions. One of the novelties (or advantages) of the GVAR model consists in the inclusion of a temporal dimension within the spatial dependence process. Some authors consider only contemporaneous spatial processes (Longhi and Nijkamp 2007; Kholodilin et al. 2008), whereas others specify only a temporally lagged type of spatial dependence (Hernández-Murillo and Owyang 2006).

Regarding the approach to including neighbouring linkages between spatial units, the contributions mentioned above use spatial weights matrices. Spatial weights matrices are positive, non-stochastic and their elements show the intensity of interdependence between pairs of spatial units, that is, eventually specify the neighbouring set for each spatial unit. In this chapter, we follow the SVAR approach proposed by Beenstock and Felsenstein (2007), where traditional VAR methods and modern spatial panel data techniques are 'mixed'. Beenstock and

[2] Structural VAR models are introduced to incorporate some necessary restrictions which are not tested by statistical tools.

Felsenstein allow for both contemporaneous and serially lagged spatially correlated variables. This SVAR model is highly nonlinear, because of the contemporaneous spatial autoregressive process. Its proponents restrict the coefficient of the endogenous contemporaneous spatial lag to zero, therefore linearizing the model.

Let us consider a country divided into n regions or municipalities ($i = 1, \ldots, n$) where the values of a set of random variables are observed over time $t = 1, \ldots, T$. In general terms, if we consider p temporal lags and s spatial cross-regressive lags, we must manage n equations like the following (one for each region):

$$y_{i,t} = c_i + \sum_{p=1}^{P} \beta_{i,p} y_{i,t-p} + \sum_{s=1}^{S} \delta_{i,s} \sum_{j=1}^{n} W_{s,i,j} y_{j,t} + \sum_{s=1}^{S} \sum_{p=1}^{P} \gamma_{i,s,p} \sum_{j=1}^{n} W_{s,i,j} y_{j,t-p} + \varepsilon_{i,t}, \quad (9.1)$$

where $w_{s,i,j}$ is the value of the cell (i, j) of spatial weight matrix. W_s for the sth spatial lag. The novelty of this model is the inclusion of the spatial cross-regressive lags. They are obtained by premultiplying each temporally lagged value by the spatial weight matrix (different contiguity orders could ideally be considered). A further relevant advantage of this approach is the possibility of testing for the significance of regional spillovers by means of Granger causality test.

In the estimation stage, it is necessary to bear in mind that each spatial unit has a unique value of the spatial lag variable, and each observation has its own set of neighbouring units. Since $\sum_{j=1}^{n} w_{s,i,j}\, y_{j,t}$ and $\varepsilon_{i,t}$ are not independent, Eq. 9.1 cannot be estimated directly. The system is estimated by seemingly unrelated regressions (SUR), because a two-stage estimation procedure is necessary, where the coefficients of the contemporary spatial lags are estimated using the (spatially weighted) predicted values of the dependent variable as instruments. These predicted values are computed using the reduced form of Eq. 9.1. This is the common solution employed to solve the endogeneity issue caused by the spatial lag of the dependent variable in a simple spatial lag model.

If the order of the temporal lags is restricted to one to preserve degrees of freedom, each equation includes a constant, the lagged variable, and the contemporaneous and temporally lagged spatially correlated variables, as follows:

$$y_{i,t} = c_i + \beta_{i,1}\, y_{i,t-1} + \delta_{i,1} \sum_{j=1}^{n} W_{1,i,j}\, y_{j,t} + \gamma_{i,1,1} \sum_{j=1}^{n} W_{1,i,j}\, y_{j,t-1} + \varepsilon_{i,t} \quad (9.2)$$

where $\sum_{j=1}^{n} w_{1,i,j}\, y_{j,t}$ collects the value of contemporaneous spatially lagged variables, and $\sum_{j=1}^{n} w_{1,i,j}\, y_{j,t-1}$ the temporally and spatially lagged variables. The accuracy of this type of models in terms of forecasting errors has been previously analysed in the aforementioned papers.

The aim of this chapter is to compare this method with the one recently proposed by Patuelli et al. (2012), based on a heterogeneous-coefficients dynamic panel data model enhanced by spatial filtering (SF). This latter approach allows us to account for spatial heterogeneity and/or autocorrelation both in the levels and in the regression coefficients, among which the one of the serially lagged term.

Eigenvector-decomposition SF (Griffith 2000, 2003) is a nonparametric solution to the problem of spatial autocorrelation in regression models. The method relies on the computational formula of Moran's I (MI, Moran 1948) – the most commonly employed statistical indicator for spatial autocorrelation – which is given by:

$$I = \frac{N \sum_i \sum_j w_{ij}(x_i - \bar{x})(x_j - \bar{x})}{\left(\sum_i \sum_j w_{ij}\right) \sum_i (x_i - \bar{x})^2}. \tag{9.3}$$

In Eq. 9.3, x_i is the value of the variable X in the ith region, and w_{ij} is the (i,j) element of the spatial weights matrix **W**. After pre- and post-multiplying **W** by a projection matrix, we obtain:

$$\mathbf{C} = (\mathbf{I}_n - \mathbf{1}\mathbf{1}^T/n)\mathbf{W}(\mathbf{I}_n - \mathbf{1}\mathbf{1}^T/n), \tag{9.4}$$

where **1** is an $n \times 1$ vector of 1's. Matrix **C** can actually be used to obtain, given variable X, the numerator of MI (Eq. 9.3), and its extreme eigenvalues are approximately the extreme values of MI (Griffith 2000). Because of this mathematical relation between **C** and MI, the eigenvectors extracted from **C** represent all mutually exclusive (orthogonal and independent) spatial patterns implied by the chosen spatial weights matrix **W**. The eigenvectors $E_1 \ldots E_n$ of **C** are extracted in decreasing order of spatial autocorrelation (MI). Therefore, E_1 has the largest MI achievable, given the choice of **W**. All subsequent eigenvectors maximize MI while being orthogonal to all previously extracted eigenvectors.

When employed in a regression model framework as additional explanatory variables, such eigenvectors may account, among other things, for unobserved heterogeneity, redundant information, and spatial spillover effects, rendering regression residuals spatially uncorrelated (at least in a cross-sectional framework). A stepwise regression approach may be used to select which eigenvectors are actually significant in a specific modelling exercise. Because the number of eigenvectors increases with the cross-sectional dimension, starting with a subset of so-called 'candidate' (or 'dominant') eigenvectors is convenient. This subset is usually defined according to a threshold of 0.25 for the ratio $I(E_k)/\max_k I_k$ (for details, see Griffith 2003). The linear combination of the set of k eigenvectors resulting from the selection procedure and their estimated regression coefficients is called a 'spatial filter'.

Additionally, SF may be employed to inspect the spatial heterogeneity of regression coefficients, equivalently to what is done in geographically weighted regression (GWR, Fotheringham et al. 2002).[3] Patuelli et al. (2012) show that, in a dynamic panel modelling framework, a heterogeneous-coefficients model can be

[3] Griffith (2008) stresses that the SF-based approach to GWR actually provides superior statistical properties (e.g., with regard to multicollinearity) than the original GWR.

successfully approximated by constructing a spatial filter-representation of the vector of serial autoregressive coefficients, simultaneously allowing for improved inference in unit root testing. This may be done by interacting each candidate eigenvector (repeated T times) with the serially lagged (dependent) variable, thus constructing a set of new variables representing its spatial decomposition in orthogonal components. The regression coefficients associated to these new variables will indicate relevance of spatial patterns in adjustment processes. The same process can be applied to any other explanatory variable (e.g., to seasonal indicator variables).

When spatial filters are simultaneously applied to the serial correlation coefficient and at the intercept level, the following model is obtained:

$$y_{i,t} = c + \beta y_{i,t-1} + \sum_{m=1}^{k} \beta_m E_{i,m} y_{i,t-1} + \sum_{m'=1}^{k'} \beta_{m'} E_{i,m'} + \varepsilon_{i,t}, \tag{9.5}$$

where m and m' are counters for the selected eigenvectors, at the lagged term and intercept level, respectively. A (standard) intercept c and an average serial correlation coefficient β are still estimated. The two spatial filters obtained provide the regional deviations from these aggregate measures.

9.3 Data and Forecasting Strategy

We test the forecasting performance of the two methods presented above on two data sets relating to Spain and Switzerland. As a numerical example, we use official regional unemployment rates at the NUTS-3 level of geographical aggregation, and analyse the temporal evolution of forecasting errors and their spatial distribution.

We choose Spain and Switzerland since two desired characteristics for this comparison exercise are verified. On the one hand, both data sets have satisfactory but different temporal (T) and spatial dimensions (n). On the other hand, the geographical size of the spatial (administrative) units analysed is widely different. The average area of Spanish provinces is about 10,499 km^2 (σ = 4,699.77 km^2), whereas the same for the Swiss cantons is 1,582 km^2 (σ = 1,822.35 km^2). Therefore, the size of the Swiss cantons is lower with a high level of variability.

Unemployment data for Spain are collected through the Spanish Labour Force Survey (Encuesta de Población Activa, EPA). The data consist of quarterly unemployment rates by province (corresponding to the Spanish NUTS-3 level of geographical classification) and cover the period 1977–2008. Most studies about Spanish labour markets assert that one of its main features is the unemployment persistence from an aggregate viewpoint, but the persistence of differences in unemployment rates across provinces is highlighted as well (Blanchard and Jimeno 1995; Jimeno and Bentolila 1998). In 2006, we can find some provinces with high unemployment rates (above 14 %, like Cádiz, Badajoz, Huelva and Cordoba), whereas others (Teruel, Soria, Navarra and Guipuzcua) have rates lower than 6 %. These relevant differences are quite similar along the entire period studied.

Switzerland is a non-EU country, and its labour market can be considered to be quite different from the EU average. Although some constraints have been relaxed with regards to employment and migration regulations (Switzerland has recently eliminated immigration quotas, and started participating in the Schengen agreement), the Swiss labour market is still strictly regulated, and migration is controlled through working permits.

Unemployment data for Switzerland are given through the Unemployment Statistics of the Swiss Federal Statistical Office. The data set we employ consists on monthly unemployment rates between 1975 and 2008, collected for the 26 cantons of Switzerland (again, corresponding to the NUTS-3 geographical aggregation level). Unemployment rates in Switzerland are much lower than in Spain. From 1995 to 2010, Switzerland's unemployment rate averaged 3.38 %, reaching an historical high of 5.40 % in 1997, and a record low of 1.60 % in 2000. The difference between Switzerland's historical high and the unemployment rates in Spain is striking. The temporal evolution of unemployment rates is also quite different, and Spanish data show a higher level of volatility as it is shown in Fig. 9.1.

The two plots composing Fig. 9.1 cover comparable time periods, and employ, for each single region, a separate colour scale, based on quantiles. The underlying graphs give the evolution of the regional unemployment rate at the median. As it can be seen, most Spanish regions experienced two highs in unemployment, around 1985 and 1995, followed by a marked improvement and, ultimately, by the first signs of unemployment rise coinciding with the 2008 financial crisis. Swiss regions, instead, experienced a marked unemployment rate increase between 1990 and 1995, which lasted until about 2000. It is worth noting that, while Swiss regions all follow the aggregate trend (the lighter and darker parts of the plot are homogeneous by row), not all Spanish regions do, suggesting a possible heterogeneity in cyclical sensitivity, which could be reflected in spatial patterning of serial correlation coefficients.

As noted above, spatial interactions between spatial units are highly relevant in a labour market context, and they are often introduced by means of a spatially lagged variable. Although there are different ways to model connectivity between areas, spatial weight matrices used in this chapter are based on the contiguity criterion, where a weight of 1 is assigned to the cell (i, j) of W if the spatial units i and j share a common boundary, and 0 otherwise. The matrices are row-standardized in the case of the SVAR, and globally standardized (Tiefelsdorf et al. 1999) for SF, because of the symmetry requirement for eigenvector extraction.

Before choosing to include spatial structure in our forecasting models, we test for the existence of spatial autocorrelation in our data, using average values per year.[4] Significant positive spatial autocorrelation is found, which suggests the existence of spillovers across regions.

[4] Moran test results are available from the authors upon request.

In order to evaluate the short-run predictive power of the two compared methods, we devise a forecasting strategy based on a rolling window. For each model and data set, estimates are obtained using a fixed window of observations, between $t = 1 + g$ and $t = T - h + g$, and *ex post* forecasts of regional unemployment rates are carried out for the nearest subsequent time period, for $g = 0, \ldots, h - 1$ and h being the number of time periods covered by the forecasting window. The forecasting window moves over 2 years, therefore providing forecasts over 8 quarters for Spain and 24 months for Switzerland. Given cross-sectional dimensions, the overall number of forecasted values is (8 × 47 =) 376 for Spain and (24 × 26 =) 624 for Switzerland.

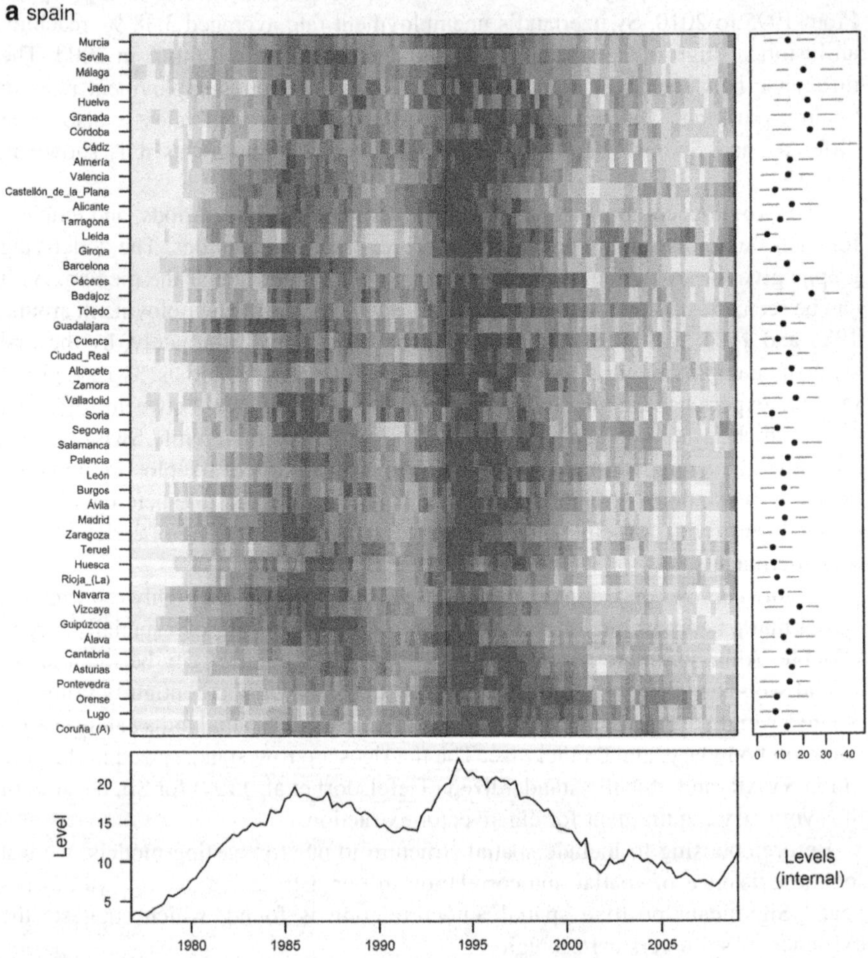

Fig. 9.1 (continued)

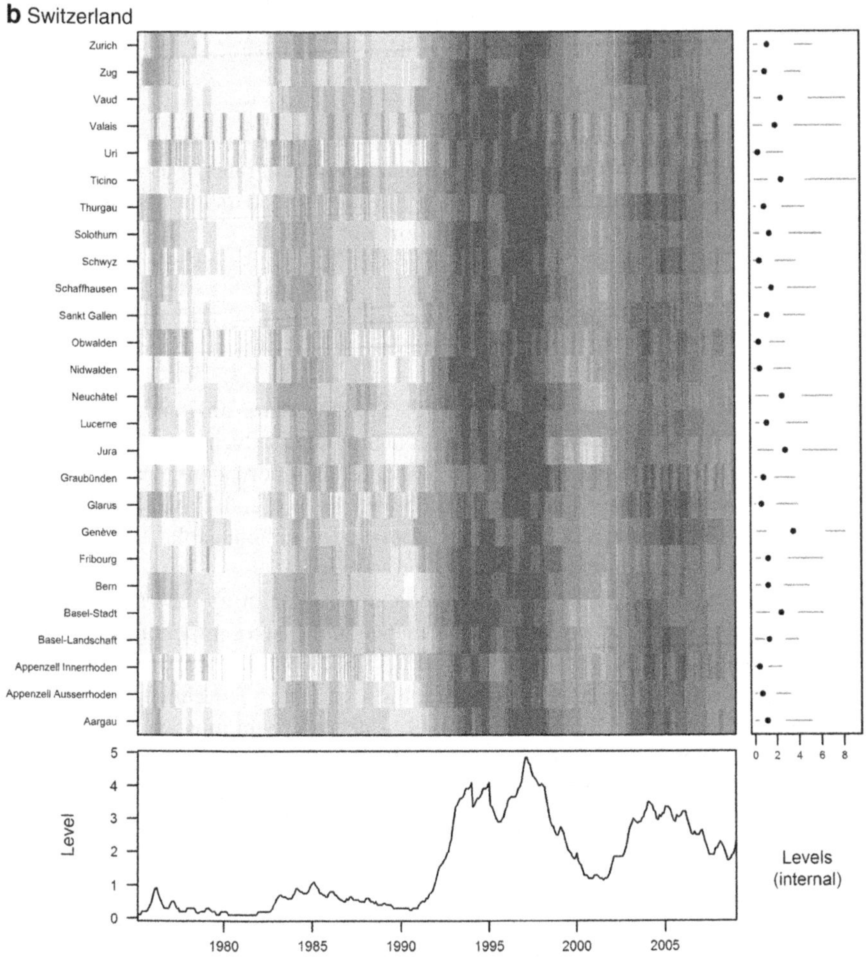

Fig. 9.1 Panel plots of regional unemployment rates for Spain (**a**) and Switzerland (**b**)

9.4 Results

The forecasting performance of the SVAR and SF methods is summarized and compared by means of statistical indicators: the mean square error (MSE), the mean absolute error (MAE) and the mean absolute percentage error (MAPE). The MSE and MAE measure deviations in absolute value from the true values, and are computed as follows:

$$MSE = \frac{1}{n}\sum_{i=1}^{n}\left(\hat{U}_i - U_i\right)^2; \tag{9.6}$$

$$MAE = \frac{1}{n}\sum_{i=1}^{n}\left|\hat{U}_i - U_i\right|. \qquad (9.7)$$

In comparing the forecasting power of competing models, it is important to take into account the scale heterogeneity in the unemployment rates of each province. It is thus convenient to consider also the MAPE, which considers forecasting errors on a percentage scale, and is given by the following:

$$MAPE = \frac{1}{n}\sum_{i=1}^{n}\frac{\left|\hat{U}_i - U_i\right|}{U_i} \times 100. \qquad (9.8)$$

Each model is tested on out-of-sample data for the years 2007 and 2008. Within this framework, the discussion on the advantages and disadvantages of *ex-ante* and *ex-post* predictions seems unnecessary, since contemporaneous spatial lags are obtained in the first stage, as described above. Angulo and Trívez (2010) avoid this debate as well, using a dynamic panel data model without explanatory variables to forecast employment levels in Spanish provinces.

On the basis of the above statistical indicators, we aim to generate inferential evidence on the relative forecasting performance of the two competing econometric models. Therefore, following Patuelli et al. (2008), we use a nonparametric test to assess if two models are equally accurate: the sign test (ST, Lehmann 1998). The sign test does not rely on the usual assumptions necessary for most comparison tests (such as the Diebold-Mariano test or the Wilcoxon test), as it does not require normal distribution or symmetry between the two vectors compared. More simply, the sign test is based on the comparison of the forecasting errors. If the methods tested present a similar forecasting performance, the number of SF (Model 2) forecasts which show a greater error than the one of SVAR (Model 1) may be expected to be 50% of the total number of forecasts obtained. Consequently, Model 1 will be considered superior to Model 2 if Model 2 has higher forecasting errors in more than 50 % of the cases. Clearly, the test does not provide insights on the error distribution, but only on comparative forecasting, pairwise. In practice, what is being tested is the hypothesis of equality in the medians. The test statistic S is computed as:

$$S = \left(C - \frac{p}{2}\right) \Big/ \frac{\sqrt{p}}{2}, \qquad (9.9)$$

where C is the number of times that Model 2 shows a higher error than Model 1's, and p is the number of forecasts carried out. The S statistic follows a normal distribution $N(0,1)$. When not standardized, C follows a binomial distribution $B(p, 0.5)$. Confidence intervals for S (C) are obtained in the standard way from the normal and binomial tests, respectively.

9.4.1 Results for Switzerland

Table 9.1 summarizes the statistical performance of SVAR and SF for the Swiss case, by means of MSE, MAE and MAPE, as we compute the cross-sectional average error for each of the 24 forecasting periods. Values in bold indicate the model with the lowest value.

From the analysis of Table 9.1, the SVAR model appears to show better forecasting performance than the SF model, although the difference between the two models is considerably reduced when the MAPE is considered. In any case, the numerical distance between the two models is rather small. A closer look at the heterogeneity of these forecasting errors is given by the inspection of the error distributions. Figure 9.2 collects the average MSE, MAE and MAPE for each estimation period and their confidence intervals. In all cases, the SF approach presents a high level of variability in comparison to the SVAR. In addition, Fig. 9.6 in the Appendix provides histograms for the aggregate distributions of the forecasting error indicators.

As a final analysis on the error indicators, the sign test is performed, depending on the forecasting errors used for comparison, along three dimensions. First, all forecasting errors are pooled (for all cross-sectional units and all forecasting periods). Then, the average forecasting errors by canton are analysed. Finally, the average forecasting errors per period are compared.

The pooled sign tests do no reject the hypothesis of equivalence in median between the SAR and SF model forecasts. The same conclusion is reached when the test is computed averaging the errors by forecasting period, but the results point out a statistically better performance of the SVAR model when the average forecasting errors by region are analysed.

Since both methods explicitly account for spatial autocorrelation, we may expect that, if spatial patterning in the data is well identified, forecasting errors should not present spatial autocorrelation. To test this hypothesis, MI is computed, for each forecasting period, on the prediction errors of both methods.

Figure 9.3 summarizes our findings for the Swiss case. For the majority of forecasting periods there is no significant spatial autocorrelation, for both SVAR and SF models. However, SVAR seems to produce spatially autocorrelated forecasting errors in a smaller number of cases. This finding, joined with the evidence above on error indicators, is not surprising, since the time dimension is much larger than the spatial dimension ($T >> n$) in the Swiss data set, therefore clearly advantaging a time-series-related method like the SVAR.

9.4.2 Results for Spain

Table 9.2 reports summary statistics for the MSE, MAE and MAPE over the eight forecasting periods (quarters) used for Spain. Our findings for Spain differ from the

Table 9.1 Summarystatistics of MSE, MAE and MAPE for Switzerland

			I	II	III	IV	V	VI	VII	VIII	IX	X	XI	XII
2007	MSE	SVAR	0.024	**0.018**	**0.015**	**0.011**	**0.011**	**0.014**	**0.005**	0.008	0.011	**0.009**	**0.008**	**0.011**
		SF	**0.015**	0.019	0.033	0.024	0.014	0.022	0.008	**0.007**	**0.010**	0.024	0.015	0.032
	MAE	SVAR	0.113	0.108	**0.103**	**0.074**	**0.090**	**0.085**	**0.056**	0.073	0.084	**0.069**	**0.070**	**0.081**
		SF	**0.092**	**0.108**	0.119	0.106	0.107	0.105	0.067	**0.067**	**0.079**	0.113	0.090	0.089
	MAPE	SVAR	4.978	4.423	4.608	**3.383**	**4.780**	**4.098**	**3.607**	4.536	4.452	**3.862**	**3.938**	**3.589**
		SF	**4.019**	**4.162**	**4.518**	5.169	5.218	4.822	3.894	**3.97**	**4.139**	6.757	4.027	3.802
2008	MSE	SVAR	0.016	**0.014**	**0.003**	**0.012**	**0.011**	**0.006**	**0.006**	**0.005**	0.007	**0.016**	**0.007**	**0.041**
		SF	**0.012**	0.021	0.015	0.017	0.013	0.012	0.008	0.006	**0.007**	0.025	0.015	0.086
	MAE	SVAR	0.091	**0.085**	**0.049**	**0.074**	**0.085**	**0.058**	**0.065**	**0.056**	0.067	**0.091**	**0.066**	**0.156**
		SF	0.087	0.091	0.063	0.089	0.095	0.072	0.074	0.066	0.066	0.111	0.083	0.198
	MAPE	SVAR	4.611	4.328	2.642	**4.057**	**4.509**	**3.925**	**4.559**	**3.655**	**4.285**	**4.879**	**3.133**	**5.861**
		SF	**4.311**	**4.076**	**2.625**	5.445	4.857	4.347	5.369	4.403	4.333	6.955	4.064	7.686

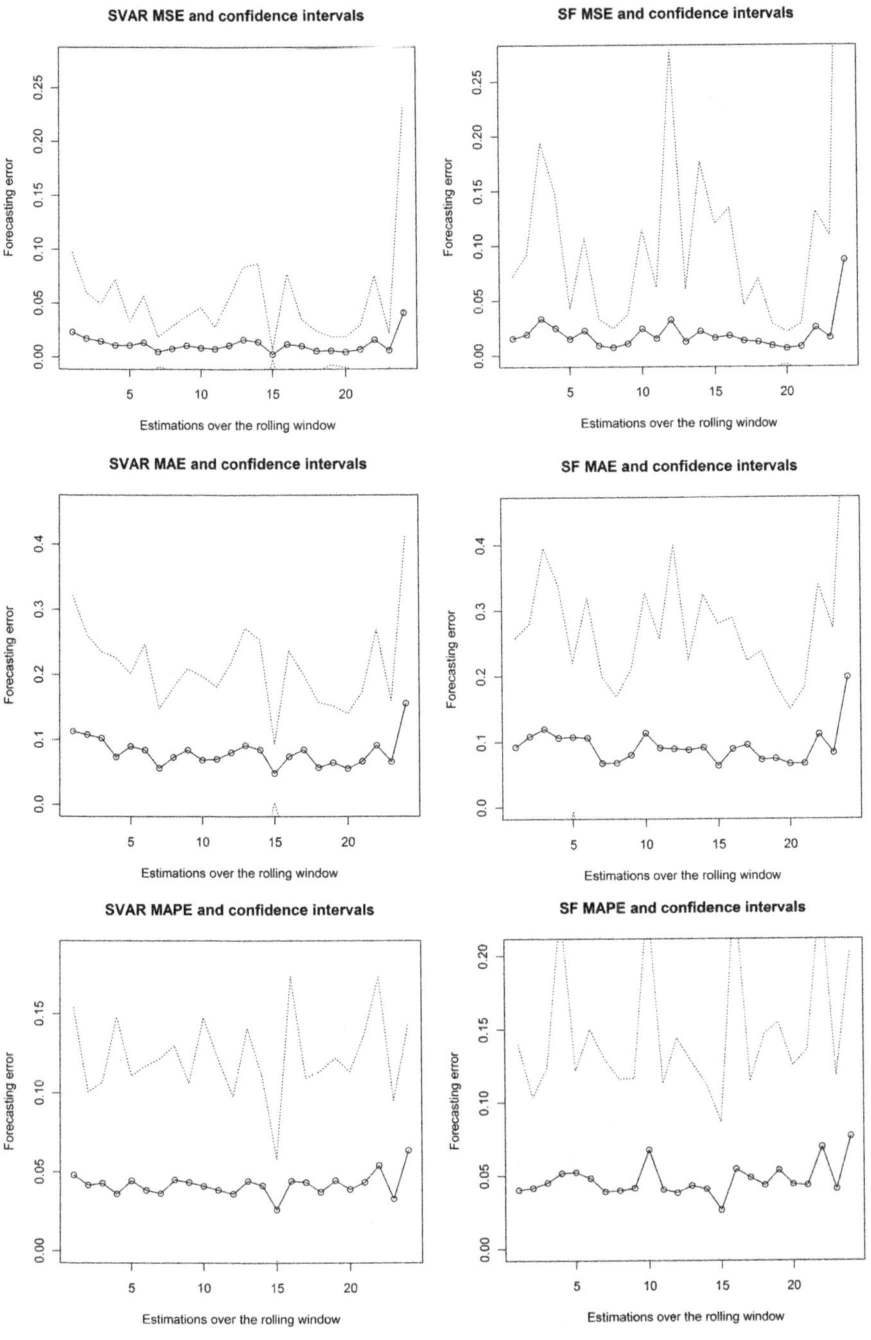

Fig. 9.2 MSE, MAE, MAPE and confidence intervals for SVAR (*left*) and SF (*right*) models for Switzerland

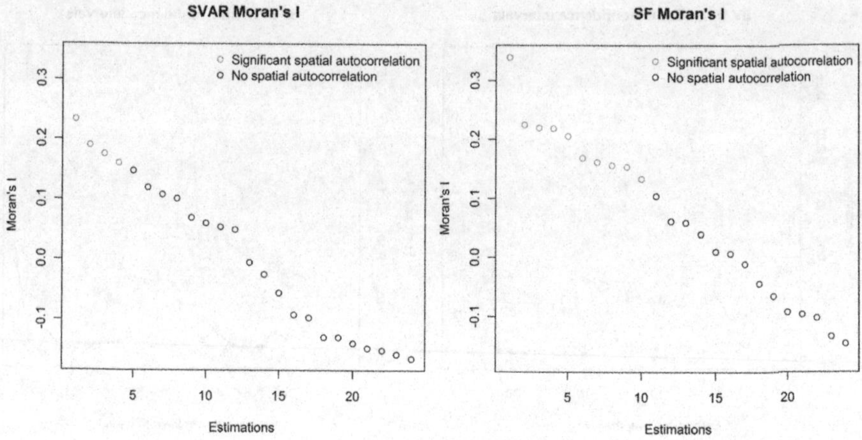

Fig. 9.3 MI of SVAR (**a**) and SF (**b**) forecasting errors for Switzerland, sorted in decreasing order

Table 9.2 Summary statistics of MSE, MAE and MAPE for Spain

			I	II	III	IV
2007	MSE	SVAR	1.883	1.344	**1.097**	**1.244**
		SF	**1.630**	**1.341**	1.279	1.850
	MAE	SVAR	1.035	**0.801**	**0.848**	**0.887**
		SF	**1.018**	0.904	0.903	1.109
	MAPE	SVAR	0.136	**0.111**	0.128	**0.115**
		SF	**0.129**	0.118	**0.121**	0.143
2008	MSE	SVAR	**1.779**	2.981	**2.459**	6.002
		SF	1.942	**2.654**	2.479	**5.035**
	MAE	SVAR	**0.942**	1.318	**1.221**	2.010
		SF	0.966	**1.258**	1.292	**1.907**
	MAPE	SVAR	0.125	0.163	0.135	0.196
		SF	**0.122**	**0.154**	**0.134**	**0.188**

ones for Switzerland in that the SF model appears to have gained in competitiveness from the different data structure (the unbalance between n and T is now of a lesser extent). In particular, the SVAR model appears to be more competitive with regard to MSE and MAE (when the error is not standardized by the level of the unemployment rates), while the SF model minimizes percentage error (MAPE), winning six out of eight comparisons.

Once again, we can inspect the heterogeneity of forecasting errors, through the plots given in Fig. 9.4. Differently from the Swiss case, it is now the SVAR model that presents a higher heterogeneity in forecasting errors. The increase in the cross-sectional dimension of the data set (from the 26 Swiss cantons to the 47 Spanish provinces) may be one reason for this finding, as spatial contiguity relationships become more meaningful in a richly disaggregated dataframe. Also noteworthy is the generalized increase in forecasting errors over time and in particular at the last

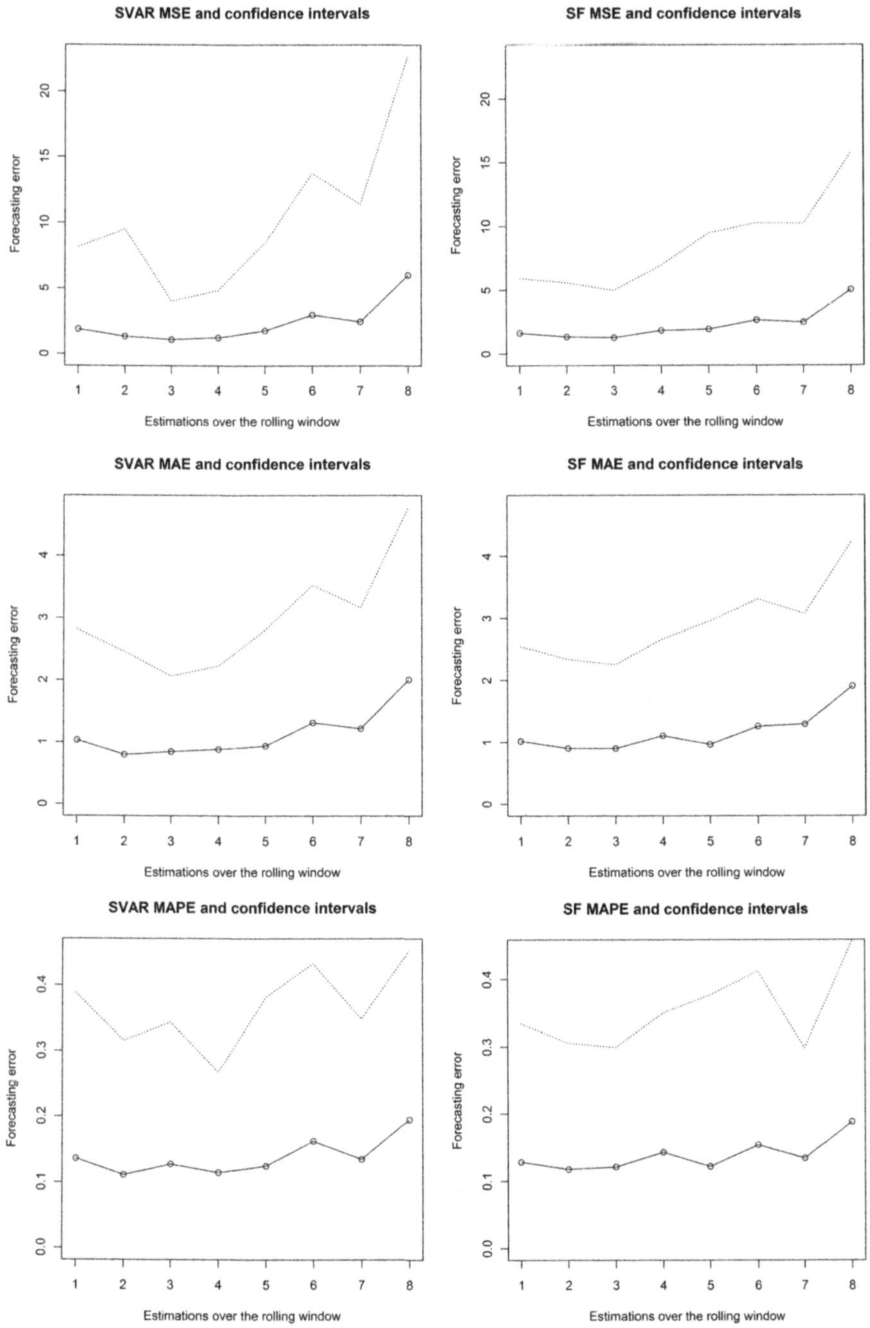

Fig. 9.4 MSE, MAE, MAPE and confidence intervals for SVAR (*left*) and SF (*right*) models for Spain

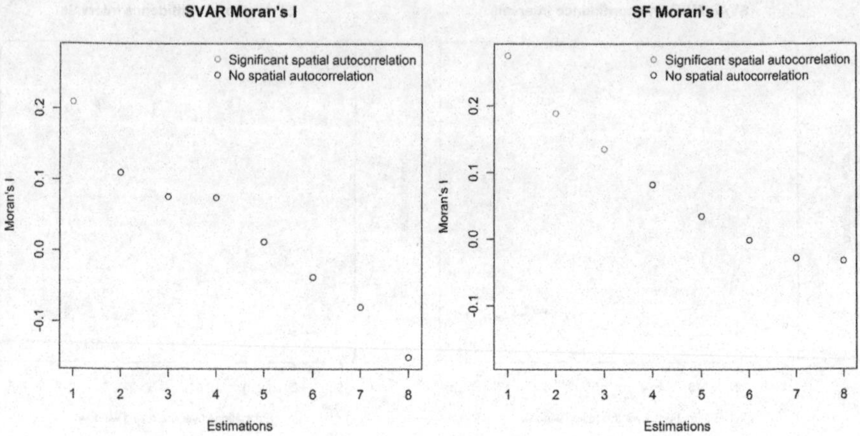

Fig. 9.5 MI of SVAR (**a**) and SF (**b**) forecasting errors for Spain, sorted in decreasing order

two quarters, coinciding with the 2008 financial crisis, which had a strong labour market impact on the Spanish labour market. Figure 9.7 in the Appendix provides further evidence on the pooled statistical distribution of the forecasting errors obtained for Spain.

We compute the ST to assess whether our two methods can be considered as equally accurate when forecasting Spanish unemployment rates. The main difference with the Swiss case lies in the temporal and cross-sectional dimension, since we now deal with moderate values of both n and T. We calculate the test using all forecasting errors (pooled test), the average forecasting errors by region, and the average forecasting errors by quarter. In all cases, the tests are not significant, suggesting an overall equivalence between the SVAR and the SF models.

Finally, Fig. 9.5 plots the MI statistic computed for each (of the eight) forecasting periods. As for the case of Switzerland, both competing methods produce spatially uncorrelated forecasting errors in most cases, but the SVAR model appears to account better for the true spatial correlation in the dependent variable. In any case, the levels of spatial autocorrelation of forecasting errors, when significant, are very low.

9.5 Rejoinder and Conclusions

The aim of this chapter was to analyse the short-run forecasting performance of two competing spatial models: a spatial vector-autoregressive model (SVAR) and a dynamic panel data model employing spatial filtering (SF). We carried out a sensitivity analysis in order to test how different number and size of the spatial units (actually administrative areas) and varying extent of the temporal dimension influence the relative forecasting performance of these methods.

Our empirical application used regional unemployment rates at the NUTS-3 level of geographical aggregation for two countries: Spain and Switzerland. Switzerland has a low number of NUTS-3 regions (26), which are also much smaller in areal extension than their (47) Spanish equivalents. Moreover, although the data in both data sets are collected for a similar number of years, Swiss unemployment data have a monthly frequency, while Spanish data are quarterly, resulting in rather different data structures: for Switzerland, $T >> n$, while for Spain this data unbalance is smaller, as there is some level of convergence between T and n.

We carried out one-period-ahead *ex post* forecasts for the years 2007 and 2008, using a fixed rolling window to estimate both models. Our results were evaluated by means of statistical indicators (MSE, MAE and MAPE), as well as a forecast equivalence test (sign test).

From an empirical viewpoint, the aforementioned differences in data structure between the Swiss and Spanish data sets appear to be a discriminating factor in terms of forecasting accuracy. The SVAR model seems slightly preferable on the SF model when the temporal dimension is much greater than the spatial dimension and the spatial units have smaller size and greater degree of variability (i.e., the Swiss data). Although not on a consistent basis, sign test results support this finding.

When moderate cross-sectional and time dimensions were used (i.e., the Spanish data), we did not find stable significance differences between the two competing methods. The SVAR models appear to minimize errors on the scale of the unemployment rates (MSE and MAE), while the SF model is preferable when percentage error is considered (MAPE). This finding is justified by methodological aspects, as the SF model computes a geographical approximation of both the autoregressive coefficients and of the fixed/random effects usually employed in dynamic panel data models. As such, it may be less efficient in estimating outliers (e.g., change in high unemployment areas), while it may be expected to provide smoother findings on the spatial patterning of coefficients.

Finally, we investigated whether the spatial patterns of the data were well-identified in both methods, testing the existence of spatial autocorrelation in the forecasting errors. The SVAR model shows a lesser number of spatially autocorrelated errors for both the Swiss and the Spanish data sets, although most estimations produced uncorrelated errors for both methods.

In summary, the SVAR models showed somehow superior performance when the time dimension was clearly dominant on the cross-sectional dimension, consistently with the underlying time-series framework of VAR models. Moving instead to moderate cross-sectional and temporal dimensions, no clear difference can be drawn between the SVAR and SF estimation frameworks. It was not possible, within this study, to test a data structure opposite to the one of Switzerland (e.g., German NUTS-3 unemployment data, which were available to the authors, and for which $n >> T$), as VAR models cannot be estimated in such case.

Further research should complement this preliminary empirical investigation. From a methodological viewpoint, it is desirable to expand our study to include more (spatial and non-spatial) econometric models (e.g., spatial panel data models, *a la* Elhorst). From an empirical viewpoint, out findings should be verified through a simulation study, which could allow for a greater variability in cross-sectional and temporal size, to provide a more complete 'map' of the comparative performance of spatial models in forecasting. Finally, expanding the forecasting horizon would allow us to test whether the relative forecasting relationships found in this chapter are preserved for longer-periods forecasts.

Appendix

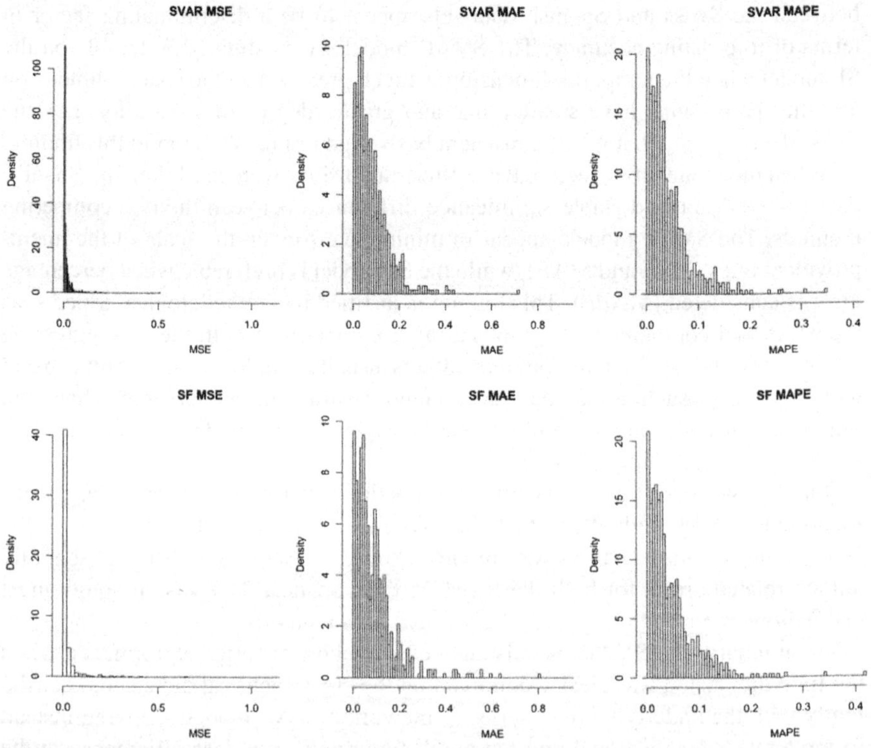

Fig. 9.6 Histograms of MSE, MAE and MAPE of SVAR (*above*) and SF (*below*) models for Switzerland

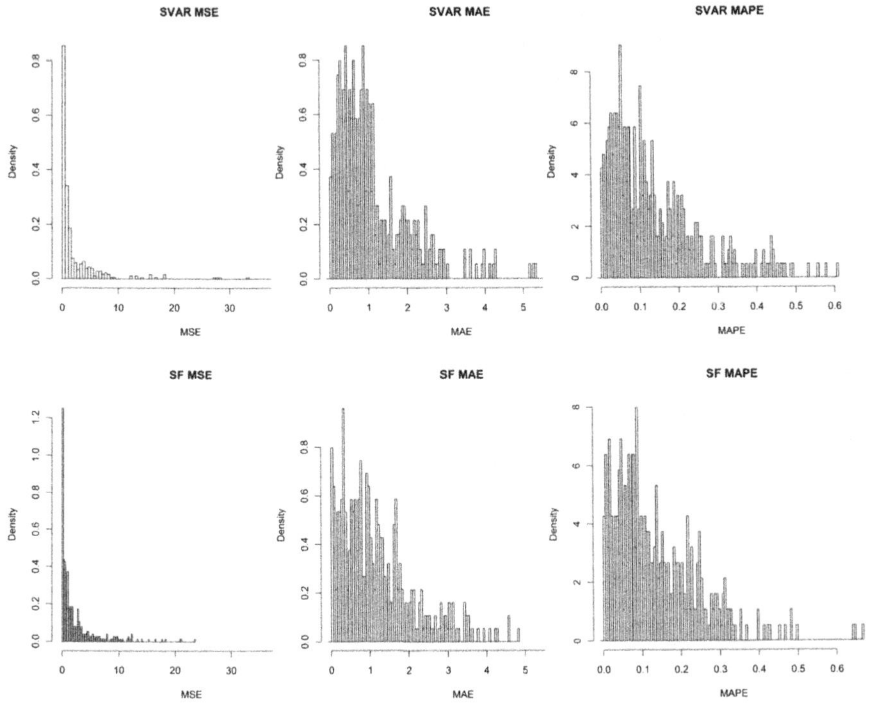

Fig. 9.7 Histograms of MSE, MAE and MAPE of SVAR (*above*) and SF (*below*) models for Spain

References

Angulo A, Trívez FJ (2010) The impact of spatial elements on the forecasting of Spanish labour series. J Geogr Syst 12(2):155–174

Baltagi BH, Li D (2004) Prediction in panel data model with spatial correlation. In: Anselin L, Florax RJGM, Rey SJ (eds) Advances in spatial econometrics: methodology, tools and application. Springer, Berlin/Heidelberg, pp 283–295

Baltagi BH, Li D (2006) Prediction in the panel data model with spatial correlation: the case of liquor. Spat Econ Anal 1(2):175–185

Beenstock M, Felsenstein D (2007) Spatial vector autoregressions. Spat Econ Anal 2(2):167–196

Blanchard O, Jimeno JF (1995) Structural unemployment: Spain versus Portugal. Am Econ Rev 85 (2):212–218

Di Giacinto V (2003) Differential regional effects of monetary policy: a geographical SVAR approach. Int Reg Sci Rev 26(3):313–341

Elhorst JP (1995) Unemployment disparities between regions in the European union. In: Armstrong HW, Vickerman RW (eds) Convergence and divergence among European unions. Pion, London, pp 209–221

Ertur C, Le Gallo J (2003) An exploratory spatial data analysis of European regional disparities, 1980–1995. In: Fingleton B (ed) European regional growth. Springer, Berlin/Heidelberg/New York, pp 55–98

Fingleton B (2009) Prediction using panel data regression with spatial random effects. Int Reg Sci Rev 32(2):195–220

Fotheringham AS, Brunsdon C, Charlton M (2002) Geographically weighted regression: the analysis of spatially varying relationships. Wiley, Chicester

Giacomini R, Granger CWJ (2004) Aggregation of space-time processes. J Econom 118 (1–2):7–26

Griffith DA (2000) A linear regression solution to the spatial autocorrelation problem. J Geogr Syst 2(2):141–156

Griffith DA (2003) Spatial autocorrelation and spatial filtering: gaining understanding through theory and scientific visualization. Springer, Berlin/New York

Griffith DA (2008) Spatial-filtering-based contributions to a critique of geographically weighted regression (GWR). Environ Plann A 40(11):2751–2769

Hernández-Murillo R, Owyang MT (2006) The information content of regional employment data for forecasting aggregate conditions. Econ Lett 90(3):335–339

Jimeno JF, Bentolila S (1998) Regional unemployment persistence (Spain, 1976–1994). Lab Econ 5(1):25–51

Kholodilin KA, Siliverstovs B, Kooths S (2008) A dynamic panel data approach to the forecasting of the GDP of German Länder. Spat Econ Anal 3(2):195–207

Kuethe TH, Pede VO (2011) Regional housing price cycles: a spatio-temporal analysis using US state-level data. Reg Stud 45(5):563–574

Lehmann EL (1998) Nonparametrics: statistical methods based on ranks (rev. ed.). Upper Saddle River: Prentice Hall

Longhi S, Nijkamp P (2007) Forecasting regional market developments under spatial autocorrelation. Int Reg Sci Rev 30(2):100–151

López-Bazo E, del Barrio T, Artis M (2002) The regional distribution of Spanish unemployment: a spatial analysis. Pap Reg Sci 81(3):365–389

Molho I (1995) Spatial autocorrelation in British unemployment. J Reg Sci 35(4):641–658

Moran P (1948) The interpretation of statistical maps. J R Stat Soc B 10:243–251

Overman HG, Puga D (2002) Unemployment clusters across Europe's regions and countries. Econ Pol 17(34):115–148

Pan Z, LeSage JP (1995) Using spatial contiguity as prior information in vector autoregressive models. Econ Lett 47(2):137–142

Patacchini E, Zenou Y (2007) Spatial dependence in local unemployment rates. J Econ Geogr 7(1): 169–191

Patuelli R (2007) Regional labour markets in Germany: statistical analysis of spatio-temporal disparities and network structures. Unpublished Ph.D. thesis, Department of Spatial Economics, VU University Amsterdam, Amsterdam

Patuelli R, Longhi S, Reggiani A, Nijkamp P (2008) Forecasting regional employment in Germany by means of neural networks and genetic algorithms. Env Plann B 35(4):701–722

Patuelli R, Schanne N, Griffith DA, Nijkamp P (2012) Persistence of regional unemployment: application of a spatial filtering approach to local labour markets in Germany. J Reg Sci 52(2): 300-323

Pesaran MH, Schuermann T, Weiner SM (2004) Modeling regional interdependencies using a global error-correcting macroeconometric model. J Bus Econ Stat 22(2):129–162

Schanne N, Wapler R, Weyh A (2010) Regional unemployment forecasts with spatial interdependencies. Int J Forecast 26(4):908–926

Sims CA (1980) Macroeconomics and reality. Econometrica 48(1):1–48

Taylor J, Bradley S (1997) Unemployment in Europe: a comparative analysis of regional disparities in Germany, Italy and the UK. Kyklos 50(2):221–245

Tiefelsdorf M, Griffith DA, Boots BN (1999) A variance stabilizing coding scheme for spatial link matrices. Enviro Plann A 31:165–180

Chapter 10
Local Weighting Matrices or the Necessity of Flexibility

Jesús Mur and Antonio Paez

10.1 Introduction

The difficulties caused by the lack of stability in the parameters of an econometric model are well known: biased and inconsistent estimators, misleading tests and, in general, wrong inference. The list of consequences explains the attention that the literature on mainstream Econometrics has dedicated to the problem. The first formal test of parameter stability was due to Chow (1960), considering only one break point known a priori. Since then, the detection of structural breaks has received substantial attention in the discipline.

The discussion about parameter instability quickly took on a spatial context. The work of Casetti (1972, 1991) is notable because of its focus on modelling instabilities through the use of contextual variables. An emerging interest in the 1990s, was the identification of pockets of local nonstationarity. This gave rise to an influential literature dedicated to the analysis of spatial autocorrelation from a local perspective (Getis and Ord 1992; Anselin 1995). Developments in this literature encouraged the development of non-parametric procedures for analyzing heterogeneous spatial data (McMillen 1996; McMillen and McDonald 1997). The most popular approach to deal with the problem of instability in spatial models is what Brunsdon et al. (1996) call Geographically Weighted Regressions (GWR in what follows), an approach that now occupies a prominent role in the spatial econometrics toolbox. This strand of literature is related to the Locally Weighted Regressions approach, introduced by Cleveland (1979) and Cleveland and Devlin (1988). In all

J. Mur (✉)
Department of Economic Analysis, University of Zaragoza. Building 'Lorenzo Normante', room 3.04. Campus 'Rio Ebro' 50018, Zaragoza
e-mail: jmur@unizar.es

A. Paez
School of Geography & Earth Sciences, 1280 Main Street West, Hamilton, Ontario L 8S 4L8, Canada

E. Fernández Vázquez and F. Rubiera Morollón (eds.), *Defining the Spatial Scale in Modern Regional Analysis*, Advances in Spatial Science,
DOI 10.1007/978-3-642-31994-5_10, © Springer-Verlag Berlin Heidelberg 2012

these papers there is constant movement from global to local perspectives, using parametric and non parametric techniques.

The benefits of a local approach are clear when strong heterogeneity and/or autocorrelation is displayed by the data. If the cause of the problem is a poor selection of the functional form, a locally linear approximation may be a second-best solution, useful to alleviate the consequences of the misspecification (McMillen 2003; Paez et al. 2008). The GWR algorithm has also been extended to directly deal with spatial correlation (Brunsdon et al. 1998a; Páez et al. 2002b). Pace and Lesage (2004) and Mur et al. (2009), among others, continue in this line relating the concepts of spatial dependence and spatial heterogeneity.

This paper contributes to the literature on GWR. Particularly, we focus on three different but related questions: (a) the development of a GWR test to compare local versus global estimates; (b) the definition of the bandwidth, that we solve in a fully adaptative framework and (c) the non-uniqueness of the GWR estimates, which follows from the uncertainty in relation to the selection of the kernel. The structure of the paper is as follows. Section 10.2 is a brief introduction to the GWR methodology. In the third section a simple GWR test is introduced. In Sect. 10.4 we advocate for a more flexible use of the concept of neighborhood. Section 10.5 discusses the problem of selecting a weighting matrix from among a finite set of alternatives. Section 10.6 contains the main conclusions.

10.2 Geographically Weighted Regressions

The GWR consists in the estimation of a model, usually linear, in each point of the sampling space using only local information (Brunsdon et al. 1998a, b; Fotheringham et al. 1999). The idea is simple. Let us assume the following model:

$$y = x\beta + u; u{:}ii\,dN(0, \sigma^2 I) \tag{10.1}$$

wherey is the $(R \times 1)$ vector of the observations of the endogenous variable, x is an $(R \times k)$ matrix of observations of the k explanatory variables, u is a white noise random vector normally distributed. For simplicity, the equation does not include spatial interaction terms. The model of Eq. 10.1 has been specified under the assumption of homogeneity. But as, indicated by McMillen (2004, p. 232): 'Spatial relationships are typically (…) complicated. Statistical tests based on simple functional forms often reveal that coefficients vary over space'. This is the issue that GWR aims to address.

The coefficients may vary over space for multiple reasons: unobserved heterogeneity, spatial instability, misspecification of the model, etc. The equation should reflect this fact:

$$y = x\beta_i + u; u{:}iidN(0, \sigma^2 I); i = 1, 2, \ldots, R \tag{10.2}$$

The systematic part of the equation is flexible so it can reflect local aspects of the process that might otherwise be overlooked. For simplicity the error term is assumed to be white noise. This is the preferred GWR scenario. The estimation algorihtm is simply replicated at different locations as desired, and results in a set of local estimates, obtained from the sequence of local regressions:

$$\hat{\beta}_i = \left[X'W_iX\right]^{-1}\left[X'W_iy\right];\ i = 1,2,\ldots,R \tag{10.3}$$

W_i is a diagonal weighting matrix that selects the observations that intervene in the estimation of the local coefficients, $\hat{\beta}_i$, in point i:

$$W_i = \begin{bmatrix} \alpha_{i1} & 0 & 0 & \ldots & 0 \\ 0 & \alpha_{i2} & 0 & \ldots & 0 \\ 0 & 0 & \alpha_{i3} & \ldots & 0 \\ \ldots & \ldots & \ldots & \ldots & \ldots \\ 0 & 0 & 0 & \ldots & \alpha_{iR} \end{bmatrix} \tag{10.4}$$

The local weights $\{\alpha_{ir}; r = 1,2,\ldots,R\}$ are constrained to the unit interval: $0\ \alpha_{ir}\ 1$, with the purpose of standardizing the influence of each observation. Generally, they are specified a priori by the user.

One of the main problems of the GWR methodology is that there are too many breaks for a limited amount of sampling information. The solution is attractive because of its simplicity: under normal circumstances, the points surrounding observation i should contain the most valuable information in relation to the behaviour of the Eq. 10.1 in point i. Consequently, the proposal is to use only these observations for obtaining the corresponding local estimate.

The specification of the local weights (for every point i!) is a complicated task in the GWR methodology. The literature suggests specifying the α's weights according to some simple function of the distance; for example:

$$\left.\begin{aligned} \alpha_{ir} &= \begin{cases} 1\ d_{ir} < d \\ 0\ d_{ir} d \end{cases} \\ \alpha_{ir} &= \exp\left[-\frac{1}{2}\left(\frac{d_{ir}}{d}\right)^2\right] \\ \alpha_{ir} &= \begin{cases} \left[1-\left(\frac{d_{ir}}{d}\right)^2\right]^2 d_{ir} < d \\ 0\ d_{ir} d \end{cases} \end{aligned}\right\} \tag{10.5}$$

where d is the bandwidth, a threshold distance beyond which it is assumed that the corresponding observations do not contain useful information in relation to the local coefficients of i.

Accordingly, there remain two decisions to be taken: the definition of the bandwidth and the selection of the functional form for the weights. There are several alternatives to treat both questions. The most popular is the so-called cross-validation score, which amounts to choosing the value of d that, globally, minimizes the (mean square) prediction error of the GWR estimation. In other cases, d may refer to the number of neighbors used in the local estimation (Farber and Paez 2007):

$$\mathrm{Min}_d\, CV(d) = \sum_{i=1}^{R} \left(y_i - \hat{y}_{i,N(i)}\right)^2 \tag{10.6}$$

being y_i the value of variable y observed in point i and $\hat{y}_{i,N(i)}$ the predicted value for this point by a linear model estimated using the $N(i) = d$ nearest neighbors of point i. In all cases, observation i is excluded from the local subsample.

The discussion in relation to the functional form of the weights has been, to our knowledge, very informal and the final decision depends of the preferences of the user. In any case, the crucial decision appears to be the bandwidth whereas the selection of the kernel plays a minor role (Wang et al. 2008; Páez et al. 2011).

A paradoxical aspect of the approach is that, since the purpose of GWR is to deal with parameter instability, the estimates will be unbiased only if the assumption of global homogeneity, implicit in Eq. 10.1, is true. In that case, it does not matter if we estimate the model of Eq. 10.1 using the whole set of R observations or only a subset of them. The bandwidth or the selection of a functional form for the weights are secondary issues given that they will affect only to the property of efficiency, through the variance of the estimates, but not to the quality of unbiasedness. On the contrary, if the equation of (10.1) is not stable, the GWR algorithm will, at the best, reduce least squares (LS) bias, but the estimates will still be biased. It is common to find in this literature the idea of a continous hyper surface of parameters that is evolving over space (Fotheringham et al. 2002) from which we have a discrete sample, in the R sampling points. If this assumption is true, the local estimation will be, probably, less problematic that the global LS estimation (Paez et al. 2002a).

From our point of view, two of the most important problems that limit the GWR methodology are the definition of what is ‘local information’ and the testing of the results. In the following sections we discuss some solutions.

10.3 A GWR Test

One of the conditions to guarantee efficient and unbiased LS estimation of Eq. 10.1 is stability of the equation over the space. GWR also produces unbiased estimates in this case, and a smaller bias if the coefficients are not stable. Despite this advantage, there are reasons why GWR must be applied judiciously. If the process is stable, GWR estimates are not efficient, whereas LS estimates are efficient, and simpler to obtain. Further, if the process is stable, GWR produces vast amounts of information that may be difficult to interpret. On the other hand, if the process is not stable, GWR must still be used with caution: spurious correlations among local parameters have been detected, especially in the case of small samples (Wheeler and

Tiefelsdorf 2005; Páez et al. 2011), something that brings into question the results if the symptoms of instability are not clear. Now, what precisely would be 'very clear symptoms'? The answer to this question may become a problem because, to our knowledge, there are few specific test for testing the adequacy of the GWR estimates (e.g. Leung et al. 2000). A proposal is as follows:

Let us assume a simple linear model like that of Eq. 10.2, where there is instability. The objective is to estimate a $(k \times 1)$ vector of parameters in each of the R sampling points. Obviously, this is not possible because we only have R observations to estimate a total of Rk parameters in R different equations. The GWR solution can be seen as a way of solving this deficit by combining pure sampling information with a priori theoretical information (in the form of a specific sequence of local breaks given by the selected kernel and bandwidth). Then, the resulting R GWR equations are evaluated independently. However some gains can be attained if the GWR estimation is solved in an extended model.

In order to achieve this, the overall design of the GWR algorithm must be modified. Let us define Λ, a $(R^2 \times R^2)$ diagonal matrix, as:

$$\Lambda = \begin{bmatrix} W_1 & 0 & 0 & \cdots & 0 \\ 0 & W_2 & 0 & \cdots & 0 \\ 0 & 0 & W_3 & \cdots & 0 \\ \cdots & \cdots & \cdots & \cdots & \cdots \\ 0 & 0 & 0 & \cdots & W_R \end{bmatrix} \tag{10.7}$$

where the W_i matrices have been defined in Eq. 10.4. Moreover X is the $(R \times Rk)$ matrix obtained as R copies of the original matrix x:

$$X = \begin{bmatrix} x & x & x & \cdots & x \end{bmatrix} \tag{10.8}$$

l is a $(R \times 1)$ vector of ones that we use to copy, vertically, matrix X and vectors y and u:

$$\begin{aligned} X &= l \otimes X \\ Y &= l \otimes y \\ U &= l \otimes u \end{aligned} \tag{10.9}$$

$\otimes$ indicates the Kronecker product. The first, X, is a $(R^2 \times Rk)$ matrix; the other two elements, Y and U, are $(R^2 \times 1)$ vectors. The three terms are related by means of a vector of parameters and by a matrix of local indicators. The first is a $(Rk \times 1)$ vector: $B = [\beta_1; \beta_2; \beta_3; \cdots; \beta_R]'$, being β_i the vector of local coefficients corresponding to point i. The second are $(R^2 \times R^2)$ matrices, similar to that of Eq. 10.7, which select the observations that intervene in each local estimation: $\bar{\Lambda} = \{\bar{\lambda}_{rs} = I(\lambda_{rs} > 0)\}$, being $I(-)$ the indicator function. The extended linear equation is the following:

$$\bar{\Lambda}Y = \Lambda XB + \bar{\Lambda}U \Rightarrow Y^* = X^*B + U^* \tag{10.10}$$

The error term of the equation, U^*, is not estandar. Depending on the specification of the local weights of Eq. 10.4, this macro vector will be heteroskedastic and cross-correlated:

$$U^* = \bar{\Lambda}U = \bar{\Lambda}(1 \otimes u) \rightarrow$$

$$\left\{\begin{array}{l} E[U^*] = 0 \\ V[U^*] = \begin{bmatrix} \bar{W}_1\bar{W}_1 & \bar{W}_1\bar{W}_2 & \cdots & \bar{W}_1\bar{W}_R \\ \bar{W}_2\bar{W}_1 & \bar{W}_2\bar{W}_2 & \cdots & \bar{W}_2\bar{W}_R \\ \cdots & \cdots & \cdots & \cdots \\ \bar{W}_R\bar{W}_1 & \bar{W}_R\bar{W}_2 & \cdots & \bar{W}_R\bar{W}_R \end{bmatrix} = \Omega \end{array}\right. \tag{10.11}$$

where $\bar{W}_s$ is the binary transformation of the weighing matrix W_s. Each term $(\bar{W}_i\bar{W}_j)$ is a diagonal matrix with values 1 or 0 . The LS estimate of Eq. 10.10 produces unbiased estimates of vector B, $\hat{B} = \left[X^{*\prime}X^*\right]^{-1}X^{*\prime}Y^*$. The efficient estimators are the GLS $\hat{B}_G = \left[X^{*\prime}\Omega^{-1}X^*\right]^{-1}X^{*\prime}\Omega^{-1}Y^*$. More important, this general framework allows us to test the GWR hypothesis:

$$\begin{array}{l} H_0 : \beta_1 = \beta_2 = \beta_3 = \cdots = \beta_R = \beta \\ H_A : \text{No } H_0 \end{array} \tag{10.12}$$

The traditional F test, in a Chow-like approach, can be used with the GLS or with the LS estimates. In the first case, the F statistic is an exact test, whereas in the second we have to use an asymptotic approximation:

$$\begin{array}{l} F_{GWR}^{GLS} = \dfrac{(R\hat{B}_G - RB)'\left(R\left[X^{*\prime}\Omega^{-1}X^*\right]^{-1}R'\right)(R\hat{B}_G - RB)}{r\hat{\sigma}_G^{*2}} : F(r, R^2 - Rk) \\ F_{GWR}^{LS} = \dfrac{(R\hat{B} - RB)'\left(R\left[X^{*\prime}X^*\right]^{-1}R'\right)(R\hat{B} - RB)}{r\hat{\sigma}^2} \underset{as}{:} \chi(r) \end{array} \tag{10.13}$$

being r the number of restrictions in Eq. 10.12: $r = (R - 1)k$.

In continuation we present Monte Carlo results in relation to the behaviour of the F_{GWR}^{LS} statistic of Eq. 10.13. We are going to simulate a basic linear model:

$$y_i = \alpha + x_i\beta_i + u_i;\ u_i{:}iidN(0, \sigma^2);\ i = 1, 2, \ldots, R \tag{10.14}$$

where the intercept is constant and equal to 1, $\alpha = 1$, but the slope, β, may vary from point to point.

Two sample sizes have been used, $R = 50$ and $R = 100$ (there is a computational burden associated to the sample size). The first step of each experiment consists in obtaining the set of coordinates, $\{(xc_i, yc_i); i = 1, 2, \ldots, R\}$, of the R sampling points.

Each coordinate has been randomly generated in the interval $[-1;1]$, meaning that the R sampling points are scattered in a square of length 2. The error term, according to Eq. 10.14, is a white noise normally distributed with unit variance, $\sigma^2 = 1$. The data of the x variable come from a continuous uniform distribution $U(0,1)$.

Under the null hypothesis, the slope is constant and equal to $\beta = 3,5$, which assures an average R^2 coefficient of 0, 5 for the LS estimation of the equation. Under the alternative, we have simulated three different types of breaks in the slope:

1. A random continuous break: $\beta_i : iidU(5,9)$. The slopes are randomly assigned to the set of spatial coordinates (xc_i, yc_i). The expected slope for a given coordinate is $E[\beta_i] = 7$ and the variance is $V[\beta_i] = \frac{4}{3}$. This case corresponds, roughly, to the random coefficients of the Swamy model (Swamy 1970). The *F* tests of Eq. 10.13 are not designed to detect this type of break; they should not react.
2. A regular continuous break according to the function: $\beta_i = abs(xc_i^a + yc_i^a)$ where a is a parameter that expands the symptoms of the break, with values $\{a = 1,2,3,4,5\}$; a higher value implies a stronger signal. Other functions, different from the power of the spatial coordinates, did not change essentially our results.
3. A regular discrete break according to the following regime:
   - Z0: $\beta_i = 7$ if $i \notin N1$ and $i \notin N2$; this is the cluster N0.
   - Z1: $\beta_i = 7 - b$ if $i \in N1$ where N1 refers to the set of the 20 % observations closer to the coordinate $(0,5;0,5)$.
   - Z2: $\beta_i = 7 + b$ if $i \in N2$ where N2 refers to the set of the 20 % observations closer to the coordinate $(-0,5;-0,5)$.

The first regime contains 60% of the observations and the slope is 7 there. The second and the third contain, each one, 20% of the sample. These observations are spatially clustered around the coordinates $(0,5;0,5)$ and $(-0,5;-0,5)$ respectively. The slopes in the N0 cluster are smaller than in N2 and higher than in N1. b is a parameter that measures the intensity of the break; three cases have been used here: $b = 1,2,3$.

Figure 10.1 presents some typical cases of the different breaks used in these part of the experiment.

Another important element is the bandwidth, d, the threshold that intervenes in the indicator function $I(-)$ of matrix $\bar{\Lambda}$ in expression Eq. 10.10. We have defined this parameter as a percentage of the sampling size with the following values: $\{d = 10\%; 30\%; 50\%; 70\%; 80\%; 90\%; 95\%\}$. In other terms, the sequence of LS local estimates have been obtained using the d% of the nearest-neighbors observations. 10% is a very low value (in the case of $R = 50$ to use only five observations to solve the LS estimation of the model of Eq. 10.14 in every coordinate, there are only three degrees of freedom). The other extreme, 95%, is too high because it implies copying almost exactly the same sample in every coordinate which results in severe multicollinearity problems for the estimation of Eq. 10.10.

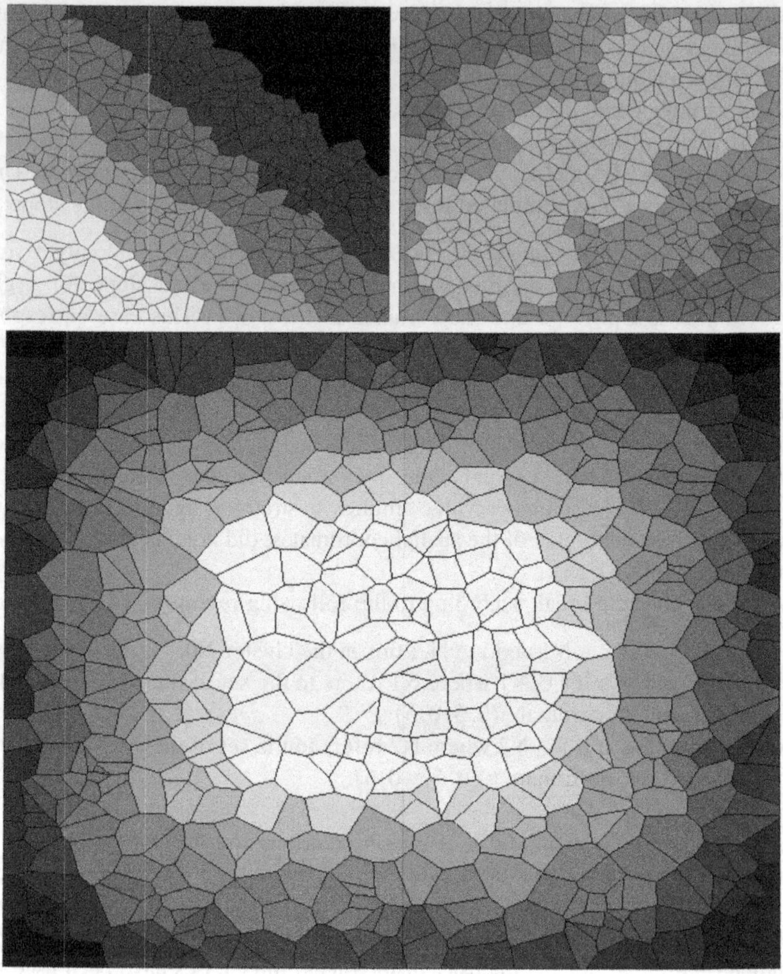

Color scale: WHITE identifies the lowest values of the coefficient and BLACK the highest.

Greater intensity of the GREY color indicates increasing values of the coefficient

Fig. 10.1 Examples of a random break (*upper left*), a discrete break (*upper right*) and a continuous break (*down*)

The more significant results of the simulation are presented in Table 10.1. Overall, the results of the F_{GWR}^{LS} test seem to be satisfactory. We would like to highlight the following aspects:

1. The test appears to be undersized. The problem is clearer for the small sample case, $R = 50$. The use of a higher bandwidth diminishes, slightly, the bias. However, the tendency is not uniform: very high values of d also produce strong biases.

Table 10.1 Size and power of the GWR test

	Size						
	10 %	30 %	50 %	70 %	80 %	90 %	95 %
R = 50	0.021	0.015	0.038	0.039	0.035	0.022	0.011
R = 100	0.033	0.034	0.042	0.045	0.041	0.031	0.014
	Power: continuous random break (m1)						
	10 %	30 %	50 %	70 %	80 %	90 %	95 %
R = 50	0.016	0.014	0.095	0.098	0.036	0.028	0.022
R = 100	0.056	0.025	0.051	0.117	0.089	0.048	0.021
	Power: continuous regular break (m2)						
	10 %	30 %	50 %	70 %	80 %	90 %	95 %
R = 50; a = 1	0.900	0.930	1.000	1.000	0.830	0.540	0.150
R = 50; a = 2	0.900	0.980	1.000	1.000	0.830	0.620	0.140
R = 50; a = 3	0.980	1.000	1.000	1.000	0.880	0.610	0.170
R = 50; a = 4	1.000	1.000	1.000	1.000	0.880	0.690	0.210
R = 50; a = 5	1.000	1.000	1.000	1.000	1.000	0.720	0.220
R = 100; a = 1	0.920	0.960	1.000	1.000	0.840	0.660	0.220
R = 100; a = 2	0.930	1.000	1.000	1.000	0.880	0.720	0.200
R = 100; a = 3	1.000	1.000	1.000	1.000	0.890	0.650	0.210
R = 100; a = 4	1.000	1.000	1.000	1.000	0.970	0.870	0.330
R = 100; a = 5	1.000	1.000	1.000	1.000	1.000	0.920	0.390
	Power: discrete regular break (m3)						
	10 %	30 %	50 %	70 %	80 %	90 %	95 %
R = 50; C1	0.040	0.170	0.320	0.420	0.240	0.130	0.050
R = 50; C2	0.090	0.790	0.640	0.610	0.530	0.370	0.160
R = 50; C3	0.280	0.970	0.990	0.980	0.990	0.900	0.630
R = 100; C1	0.380	0.710	0.790	0.740	0.670	0.360	0.240
R = 100; C2	0.460	0.660	0.940	0.930	0.850	0.480	0.450
R = 100; C3	0.630	0.840	1.000	1.000	1.000	0.950	0.740

Note: C1, C2 and C3 refer to the expected values of the parameters of the discrete break, C1 = [6; 7; 8]; C2 = [5; 7; 9]; C3 = [4; 7; 10]

2. A random break in the parameters results in problems of heteroskedasticity and endogeneity (if the regressors are not exogenous) but not necessarily creates parameter instability. This is reflected in Table 10.1, given that the estimated percentage of rejections of the null hypothesis of the F_{GWR}^{LS} test for this case is rather low and stable with d.
3. The behaviour of the test is acceptable for the case of a continuous regular break. The estimated power is at its highest even with small sample sizes, $R = 50$, very narrow bandwidths ($d = 10\%$) and moderate to strong symptoms of break ($a = 3, 4$ and 5). The optimal value of the bandwidth appears to be around 50% of the sample. In any case, it is recommendable to use a value below 70%; higher values result in power loses.
4. The estimated power of the F_{GWR}^{LS} test is lower in the case of a discrete regular break. The sample size plays a more important role, as can be seen by the differences between the cases of $R = 50$ and $R = 100$. Also the bandwidth

must be larger and, preferably, not smaller than 50% of the sample in any case. The upper threshold of 70% still is in use.

10.4 Criteria to Define the 'Local Neighbors'

The preceding test can assist with the decision of whether to use GWR. Once there is spatial instability in the coefficients, we must move to the local level. The key issue now is the exact meaning of "local neighbors", for each point in the space. The common solution consists in the introduction of a set of local spatial weights, as in Eq. 10.5. These weights are specified a priori by the user in a practice similar to the use of kernels in time series (Robinson 1983; Härdle and Vieu 1992; Lütkepohl 2005; Li and Racine 2007).

However, space may be much more heterogeneous than time and some flexibility may be beneficial. For example, why a unique kernel? Depending on the heterogeneity, it could be preferable to allow the moving spatial window to adjust locally. If the symptoms of instability in a certain part of the space are weak the kernels should be large there to gain efficiency. On the contrary, if the instability dominates in other areas, the size of the kernels should be smaller in order to reduce bias. Even the shape of the kernels may vary depending on how evolves the heterogeneity.

For this to happen, we need to identify the patterns of instability present in the data. The scanning window of the Scan statistic can be useful here (Kulldorff 2001; Kulldorff et al. 2009). The Scan statistics was developed to find spatial clusters of a given variable. By spatial cluster we mean a group of spatially clustered observations which have either a larger or smaller mean than the other observations.

The null hypothesis of the Scan statistic, for this case, is that all the observations come from the same distribution (there is no instability). Assuming normality, the likelihood corresponding to any window centred on, for example, observation i may be written as:

$$L_{0i} = \left(2\pi\sigma_0^2\right)^{-R2} \exp - \frac{\sum_{r=1}^{R} (x_r - \mu)^2}{2\sigma_0^2} \tag{10.15}$$

Under the alternative, there are differences in the mean, μ and λ, for the observations inside and outside the cluster, respectively. The likelihood of Eq. 10.15 becomes:

$$L_{Ai} = \left(2\pi\sigma_A^2\right)^{-R2} \exp - \frac{1}{2\sigma_A^2} \left[\sum\nolimits_{r\in N(i)} (x_r - \mu)^2 + \sum\nolimits_{r\notin N(i)} (x_r - \lambda)^2\right] \tag{10.16}$$

where $r \in N(i)$ means that observation r pertains to the local neighborhood of i. The estimates of the variances, σ_0^2 and σ_A^2, correspond to the respective hypothesis. The decision rule is to select the cluster that maximizes the difference between both likelihoods:

$$\mathrm{Max}\left(\frac{L_{Ai}}{L_{0i}}\right) = \left(\frac{\sigma_0^2}{\sigma_A^2}\right)^{R2} \tag{10.17}$$

The significance of the cluster can be evaluated using Monte Carlo hypothesis testing, randomly permuting the observed values x_r and their corresponding locations r.

Our interest focuses on the stability of the regression coefficients that relates the two variables, which is a slightly different question.

Let us assume a bivariate relation between variables y and x and a bivariate normal distribution:

$$f(y,x) = \frac{1}{\sqrt{2\pi}} \times |V| \exp\left\{-\frac{1}{2}\left[\left(y-\mu_y\right);\ \ \left(x-\mu_x\right)\right]V^{-1}\left[\left(y-\mu_y\right);\ \ \left(x-\mu_x\right)\right]'\right\} \tag{10.18}$$

$$V = \begin{bmatrix} \sigma_y^2 & \gamma_{yx} \\ \gamma_{yx} & \sigma_x^2 \end{bmatrix}$$

The regression coefficient depends on the covariance between the two variables: $\beta = \left(\gamma_{yx}\sigma_x^2\right) = \rho_{yx}\left(\sigma_y\sigma_x\right)$, where ρ_{yx} is the correlation coefficient. Assuming stability in all of the parameters of the relation, the null model can be summarized by the likelihood:

$$L_{0i} = (2\pi)^{-R2}\frac{\left(1-\rho_{yx}^2\right)^{-R2}}{\left(\sigma_y^2\sigma_x^2\right)^{R2}}\exp\left\{-\frac{1}{2\left(1-\rho_{yx}^2\right)}\left[\sum_{r=1}^{R}\left(\frac{y_r-\mu_y}{\sigma_y}\right)^2 + \sum_{r=1}^{R}\left(\frac{x_r-\mu_x}{\sigma_x}\right)^2 - 2\rho_{yx}\sum_{r=1}^{R}\left(\frac{y_r-\mu_y}{\sigma_y}\right)\left(\frac{x_r-\mu_x}{\sigma_x}\right)\right]\right\} \tag{10.19}$$

Under the alternative hypothesis we allow for some instability in the regression coefficient β in the neigborhood of point i. The covariance or, what the same, the correlation coefficient between the two variables changes: ρ_{yx}^{i} is the value around point i and $\rho_{yx}^{(i)}$ in other parts of the space. The likelihood function is:

$$L_{Ai} = (2\pi)^{-R2}\left[\frac{\left(1-\rho_{yx}^{(i)2}\right)^{-(R-|N(i)|)2}\left(1-\rho_{yx}^{i2}\right)^{-|N(i)|2}}{\left(\sigma_y^2\sigma_x^2\right)^{R2}}\right] \tag{10.20}$$

$$\exp\left\{-\frac{1}{2}\left[\sum\nolimits_{r=1}^{R}\left(\frac{y_r-\mu_y}{\sigma_y}\right)^2+\sum\nolimits_{r=1}^{R}\left(\frac{x_r-\mu_x}{\sigma_x}\right)^2\right.\right.$$

$$\left.\left.-2\rho_{yx}^{(i)}\sum\nolimits_{r\notin N(i)}\left(\frac{y_r-\mu_y}{\sigma_y}\right)\left(\frac{x_r-\mu_x}{\sigma_x}\right)-2\rho_{yx}^{i}\sum\nolimits_{r\in N(i)}\left(\frac{y_r-\mu_y}{\sigma_y}\right)\left(\frac{x_r-\mu_x}{\sigma_x}\right)\right]\right\}$$

where $|N(i)|$ means the cardinality of the set $N(i)$. The likelihood ratio that we have to maximize is:

$$\mathrm{Max}\left(\frac{L_{Ai}}{L_{0i}}\right) = \left[\frac{1-\rho_{yx}^2}{\left(1-\rho_{yx}^{(i)2}\right)^{1-|N(i)|R}\left(1-\rho_{yx}^{i2}\right)^{|N(i)|R}}\right]^{R2} \tag{10.21}$$

The 'local neighborhood' of point i corresponds to the area around point i that produces the greatest difference between the two correlation coefficients, ρ_{yx}^{i} and $\rho_{yx}^{(i)}$. The next question is testing for the significance of the cluster i in relation to parameter β for which the permutation approach seems adequate.

The same as in the previous section, we include a small Monte Carlo with the purpose of checking the use of the Scan test in the GWR estimation. We maintain the design of the experiment: (a) the spatial coordinates $\{(xc_i, yc_i); i = 1, 2, \ldots, R\}$ are randomly distributed in the square $(-1,-1); (-1,1); (1,1); (1,-1)\}$; (b) we have simulated the model of Eq. 10.14 where the coefficient β varies from point to point; (c) a GWR estimation of this model is obtained using the binary kernel function of Eq. 10.5. The problem of this section is to determine the value of d and our proposal consists of using the generalization of the Scan test of Eq. 10.21. In short, we are going to obtain a specific bandwidth for every point in the sample.

Because of the computational burden, we only present the results corresponding to the case of $R = 100$. We have used two different types of break: continuous and discrete, with one or four inverse peaks in the first case and three or five different regimes in the second. Figure 10.2 shows some examples of the type of breaks used in these part of the experiment.

Table 10.2 presents the average Mean Squared Error, MSE, of the GWR estimation of parameter β after 1,000 simulations:

$$\mathrm{MSE}(d) = \frac{\sum_{i=1}^{R}\left(\beta_i-\hat{\beta}_{i,N(i)}\right)^2}{1{,}000} \tag{10.22}$$

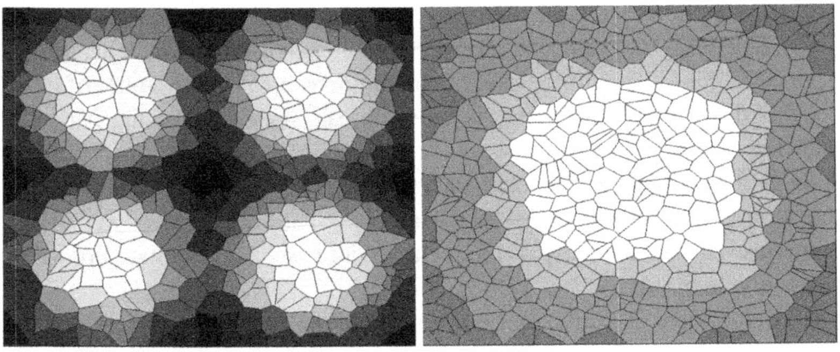

Color scale: WHITE identifies the lowest values of the coefficient and BLACK the highest.

Greater intensity of the GREY color indicates increasing values of the coefficient

Fig. 10.2 Examples of continuous (*left*) and discrete (*right*) breaks

Table 10.2 GWR estimates under scan and cross-validation criterion

	Continuous regular break: 1 peak					Discrete regular break: 3 regimes				
	a = 1	a = 2	a = 3	a = 4	a = 5	C1	C2	C3	–	–
Scan										
MSE	0.055	0.047	0.064	0.087	0.125	0.081	0.099	0.184	–	–
% cells[a]	32.1	31.5	26.4	22.1	17.6	19.5	20.8	18.4	–	–
% signif.[b]	67.5	72.1	77.1	89.6	92.5	88.6	91.4	97.6	–	–
CV										
MSE	0.021	0.039	0.068	0.116	0.193	0.054	0.087	0.187	–	–
% of cells[a]	25.2	21.7	23.5	19.8	22.4	22.6	20.4	17.5	–	–
	Continuous regular break: 4 peaks					Discrete regular break: 5 regimes				
	a = 1	a = 2	a = 3	a = 4	a = 5	C1	C2	C3	C4	C5
Scan										
MSE	0.121	0.117	0.156	0.198	0.301	0.071	0.084	0.086	0.091	0.116
% cells[a]	18.6	15.2	16.7	21.1	13.4	13.6	14.1	12.4	11.6	12.7
% signif.[b]	59.4	68.4	67.2	76.4	83.6	84.5	91.1	91.4	93.5	98.7
CV										
MSE	0.102	0.136	0.189	0.274	0.411	0.062	0.088	0.103	0.147	0.163
% cells[a]	26.8	29.7	21.6	22.4	27.8	17.6	19.4	18.5	17.5	19.1

C1, C2, C3, C4 and C5 refer to the expected values of the parameters used in the discrete break: C1 = [6;7;8]; C2 = [5;7;9]; C3 = [4;7;10], C4 = [3;7;10]; C5 = [2;7;11]

[a]Average percentage of cells in the local estimates.

[b]Average percentage of cases where the test of Eq. (10.21) rejects the null of a non significant break in the point.

The estimation error depends on the value of d determined by each approach, globally as in Eq. 10.22 of locally as in Eq. 10.21.

We would like to stress the following aspects from this table:

1. The Scan approach uses, in average, less observations than the CV criterion. The difference is more important when the break is of a discrete type.

2. The number of observations determined by the Scan criterion depends on the heterogeneity of the data. The relation is: more heterogeneity, smaller value of d. The relation is evident in the case of a continuous break: the number of neighbors decided decreases as the intensity of the break, measured by the parameter a, increases.
3. A high proportion of the local tests of instability of Eq. 10.21 leads to the rejection of the null hypothesis of stability.[1] The estimated power of the test increases with the intensity of the break. In all cases, the estimated power of the test is above 50 %.
4. The Scan approach is a useful alternative to the global cross-validation score. The second criterion outperfoms the Scan criteria, in terms of minimizing the MSE, when there is a break of weak to moderate intensity. The local determination of the bandwidth seems preferable in cases where the violence of the break is strong or the number of regimes increases.

10.5 Selection of the 'Local Weights'

The algorithm based on the Scan test allows us to answer several important questions: how many breaks are there, where they are, and which parameters are affected. However, there is another question that also needs to be addressed: what is the best form to model the breaks? Returning to the GWR terminology, this amounts to specify the sequence of local GWR weighting matrices that captures the evolution of the breaks.

At this point it is worth to restate the following:

1. We can construct GWR weighting matrices in different ways using different hypothesis. Each hypothesis results in a different local weighting matrix and, accordingly, in a different set of spatial regressors. Different regressors result in different models. These implications are clear from expressions (10.7), (10.8, (10.9, and (10.10).
2. There are some general guidelines in the spatial econometrics literature devoted to the specification of a global weighting matrix, using concepts like nearness, accessibility, influence, etc. (Kooijman 1976; Bavaud 1998; Anselin 2002; Folmer and Oud 2008, among others).
3. The two problems (specifying a global or a set of GWR weighting matrices) are similar: there is little information in relation to the correct matrix and, usually, the researcher has some priors to guide the specification.

[1] In order to solve the test, (1) the pairs of values $\{(x_i, y_i); i = 1, 2, \ldots, R\}$ have been randomly permuted $(N-1)$ times; (2) for each permutation a value of the likelihood ratio of Eq. (10.21) has been obtained; (3) the sequence of $(N-1)$ likelihood ratios has been ordered from lowest to highest; (4) finally, we reject the null if the observed value pertains to the five highest values of the sequence.

Let us assume that we have a set of N linearly independent GWR weighting matrices, $\mathrm{Y} = \{\mathrm{W}^1; \mathrm{W}^2; \ldots; \mathrm{W}^N\}$. Usually N corresponds to a small number of different competing matrices but in some cases this number may be quite large, reflecting a situation of great uncertainty. They are general matrices which correspond to different interaction hypothesis and any of these matrices can be used to tackle the breaks. As said, each matrix generates a different set of spatial regressors and, consequently, a different spatial model. These matrices are linearly independent and they can be nested or non-nested. This is important because, at the end, we have a problem of model selection, between nested or non nested models.

Two GWR weighting matrices may be nested if, for example, binary weights are used: the weights of the first matrix correspond to the m nearest neighbors of point i (a 1 indicates that the point is one of the m nearest neighbors), whereas the weights of the second matrix correspond to the $(m + s)$ nearest neighbors $(s > 0)$. The weights of the first matrix are contained in the second matrix plus some other non-zero weights. Discriminating between the two matrices is not difficult using standard techniques of model selection (Herrera et al. 2011, develop a simple Lagrange Multiplier).

For the case of non-nested matrices, we propose the use of the J-test of Davidson and McKinnon (1981, 1982), adapted to a spatial context by Leenders (2002). Let us assume that there are two competing GWR spatial weighting matrices for the same basic model:

$$\Lambda^1 = \begin{bmatrix} W_1^1 & 0 & 0 & \cdots & 0 \\ 0 & W_2^1 & 0 & \cdots & 0 \\ 0 & 0 & W_3^1 & \cdots & 0 \\ \cdots & \cdots & \cdots & \cdots & \cdots \\ 0 & 0 & 0 & \cdots & W_R^1 \end{bmatrix} \quad \Lambda^2 = \begin{bmatrix} W_1^2 & 0 & 0 & \cdots & 0 \\ 0 & W_2^2 & 0 & \cdots & 0 \\ 0 & 0 & W_3^2 & \cdots & 0 \\ \cdots & \cdots & \cdots & \cdots & \cdots \\ 0 & 0 & 0 & \cdots & W_R^2 \end{bmatrix} \tag{10.23}$$

where the W_i^j matrices have been defined in Eq. 10.4. As said, each matrix results in a different GWR spatial model that we may compare in a classical decision problem.

$$\left.\begin{array}{l} H_0 : \bar{\Lambda}^1 Y = \Lambda^1 XB + \bar{\Lambda}^1 U \Rightarrow Y^{1*} = X^{1*}B + U^{1*} \\ H_A : \bar{\Lambda}^2 Y = \Lambda^2 XB + \bar{\Lambda}^2 U \Rightarrow Y^{2*} = X^{2*}B + U^{2*} \end{array}\right\} \tag{10.24}$$

As it is well-known, the J-test uses an augmented regression like the following:

$$Y^{1*} = (1 - \theta)\left[X^{1*}B\right] + \theta\left[X^{2*}\hat{B}\right] + v \tag{10.25}$$

being $\hat{B}$ the corresponding maximum-likelihood estimates of the respective parameters on a separate estimation of the model of H_A. In order to attain a decision it is recommendable to calculate a second J test reversing the order of the two matrices in the null and alternative hypotheses of Eq. 10.24. Leenders (2002) and Kelejian (2008) show that the J test can be extended to the comparison of a null model against $(N - 1)$ different models. In this case, the reversion would not be necessary.

Table 10.3 The J test and the selection of the Kernel

$R = 100$	$H_0 = \Lambda^A$ $H_A = \Lambda^B$	$H_0 = \Lambda^B$ $H_A = \Lambda^A$	$H_0 = \Lambda^A$ $H_A = \Lambda^C$	$H_0 = \Lambda^C$ $H_A = \Lambda^A$	$H_0 = \Lambda^B$ $H_A = \Lambda^C$	$H_0 = \Lambda^C$ $H_A = \Lambda^B$
	No break: constant slope					
	0.048	0.064	0.071	0.088	0.062	0.094
	Continuous regular break					
a = 1	0.782	0.816	0.779	0.836	0.935	0.921
a = 3	0.796	0.857	0.826	0.901	0.924	0.951
a = 5	0.908	0.924	0.897	0.893	0.972	0.948
	Discrete regular break					
Cl regime	0.171	0.984	0.159	0.962	0.762	0.863
C2 regime	0.126	0.926	0.118	0.943	0.846	0.841
C3 regime	0.109	0.967	0.086	0.953	0.892	0.927

Note: C1, C2 and C3 refer to the expected values of the parameters used in the discrete break, C1 = [6;7;8]; C2 = [5;7;9]; C3 = [4;7;10]

The Monte Carlo of this section refers to the behaviour of the J test, of Eqs. 10.24 and 10.25, as a method of helping in the selection of the most adequate spatial kernel, matrix Λ^i. We maintain the design of the previous sections in relation to the random distribution of the spatial coordinates, the model simulated (that of Eq. 10.14) and the sample size (R = 100). Moreover, we use the three kernels of Eq. 10.5 which result in the matrices Λ^A, with weights $\{\lambda_{rs} = 1\ d_{ir} < d; 0\ d_{ir}\ d\}$, Λ^B, with weights $\left\{\lambda_{rs} = \exp\left[-\frac{1}{2}\left(\frac{d_{ir}}{d}\right)^2\right]\right\}$ and Λ^C, with bisquare weights $\left\{\lambda_{rs} = \left[1 - \left(\frac{d_{ir}}{d}\right)^2\right]^2 d_{ir} < d; 0\ d_{ir} d\right\}$. The bandwidth, d, has been fixed in such a way that each local estimation contains 50% of the observations.

Three different sets of β coefficients have been used:

1. A constant slope equal to 3, 5. The purpose is to check the behaviour of the J tests under the assumption of stability.
2. A regular continuous break according to the function: $\beta_i = 7 \times \text{abs}(xc_i^a + yc_i^a)$ with three different values for a, $\{a = 1, 3, 5\}$.
3. A regular discrete break in three different regimes which, as described in Sect. 10.3, contain, respectively, 60% 20% and 20% of the observations in the sample.

Table 10.3 shows the percentage of rejections of the model of the null in favour of the model of the alternative hypothesis, as indicated in the second row. The results are somewhat puzzling although some general guidelines can be extracted:

1. The simultaneous acceptance of both null hypotheses (after reversion) is a signal of stability in the coefficients. The use of another, different kernel cannot improve the results of the first kernel because the two are unbiased. The J test does not assist in the decision.

2. The simultaneous rejection of both null hypotheses is a signal of misspecification, as it is evident in the case of a continuous regular break. It is important to remind that the use of kernels in the GWR estimation does not correct the specification, which is unstable per se. At best, a wise selection of the kernel will reduce the estimation bias, but not totally.
3. A discrete regular break fits better into a binary kernel scheme as reflected in Λ^A. This kernel is the preferred alternative almost 90% of times, both against Λ^B and the bisquare Λ^C. The use of the last two kernels in the J test point, once more, to a misspecified model.

10.6 Conclusions and Further Research

Geografically weighted regressions are a widely used technique in the analysis of spatial data, with a large specialized literature and an increasing number of empirical applications. Conventional analysis is based on a global perspective; however spatial location matters. This implies that heterogeneity, in variables and/or in their marginal effects, is a question of concern.

Current GWR techniques are more consistent, computational restrictions have been reduced, regression output are more friendly presented, etc. However, there remain some weaknesses that need to be addressed. We focus on three points.

The first takes up the suggestion of Wheeler and Tiefelsdorf (2005, p. 163): 'A new class of GWR models would be required (...). This class would need to connect all location specific GWR regression equations simultaneously into a seemingly unrelated regression model, which allows jointly testing across several local models and a nested specification of a global model with local parameter variations'. Fotheringham et al. (1997) approached this question through Monte Carlo techniques whereas Leung et al. (2000) developed a goodness-of-fit test that compares GWR with global LS. We propose a simple test based on a general nesting equation which combines sampling with theoretical information. The LS version of the resulting GWR test appears to be a slightly undersized but it has good power against continuous or discrete breaks. There is a computational burden in obtaining the test, because $(R^2 \times R^2)$ matrices are involved, although a sampling scheme can alleviate the problem. We are working in this direction.

The bandwidth is another point of concern in the GWR approach. There are several proposals in the literature to fix this value, among which the cross-validation score is the most popular. We explore a different solution based on an adaptative weighting scheme: the size, even the shape, of the bandwidth should reflect the density of the data and, where possible, also their heterogeneity. The scanning window of the Kulldorf test allows us to tackle this problem by defining a specific bandwidth for each local estimation. The Scan statistic emerges as a useful complement to the cross-validation score specially when there is a high instability in the data. Our simulationd show that, for these cases, the Scan criterion tends to determine a narrow bandwidth thus minimizing the estimation bias. On the negative side, we must acknowledge the computational burden of the technique and the requirement of normality.

The third question refers to the non uniqueness of the GWR estimates: different sets of parameters can be obtained using different bandwidths and different weighting schemes. The selection of the kernel appears to have less impact but, even in this case, it is somewhat surprising the lack of guidelines in the literature. We argue that different local weights amount to different regressors and, thus, the problem of choosing the kernel is equivalent to a model selection problem. Our simulation with the J test is a first attempt in this direction. The test works properly in cases of a discrete break in a finite number of regimes, although is not conclusive for breaks of a continuous type. Unless the weighting scheme that generated the data is in the set of kernels selected (a not very reasonable hypothesis), the J test points towards a misspecification problem by rejecting the different alternatives of the null.

There are indications that the size of the sample may have an important effect on estimation and identification results (Páez et al. 2011). The simulations conducted in this paper were conducted with relatively small samples. A direction for further research is to expand these experiments to assess the effect of sample size on the various effects investigated.

References

Anselin L (1995) Local indicators of spatial association – LISA. Geogr Anal 27:93–115

Anselin L (2002) Under the hood: issues in the specification and interpretation of spatial regression models. Agric Econ 17:247–267

Bavaud F (1998) Models for spatial weights: a systematic look. Geogr Anal 30:153–171

Brunsdon C, Fotheringham S, Charlton M (1996) Geographically weighted regression: a method for exploring spatial nonstationarity. Geogr Anal 28:281–298

Brunsdon C, Fotheringham A, Charlton M (1998a) Geographically weighted regression – modelling spatial non-stationarity. Statistician 47:431–443

Brunsdon C, Fotheringham S, Charlton M (1998b) Spatial nonstationarity and autoregressive models. Environ Plann A 30:957–973

Casetti E (1972) Generating models by the expansion method. Application to geographical research. Geogr Anal 4:81–91

Casetti E (1991) The investigation of parameter drift by expanded regressions: generalities and a family planning example. Environ Plann A 23:1045–1051

Chow G (1960) Tests of equality between sets of coefficients in two linear re-gressions. Econometrica 28:591–605

Cleveland W (1979) Robust locally weighted regression and smoothing scatterplots. J Am Stat Assoc 74:829–836

Cleveland W, Devlin S (1988) Locally weighted regression: an approach to re-gression analysis by local fitting. J Am Stat Assoc 83:596–610

Davidson R, MacKinnon J (1981) Several tests for model specification in the presence of alternative hypotheses. Econometrica 49:781–793

Davidson R, MacKinnon J (1982) Some non-nested hypothesis tests and the relations among them. Rev Econ Stud XLIX:551–565

Farber S, Paez A (2007) A systematic investigation of cross-validation in GWR model estimation: empirical analysis and Monte Carlo simulations. J Geogr Syst 9:371–396
Folmer H, Oud J (2008) How to get rid of W? A latent variable approach to modeling spatially lagged variables. Environ Plann A 40:2526–2538
Fotheringham A, Charlton M, Brunsdon C (1997) Two Techniques for exploring non-stationarity in geographical data. Geogr Syst 4:59–82
Fotheringham A, Charlton M, Brunsdon C (1999) Geographically weighted regression a natural evolution of the expansion method for spatial data analysis. Environ Plann A 30:1905–1927
Fotheringham AS, Brunsdon C, Charlton M (2002) Geographically weighted regression: the analysis of spatially varying relationships. Wiley, West Sussex
Getis A, Ord JK (1992) The analysis of spatial association by use of distance statistics. Geogr Anal 24:189–206
Härdle W, Vieu P (1992) Kernel regression smoothing of time series. J Time Ser Anal 13:209–232
Herrera M, Mur J, Ruiz M (2011) A study on the criteria for selecting the most adequate spatial weighting matrix. Working paper, Department of Economic Analysis, University of Zaragoza
Kelejian H (2008) A spatial J-test for model specification against a single or a set of non-nested alternatives. Lett Spat Resour Sci 1:3–11
Kooijman S (1976) Some remarks on the statistical analysis of grids especially with respect to ecology. Ann Syst Res 5:113–132
Kulldorff M (2001) Prospective time-periodic geographical disease surveillance using a scan statistic. J Roy Stat Soc Ser A 164:61–72
Kulldorff M, Huang L, Konty K (2009) A scan statistic for continuous data based on the normal probability model. Int J Health Geogr 8:1–9
Leenders R (2002) Modeling social influence through network autocorrelation: constructing the weight matrix. Soc Networks 24:21–47
Leung Y, Mei C, Zhand W (2000) Statistical tests for spatial nonstationarity based on the geographically weighted regression model. Environ Plann A 32:9–32
Li Q, Racine J (2007) Nonparametric econometrics: theory and practice. Princeton University Press, Princeton
Lütkepohl H (2005) New introduction to multiple time series analysis. Springer, Berlin
McMillen D (1996) One hundred fifty years of land values in chicago: a non-parametric approach. J Urban Econ 40:100–124
McMillen D (2003) Spatial autocorrelation or model misspecification? Int Reg Sci Rev 26:208–217
McMillen D (2004) Employment densities, spatial autocorrelation, and subcenters in large metropolitan areas. J Reg Sci 44:225–244
McMillen D, McDonald J (1997) A nonparametric analysis of employment density in a policentric city. J Reg Sci 37:591–612
Mur JF Lopez, Angulo A (2009) Testing the hypothesis of stability in spatial econometric models. Papers in Regional Science. 88:409–444
Pace K, Lesage J (2004) Spatial autoregressive local estimation. In: Getis A, Mur J, Zoller H (eds) Spatial econometrics and spatial statistics. Palgrave, Londres, pp 31–51
Páez A, Uchida T, Miyamoto K (2002a) A general framework for estimation and inference of geographically weighted regression models. 1: location-specific kernel bandwidth and a test for locational heterogeneity. Environ Plann A 34:733–754
Páez A, Uchida T, Miyamoto K (2002b) A general framework for estimation and inference of geographically weighted regression models. 2: spatial association and model specification tests. Env Plann A 34:883–904
Paez A, Long F, Farber S (2008) Moving window approaches for hedonic price estimation: an empirical comparison of modelling techniques. Urban Stud 45:1565–1581

Páez A, Farber S, Wheeler D (2011) A simulation-based study of geographically weighted regression as a method for investigating spatially varying relationships. Environ Plann A 43:2992–3010

Robinson P (1983) Nonparametric estimators for time series. J Time Ser Anal 4:185–207

Swamy P (1970) Efficient inference in a random coefficient regression model. Econometrica 38:311–323

Wang N, Mei C, Yan X (2008) Local linear estimation of spatially varying coefficient models: an improvement on the geographically weighted regression technique. Environ Plann A 40:986–1005

Wheeler D, Tiefelsdorf M (2005) Multicollinearity and correlation among local regression coefficients in geographically weighted regression. J Geogr Syst 7:161–187

Part III
Applications of Spatial Analysis with Small Area Observations

Chapter 11
The Estimation of Urban Premium Wage Using Propensity Score Analysis: Some Considerations from the Spatial Perspective

Dusan Paredes, Marcelo Lufin, and Patricio Aroca

11.1 Introduction

The urban economics literature supports that thick labor markets pay higher wage levels than thin labor markets. Glaeser and Mare (2001) estimate the elasticity wage-city size larger than one million inhabitants around of 36 % higher than smaller areas, while Glaeser and Resseger (2010) identify a elasticity of 45 % for the case of skilled workers. This positive relation also exists within industries, but with an uneven impact (Elvery 2010). In spite of the extensive empirical evidence, the most of the applications have been focused on North American, European and Asian contexts. In this chapter we extend the analysis toward the Latin American case, where the ONU-Wider has strongly recommended focusing on "increasing inequalities partly as a consequence of the uneven impact of trade openness and globalization" (Kanbur et al. 2005). We use the Chilean case and provide a first estimation of wage differentials between thick and thin labor markets. Although the extension toward new contexts could be considered a contribution as itself, the particular scenario of Latin American realities must be discussed.

Puga (1998, 1999), Rosenthal and Strange (2004) and Venables (2005) among other authors suggest that developing countries, such as the Chilean case, are characterized by high primacy index. This fact implies that a unique region, in the most of the cases the capital of the country, concentrates a large portion of the total economic activity and population. For example, the Metropolitan Region of Santiago concentrates around the 40 % of the total population and the 50 % of production in the 2 % of the total territory.[1] That indicates than developing countries present less thick labor markets than developed countries and even,

[1] Annual National Manufacturing Survey, Chile, 1997

D. Paredes (✉) • M. Lufin • P. Aroca
Facultad de Economia, Departamento de Economia, Universidad Católica del Norte, Avenida Angamos 0610, Antofagasta, Chile
e-mail: dparedes@ucn.cl

E. Fernández Vázquez and F. Rubiera Morollón (eds.), *Defining the Spatial Scale in Modern Regional Analysis*, Advances in Spatial Science,
DOI 10.1007/978-3-642-31994-5_11, © Springer-Verlag Berlin Heidelberg 2012

some countries could present only one dense labor market. In the Chilean context, the MR rises as a natural benchmark. Given that MR is a thick labor market, then we expect that this region pay a wage premium in comparison with any other Chilean region.[2] In spite of the simplicity of this statement, some additional discussion about the labor markets is needed.

We expect to contribute with three empirical discussions for the estimation of spatial wage differentials in the Chilean case. The first is the issue raised when the aggregated wage at regional level is used as a dependent variable. Glaeser and Mare (2001), Berry and Glaser (2005), Glaeser and Resseger (2009) and Elvery (2010) are examples of spatial wage differentials using averages wages as dependent variable. The authors add an independent variable indicating the size of the region and its coefficient identifies the wage premium. This measure can be accepted when there are not data for capturing the heterogeneity in human capital, but its use presents several problems. Firstly, the theory of spatial sorting suggests that high skills workers are concentrated in large cities (Combes et al. 2008 and Mion and Naticchioni (2009)). This theory rejects the direct comparison of wages at aggregated level due to the worker heterogeneity could also explain the wage differentials. Additionally, the skills concentrations vary across industries and economic sectors, implying a wage variation at intra-city level. Even controlling by worker characteristics, the industrial mix affects the estimation of wage differentials. Both problems put in evidence the advantages of considering the wages at micro data instead of any aggregated level. We partially overcome this problem using a rich database for the Chilean case which is disaggregated at individual level. Our survey allows identifying characteristics of the worker such as age, education (years), educational level, economic sector, occupation, type of contract, and marital status among other personal characteristics.

A second dimension is how the literature has defined the geographical scale of a thick labor market. For the Chilean case, and the most of the empirical literature, the researcher would use the administrative division of the geographical space and the MR, again, appears a natural candidate for benchmark area. However, this administrative approach is problematic. A thick labor market emerges due to the existence of economics of agglomerations that work to maintain a spatial concentration of demand and labor supply. The continuous interaction in the labor market generates increasing returns to scale and peer effect networks that increase the worker productivity and, consequently, generates higher wages. In this framework, the economics of agglomeration works in a functional space, which does not necessarily fit with the administrative division. In other words, the thick labor market in the MR could be larger that the administrative division because of the interaction MR with other contiguous regions. We cover this gap proposing a stepwise tool for spatial delineation of thick labor markets following a combination of methods. We use commuting flows among counties for building the geographical scope of labor markets. The proposed method is rooted on the spatial analysis

[2] The Chilean administrative division presents ifteen regions including a Metropolitan Region (MR).

literature (Anselin 1995; Getis and Griffith 2002) and also social networks analysis (Bonacich 1972; Green 2007). We generate a new division of the geographical space where we confirm the existence of only one thick labor market, namely the MR, but we show a larger geographical scope where some communes of other regions are absorbed by this thick labor market.

The third dimension emerges when the wage differential is considered as a causal effect of thick labor market instead of other control variables such as human capital. Two workers could present different wages under different labor markets, lets say thick and thin labor markets, but we cannot identify this causality if the workers also have different human capital. We face this problem bringing the counterfactual framework inside the regional science literature. The counterfactual framework makes a comparison just among comparable workers. For example, a thick labor market could group very high skilled workers and its comparison with any other thin labor market would be biased. In order to improve the previous literature, we use a Coarsened Exact Matching (CEM) to find identical workers between a thick and thin labor markets as a previous step to estimate the wage differential. We use CEM instead other matching approach such as Propensity Score Analysis (PSA) due to the exact matching find the similar worker, but the PSA could match two workers with similar propensity score and still with different characteristics.

According to the three empirical contributions previously discussed, we set the structure of our article. Our hypothesis is that thicker labor markets generate a wage premium in comparison with thin labor markets. To carry out our testing strategy, we require a redefinition of the geographical space to identify functional thick labor markets. We used spatial and social network analysis to identify the functional area. With this result, we are able to compare two geographical divisions: an administrative space (14 regions and the Metropolitan Region of Santiago labeled as thick labor market) and a functional division (14 regions and a new thick labor market which captures the old MR plus a portion of other two regions). Next, we estimate the spatial wage differentials using both divisions and we test if the functional division increases the wage premium in comparison with the previous spatial structure.

Our results support the established hypothesis. First, the functional approach shows a new geographical space. The administrative perspective implies 15 different markets, but our approach suggests a different configuration of the space. The Metropolitan Region of Santiago is a huge labor market, which also captures a portion of other regions. Moreover, a portion of the Metropolitan Region, particularly the rural areas, does not belong to the new thick labor market. These results suggest evidence how the geographical extent of labor markets is not defined by administrative reasons, but rather by its functionality. Using the administrative and functional geographical division, the wage differential was estimated for Chilean regions. The results show how, after a matching procedure, a similar set of workers earn higher wages in the thick labor markets comparison for almost any region. This premium is bounded between 4 % and 21 %. Alternatively, a set of three regions show a higher wage than thick market, but its difference is reduced with our functional approach. These shown results suggest the existence of spatial wage differentials in the Chilean case. This output opens new discussion for policy makers, especially for concerns about how the density of labor market affects the welfare measurements such as wages.

11.2 Methodology

Our empirical strategy implies two steps: a functional definition of thick labor markets and the estimation of the wage premium through matching estimator at micro data level. The administrative division of Chile presents three administrative divisions: 346 counties, 53 provinces and 15 regions. The densest region is the Metropolitan Region of Santiago (MR) which comprises 6 provinces and 53 counties. The MR contains the 40 % of the total population with a density of 390 habitants per square kilometers when the national average density is 19.9. These characteristics posit the MR as the largest region in Chile and a perfect spatial unit to use as a thick labor market. However, the choice of the MR as a dense urban area could be arbitrary. According to Marshall (1920) the agglomeration economics, which are the sources to argue higher wages in dense areas, operate in a geographical space defined by the extent of the labor market. Thus, several critiques appear by using the MR as a benchmark area. First, the MR could represent more than one labor market and an aggregated perspective could bias the estimation. Second, other large geographical labor markets could exist in others places of the country. Third, the labor market extent must be defined with homogeneous labor characteristics more than political-administrative areas.

This chapter contributes with an alternative methodology for building a new "space" dimension to measure the scope of economics of agglomeration. Firstly, we identify a thick labor market using functional regions based on geographic contiguity and people flow interaction among the 346 counties. This methodology keeps a geographic dimension, but it is built according to some relation-measure, leaving out the a priori spatial administrative division. Here, as a complement to flows between areas, the presence of a significant level of spatial local autocorrelation (Getis and Griffith 2002) is proposed to identify thick labor markets in order to maximize areas that behave similar in terms of their importance in the networks of workers flows between locations.

To identify the relative importance of each location in the system of flows, we propose a functional design based on the social networks literature, there the Eigen-centrality has been used to identify relevant actors in a relational structure (Bonacich 1972; Borgatti 2005), it is defined as the principal eigenvector of the matrix defining the network. Using that index, each node (location) is ranked base on its eigenvector score, in a manner in which those more centrals are adjacent to nodes that are themselves more centrals. This idea has been also explored and used in geography literature (Boots and Kanaroglou 1988; Boots 1984; Boots and

Tiefelsdorf 2000). In that fashion, locations that are similar in terms of their centrality in the system of workers' flows, but also their constitute a spatial cluster of contiguous units associated to a hotspot in terms of commuting rates, belong together to in a new functional geographical area defined as a thick labor market.

11.2.1 Eigen-Centrality Index and Functional Region Identification

This centrality measure was proposed by Bonacich (1972) as a modification of the degree centrality perspective, traditionally the centrality approach establishes that nodes (counties in our case) who have more connections are more important in the relational system, because they can affect more other actors. Bonacich argued that each node's centrality is a function of the extent of its connections (degree), but also it is depending of the number of connections that its neighbor locations hold. This new centrality perspective rest in the idea that being connected to well connected others makes an actor more central.

The eigenvector centrality was developed by Bonacich (1972) to measure of how well connected is an actor in the whole system. The method works as the factor analysis (FA), assuming that each municipality in the commuting flows matrix is a variable, FA identifies "dimensions" of the distances, based on commuting flows, among them. The location of each municipality with respect to each dimension is called an "eigenvalue," and the collection of such values is called the "eigenvector." Usually, the first dimension captures the "global" aspects of distances among actors; second and further dimensions capture more specific and local sub-structures. Higher scores indicate that actors are "more central" to the main pattern of social network distances among all of the actors, lower values indicate that actors are more peripheral.

Later, those Eigen-Centrality indexes by municipality, jointly with the proportion of total population that commutes and population density were mapped in their geographical representation, to identify thick labor markets areas a spatial cluster analysis were performed using Lisa Indicators (Anselin 1995), to test those areas three adjacency matrices based on contiguity type queen order 1, shortest distance between polygons, and six-nearest neighbors were used, with 9,999 permutations. Those municipalities systematically associated with the same hotspot were included in the same functional thick labor region.

11.2.2 Counterfactual Framework and Matching Comparison

Assume that the worker can be assigned to two exposures: T = 1 if he is at a thick labor market and T = 0 if he is not. Assume that the wages can be observed under

$T=1$ and $T=0$, with W_1 and W_0, respectively. Using the Neyman-Rubin counterfactual framework, the potential wage for worker i is:

$$W_i = T_i W_{1i} + (1 - T_i) W_{0i} \tag{11.1}$$

Note that the researcher cannot observe W_i under $T_i = 1$ and $T_i = 0$ which is known as the fundamental problem of causal inference (Holland 1986). Using the counterfactual framework, the wage premium is defined by:

$$UWP = E(W_0|T=0) - E(W_1|T=0) \tag{11.2}$$

While the $W_0|T=0$ can be easily observed (the wage of a worker in a thin labor market), the counterfactual $W_1|T=0$ (the wage of the same worker in thick labor market) is unobservable. Note that treatment T is not a random selection, and there are several literatures discussing how the workers are moved toward a particular labor market. Moreover, the treatment $T_i = 1$ or $T_i = 0$ is motivated by some observables and unobservable variables (X). We assume that we can observe X and the orthogonal condition Eq. 11.3 is satisfied.

$$(W_0, W_1) \perp T|X \tag{11.3}$$

Equation 11.3 is the Ignorable Treatment Assignment Assumption (ITAA),[3] which allows identifying the spatial wage differential. The ITAA can be perfectly assumed in the case of randomized experiments, where the treatment is absolutely orthogonal to the output, but this assumption must be carefully considered in the case of quasi-experimental methods. A second assumption to be maintained is the Stable Unit Treatment Valued Assumption (SUTVA) (Rubin 1986). This assumption imposes that W for the worker i exposed to the treatment T is the same no matter what mechanism is used to assign the treatments to the rest of workers. According to Heckman (2005), this assumption rules out social interactions and general equilibrium effects. While the ITAA can be considered with a correct selection of the set X in the propensity score framework, we must necessarily maintain the SUTVA.

Both assumptions help to establish Eq. 11.2 as a measure of wage differential, but it cannot be empirically estimated: we cannot observe the worker under both conditions simultaneously. The matching estimator is proposed as an alternative to get an estimation of Eq. 11.2. The method proposes to identify a clone worker in the MA ($E(W_1|T=0)$) with a similar set of characteristics than the worker in a region r. The workers at r are called "treated", while the counterfactual workers at thick labor markets are called "controls". According to Eq. 11.3, the search of clones, hereafter called matching, must use the set of observable X. The matching is highly

[3] This assumption has received several names in the literature: unconfoundness (Rosenbaum and Rubin 1983), selection on observables (Barnow et al. 1980), conditional independence (Lechner 1999), and exogeneity (Imbens 2004).

conditioned by the dimension of the set X. If X is a large matrix, the matching procedure can be problematic. The Coarsened Exact Matching is a new technique inside the family of matching estimators that leads the dimensionality problem. Iacus (2008) proposes a two-step technique where in a first stage the variables are portioned in blocks in order to reduce the number of potential matching workers. In a second step, an exact matching is warrantied among workers. Finally, with the new sample of workers the UWP is computed.

11.3 Data

11.3.1 Data for Functional Thick Labor Markets

A set of variables is used to identify thick labor market. Such as it was described previously; we define the new MA in order to represent the geographical area where the thick labor market is located. Firstly, we used a matrix of commuting flows between municipalities, including just workers, normalizing the row flows by the size of the total population in the origin area, this relative flows matrix was used to calculated Eigen-centrality indexes. Secondly, the population density by municipality was calculated considering total population divided by its area measured in kilometers squared. Thirdly, to represent the relative flows as a network, a particular geographical layout was setting using roads distances between municipalities, those distances were kilometers taken to reach a municipality's local capital from another, that matrix was reduced to a unique pair of vectors by metrical multidimensional scaling, and in that manner was possible to integrate the geographical space into the commuting network flows.

With the purpose to visualize the relative importance of each municipality in the commuting system of flows, two different kinds of algorithms were used. The first algorithm, the Fruchterman and Reingold (1991) layout (F-R) was used to produce a representation of the networks of flows, in which those municipalities more central were located in the core of the system.

The F-R algorithm operates under two principles for graph drawing: (1) Vertices connected by an edge should be drawn near each other, and (2) Vertices should not be drawn too close to each other. To determine how close vertices should be placed some gravitational laws analogies are used, F-R assumed that vertices behave as celestial bodies, exerting attractive and repulsive forces on one another; those forces are modeled making only vertices that are neighbors attract each other, and at the same time, all vertices repel each other. The algorithm defines a particular static equilibrium to such system, in that way be close means to be connected but also it means to play a similar role in the relational system, in opposition be far means to be played an opposite role in that system.

The second algorithm, it is following Castell's ideas about the network society as a dichotomy between the space of flows and the space of places (Castells 2006), its main purpose is to be able to mix geographical topology associated with the spatial distribution of municipalities with their role in the system of flows, indicated by their hierarchy in terms of Eigen-Centrality, it is assumed here, that thick labor markets are those areas more important in terms of flows system and to show their importance into the conventional geographical terms, it is necessary to weight their physical areas by their centrality. To mix both dimensions, the physical municipality area was weighted using its Eigen-Centrality importance using the Diffusion–Based method proposed by Gastner and Newman (2004). The diffusion process generates a "diffusion cartogram"; where areas of municipalities have been rescaled to be proportional to their Eigen-Centralities, in that manner thick labor markets appears a bigger areas in the map, in opposition thin labor markets are reduced to reflex their less importance in the system.

11.3.2 Data for Coarsened Exact Matching

This chapter uses information from the Characterization National Survey 2009 (CASEN 2009), which is available at micro data level. The selected workers are 14 and 64 years and all of them receive a wage/hour greater than zero. We do not consider the army workers to properly represent the market equilibrium of labor demand and labor supply in the regional markets. The workers with economic sectors labeled as "Unspecified sectors" were also deleted. Finally, the outliers were treated using three different procedures. We run a regression by region where the dependent variable is the logarithm hourly wage and the independent variables are described in the Table 11.1. In a second step, we estimate the studentized residuals, standard residuals and dfbeta. We choose only those observations with dfbeta smaller than one and studentized and standard errors lower than three (absolute value).[4] The Table 11.1 reports a set of descriptive statistics for worker characteristics after the treatment of missing.

Such as it was previously discussed, Chile presents 15 regions and they are presented in the first column of the Table 11.1. Given its particular geographical shape, the geographical structure of the country is similar as the column is presented: "XV: Arica and Parinacota" is the region in the extreme north of the country and "XII: Magallanes y de la AntarticaChilena" is the extreme south. The second column describes an average hourly wage comparison, where the higher wage is observed in the Metropolitan Region of Santiago (MR). This exercise shows how the data suggest that the thick labor market generates a wage premium. Only a subset of three regions seem to have a wage close to the MR: II: Antofagasta, XI: Aysen and XII: Magallanes. This three are very specialized regions where the copper (II), salmon and oil (XI, XII) are the main industries. The second and third column represents the education (years) and experience. For the case of education,

[4] The Stata® code is available upon autor request.

Table 11.1 Worker chracteristics

Región administrativa	Sxh	Esc	Exper	Head	Casado	Contract	Permanent	Alphabet	Participation	Observations
XV. ARICA Y PARINACOTA	2156.34	12.40	20.61	0.48	0.62	0.87	0.81	1.00	0.50	420
I. TARAPACÁ	1997.28	11.82	20.40	0.52	0.63	0.88	0.80	0.99	0.52	877
II. ANTOFAGASTA	2303.95	11.65	20.93	0.42	0.61	0.90	0.84	0.99	0.54	1.689
III. ATACAMA	1925.03	11.30	22.15	0.53	0.66	0.87	0.77	1.00	0.49	1.129
IV. COQUIMBO	1754.28	11.29	22.53	0.50	0.58	0.80	0.73	1.00	0.53	1.597
R.M. METROPOLITANA DE SANTIAGO	2871.26	12.29	21.34	0.46	0.61	0.87	0.87	0.99	0.57	13.647
V. VALPARAÍSO	1820.20	11.71	22.12	0.50	0.61	0.86	0.79	0.99	0.51	5.775
VI. LIBERTADOR GENERAL BERNARDO O'HIGGINS	1886.84	11.18	21.85	0.48	0.62	0.88	0.67	0.98	0.55	3.444
VII. MAULE	1945.00	10.96	23.12	0.54	0.62	0.83	0.70	0.99	0.52	2.777
VIII. BÍO-BÍO	1752.75	11.68	22.00	0.52	0.64	0.86	0.75	0.99	0.48	6.415
XIV. LOS RÍOS	1854.44	11.24	22.00	0.56	0.64	0.86	0.78	0.99	0.48	1.103
IX. LA ARAUCANÍA	1999.98	11.42	21.78	0.52	0.67	0.83	0.75	0.98	0.49	2.385
X. LOS LAGOS	1731.35	10.98	22.09	0.50	0.64	0.84	0.77	0.98	0.54	2.26
XI. AYSÉN DEL GENERAL CARLOS IBÁÑEZ DEL CAMPO	2178.00	10.68	22.67	0.52	0.66	0.89	0.74	0.99	0.58	625
XII. MAGALLANES Y DE LA ANTÁRTICA CHILENA	2362.54	11.33	21.64	0.51	0.68	0.89	0.72	1.00	0.56	487

the MR presents the highest average education with almost 12.30 years. Head household and marital status represent two socio characteristics variables. We use this variable because of the responsibilities of the workers push the labor participation and the incentive the workers to improve its human capital in order to increase the wages. The rest of the variables area standard set of human capital characteristics. Second sets of independent variables are related with the occupation of the worker: two workers can earn different wages according to the type of occupation. CASEN 2009 provides a rich data set of information and it is shown in Table 11.2.

An interesting pattern is detected in the Table 11.2: the largest proportion of scientific and professionals are spatially sorted in the MR. A 15 % of the total sample of professional scientific are concentrated in this region against a 6 % in III: Atacama. This value reveals the potential of matching techniques because a direct comparison between workers in the MR and the III region would be highly biased due the heterogeneity in human capital. Both set of workers are not comparable and, moreover, the large set of workers in the MR could be do not exist in the III region. The Table 11.3 shows a descriptive statistics to analyze the economic sector of the regionals markets.

The Table 11.3 completes the spatial sorting picture around the MR. Again; we can see a highest proportion of financial sector with a representation of the 13 %. An interesting fact is the highly mining specialization of the II and III region where around of the 18 % of the population is related with this sector. This event also supports the high level of specialization around a particular sector, which almost does not exist in the MR (1 %). This set of tables support our hypothesis about the spatial sorting of human capital and about how this sorting must be controlled in order to warranty an appropriate wage comparison

11.4 Results

11.4.1 Thick and Thin Labor Markets

In Map 11.1 is displayed the Chilean administrative political division at municipal level, considering 334 spatial units (left side). It is important to remark that Chilean territory is extremely long and narrow, near 4,300 km from North to South, and 177 km as average wide. The municipality's areas are also heterogeneous, typically those located in both extremes (south and north regions) are larger than those located in central area, this feature determines in an important manner average travelling time between municipalities, and it also affects the commuting flows among them. In that map, it is also remarked those municipalities (52) that are part of the Political Administrative Metropolitan Region of Santiago (where the Chile's capital city is located) because they are highly populated and they concentrate an important share of the total commuting flows. Using the method previously

Table 11.2 Occupations

Región administrativa	Poder Ejecutivo	Profesionales científicos	Ténicos y prof. nivel medio	Empleados de oficina	Vendedores, comerciantes	Agricultores y trab. calificado	Oficiales, operarios y artesanos	Operadores y montadores	Trabajadores no calificados
XV. ARICA Y PARINACOTA	0.05	0.09	0.13	0.06	0.21	0.02	0.14	0.10	0.19
I. TARAPACÁ	0.03	0.10	0.10	0.08	0.21	0.01	0.15	0.18	0.15
II. ANTOFAGASTA	0.02	0.08	0.11	0.07	0.18	0.01	0.22	0.14	0.17
III. ATACAMA	0.02	0.06	0.10	0.08	0.15	0.02	0.22	0.16	0.19
IV. COQUIMBO	0.02	0.07	0.09	0.07	0.17	0.04	0.19	0.10	0.24
R.M. METROPOLITANA DE SANTIAGO	0.04	0.15	0.12	0.11	0.19	0.01	0.13	0.09	0.17
V. VALPARAÍSO	0.03	0.09	0.11	0.08	0.20	0.02	0.15	0.10	0.22
VI. LIBERTADOR GENERAL BERNARDO O'HIGGINS	0.01	0.09	0.09	0.08	0.14	0.03	0.13	0.11	0.31
VII. MAULE	0.02	0.11	0.09	0.06	0.18	0.03	0.13	0.09	0.30
VIII. BÍO-BÍO	0.02	0.11	0.10	0.08	0.17	0.02	0.17	0.12	0.20
IX. LA ARAUCANÍA	0.02	0.12	0.09	0.07	0.19	0.02	0.16	0.10	0.21
XIV. LOS RÍOS	0.02	0.10	0.10	0.12	0.17	0.05	0.13	0.14	0.17
X. LOS LAGOS	0.02	0.08	0.10	0.09	0.17	0.03	0.17	0.11	0.22
XI. AYSÉN DEL GENERAL CARLOS IBÁÑEZ DEL CAMPO	0.03	0.09	0.10	0.07	0.18	0.03	0.16	0.13	0.21
XII. MAGALLANES Y DE LA ANTÁRTICA CHILENA	0.06	0.07	0.09	0.09	0.13	0.06	0.21	0.13	0.16

Table 11.3 Regional industrial mix composition

Región administrativa	Agricultura, caza y silvicultura	Explotación minas y canteras	Industria manufacturera	Electricidad, gas y agua	Construcción	Comercio/ rest. hoteles	Transporte y comunicaciones	Establecimi. Financieros	Servicios comunales y sociales
XV. ARICA Y PARINACOTA	0.05	0.07	0.09	0.00	0.13	0.24	0.06	0.08	0.28
I. TARAPACÁ	0.02	0.08	0.05	0.01	0.09	0.24	0.13	0.08	0.31
II. ANTOFAGASTA	0.02	0.17	0.09	0.02	0.11	0.18	0.11	0.07	0.24
III. ATACAMA	0.08	0.20	0.05	0.01	0.10	0.18	0.10	0.04	0.23
IV. COQUIMBO	0.14	0.09	0.05	0.01	0.13	0.20	0.08	0.05	0.24
R.M. METROPOLITANA DE SANTIAGO	0.02	0.01	0.13	0.01	0.09	0.22	0.10	0.13	0.30
V. VALPARAÍSO	0.08	0.02	0.08	0.01	0.11	0.22	0.10	0.08	0.30
VI. LIBERTADOR GENERAL BERNARDO O'HIGGINS	0.22	0.07	0.09	0.01	0.08	0.19	0.07	0.04	0.24
VII. MAULE	0.22	0.01	0.10	0.01	0.07	0.24	0.06	0.05	0.25
VIII. BÍO-BÍO	0.09	0.01	0.13	0.01	0.12	0.18	0.09	0.06	0.31
IX. LA ARAUCANÍA	0.08	0.00	0.09	0.01	0.12	0.22	0.08	0.06	0.32
XIV. LOS RÍOS	0.12	0.00	0.10	0.01	0.11	0.19	0.11	0.04	0.32
X. LOS LAGOS	0.10	0.00	0.10	0.01	0.12	0.24	0.09	0.04	0.30
XI. AYSÉN DEL GENERAL CARLOS IBÁÑEZ DEL CAMPO	0.07	0.01	0.07	0.00	0.18	0.18	0.08	0.06	0.36
XII. MAGALLANES Y DE LA ANTÁRTICA CHILENA	0.06	0.02	0.08	0.01	0.12	0.25	0.08	0.05	0.33

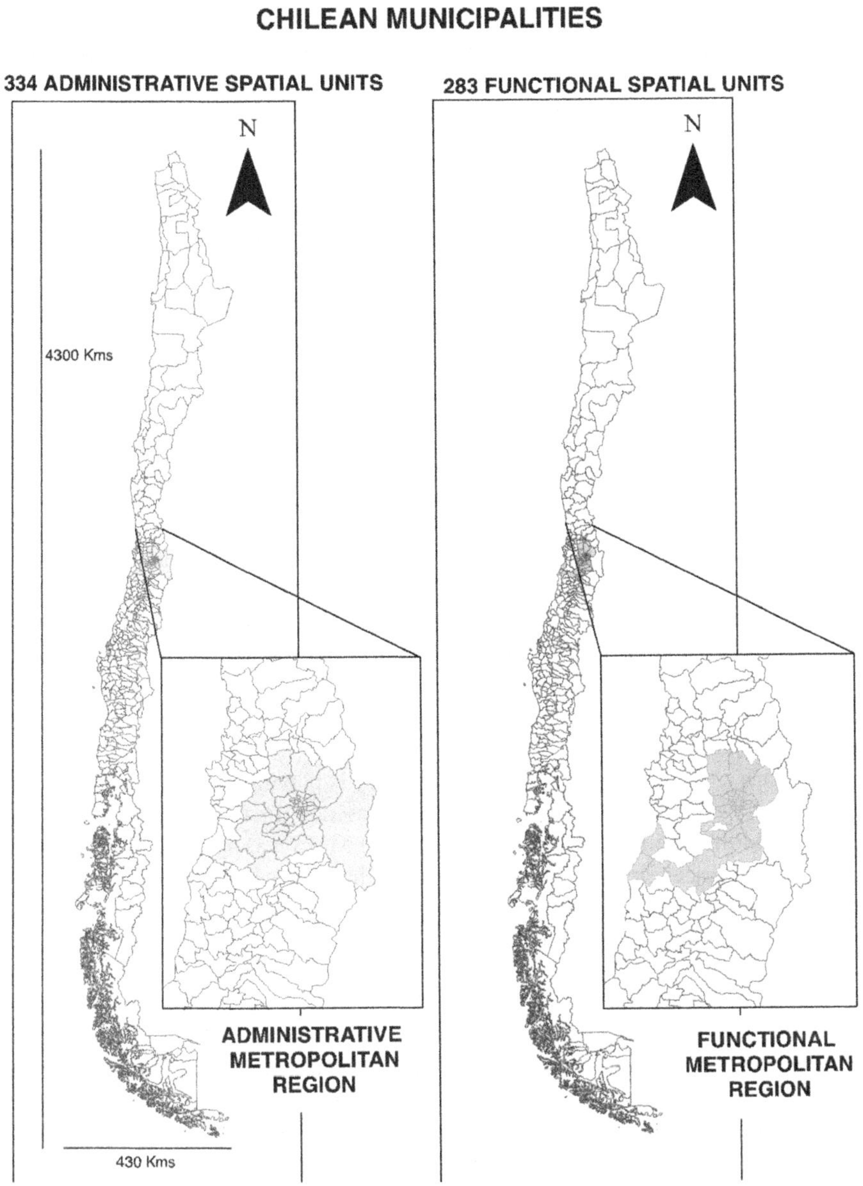

Map 11.1 Administrative RM and Functional RM

discussed a thick labor market was identified as a local spatial cluster of highly correlated values in terms of population density, high rates of commutation and higher values in terms of Eigen-Centrality indicators. In the map No 11.1 (right side), the thick functional labor market was named Functional Metropolitan Area (FMA), its presence reduces the original number of administrative counties to a new total of 283 remaining municipalities, in that manner FMA implies to collapse 51 spatial original municipalities into FMA, those municipalities come from the administrative Region of Santiago (44), Valparaiso (1) and O'higgins Region (6).

In Map 11.2 (left side) is displayed a standard deviation map of population density, there should be noticed that central municipalities are highly dense populated and smaller in terms of geographical areas, in opposition extreme regions are bigger in terms of area but also less populated. In that map (right side), the results from the "diffusion cartogram" are presented, there those bigger areas represent municipalities more relevant in terms of their hierarchy of commuting flows (greater Bonacich's Eigen-Centrality). In that particular map, the identified thick labor market area (FMA), matches exactly with the more central spatial units, and for this reason they appear with greater weighted areas, indicating that it is the Core of the commuting flows system and also the thicker labor market, in contrast the thin labor markets appear with reduced areas.

To understand why those municipalities are collapsing together into a FMA, it is necessary to describe the main characteristics of Chilean system of commuting flows. In Network 11.1, those flows are plotting in a F-R layout. That network corresponds to a valued and directed type, there nodes represent each municipality, and edges represent a connection among them as the relative commuting flow between them normalized by total population at origin place, to simplify that network connections a cutoff equal to 1 % was used. The network shows that municipalities tend to cluster into cohesive groups, implying that cross-groups commuting are rare.

To clarify the role that geographical topology is playing there, in Network 11.2, the same flows are plotted, but using a layout based on first two dimensions from a Metrical Multidimensional Scalling (MMS), the MMS was applied over the matrix of travel distances among municipalities, in that manner vertexes located close means that are geographically close and/or between them it is relatively easy to commute, fact the normally occurs when they are located into the same political administrative region or between geographically contiguous areas. There, central – and densely interlinked- area is occupied by vertexes (municipalities) located in the Chilean central regions (Valparaiso, Metropolitana of Santiago, and O'Higgins), in contrast the extreme and less connected nodes are municipalities located in extreme South and North regions, in that fashion it is clear that geographical location (or the Castellian System of Places) plays and important role to configure the System of Flows, in our case commuting interactions.

Both networks contribute to explain why the Administrative Metropolitan Region of Santiago is not an appropriate benchmark region for estimating the UWP. The identified Functional Thick Labor Market is absorbing a portion of other two regions and the rural side is left out. This new region is considered the new densest labor area of the country. We use this new geographical division (FMA) as the second scenario discussed in the introduction and also it was used for estimating a new UWP.

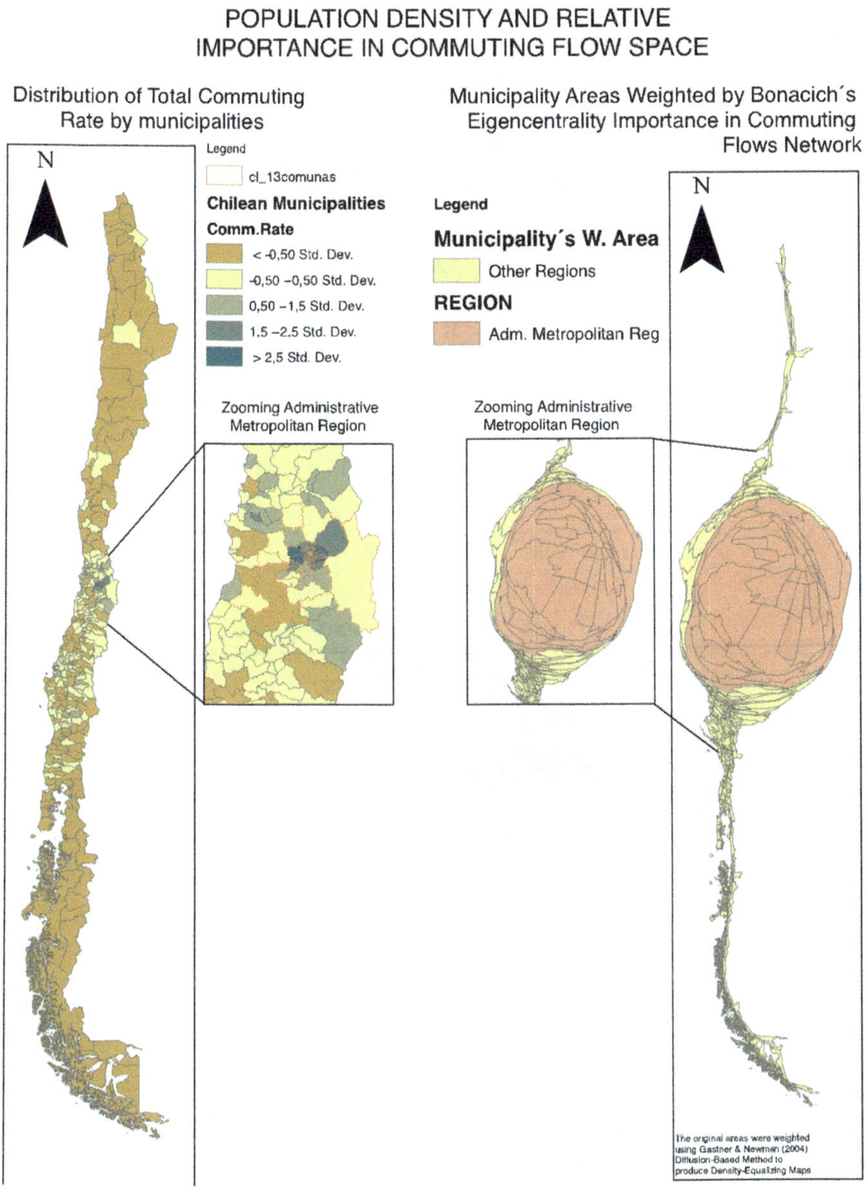

Map 11.2 Administrative RM characteristics and Functional RM relative importance

11.4.2 Estimation of UWP Using Both Geographical Divisions

Our hypothesis establishes than the wage premium is better identified when the thick labor market is defined in a functional sense. We ground our hypothesis on the

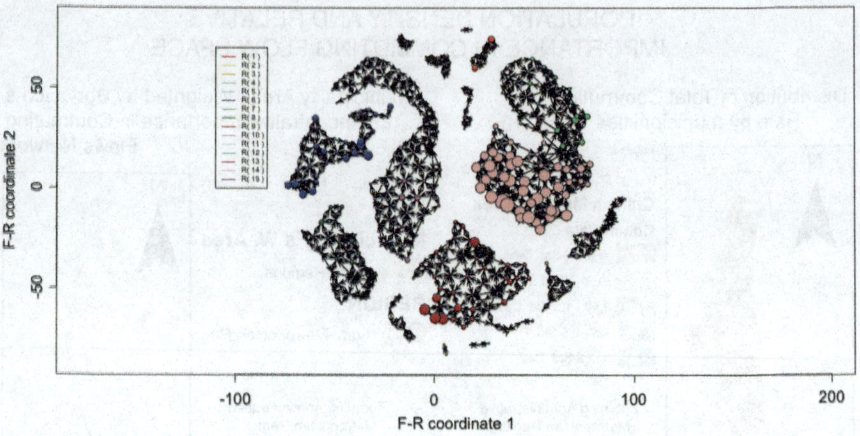

Network 11.1 Commuting flows networks representation

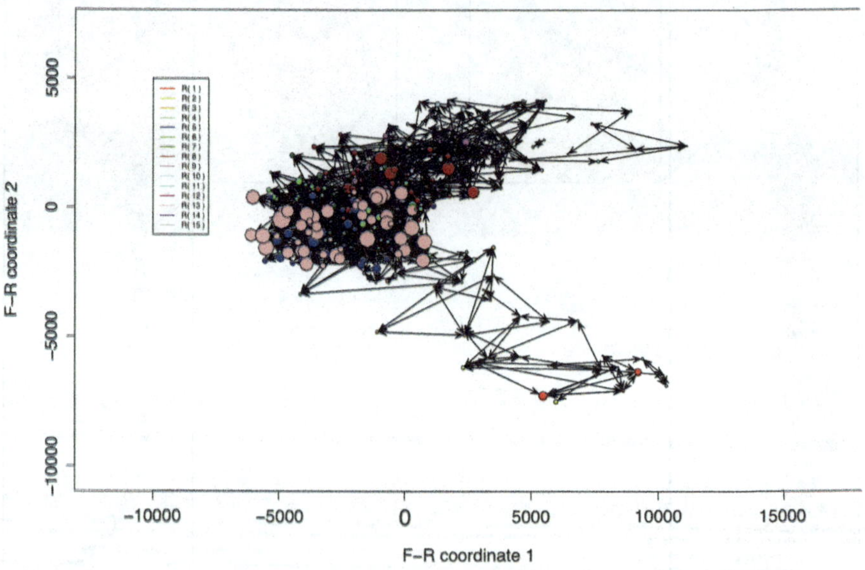

Network 11.2 Commuting flows networks and geographical space

logical of economics of agglomeration: if the spatial spillovers explain the higher productivity of workers, then they exist in a labor market were the interaction between economic agents is warranted. However, the geographical scope of the market could not coincide with the administrative division. If the administrative area does not fit with the functional, then the wage gap could not be properly identified.

In order to test our hypothesis, we follow a two-step procedure. First, we estimate the wage differential between a thin labor market (each one of the 14 Chilean political administrative regions) and the MR of Santiago (a region selected as benchmark because it is traditionally considered a thick labor market). In a second step, we repeat the same exercise, but using the FMA instead the administrative one as control.

This testing strategy rests on the critical role played by a fair comparison of wages among a set of heterogeneous workers. In this sense, we estimate the differentials using a Coarsened Exact Matching. The CEM allows a resampling where only a set of comparable workers is considered. Given that, the balance is the core of CEM, we must provide a statistical measurement to evaluate how comparable are the workers after matching. The balance between both groups of workers is evaluated with the L_1 statistics proposed by Iacus et al. (2008). Instead of considering only a mean test, this statistics considers the proximity of multivariate histogram between groups. The statistics is:

$$L_1(f, g) = \frac{1}{2} \sum_{l_1 \ldots l_k} \left| f_{l_1 \ldots l_k} - g_{l_1 \ldots l_k} \right| \tag{11.4}$$

The procedure builds a crosstab of kcovariates for treated (thin labor market) and control (thick labor market) group. The frequencies for treated $f_{l_1 \ldots l_k}$ and control $g_{l_1 \ldots l_k}$ are used to estimate the balance measure. If L_1 is close to zero, then the balance property is satisfied. Otherwise, the matching procedure does not satisfy the core assumption. Formally, if the matched frequencies are labeled g^m , then we expect :

$$L_1(f^m, g^m) < L_1(f, g). \tag{11.5}$$

Assuming the frequencies without matching is labeled as $L_1(f, g)$ then we expect $L_1(f^m, g^m) < L_1(f^c, g^c)$. We estimate the L_1 for each region and the MR with and without matching.

The Fig. 11.1 contains the Chilean regions, on the horizontal axis they are displayed in a geographical order from North to South, and the vertical axis shows the L1 statistics. Those statistics represent the heterogeneity of workers before and after matching, using the administrative division as control. The CEM finds similar workers using the complete set of variables described on Tables 11.1, 11.2 and 11.3. The black line (circles) represents the L1 statistic for each region in comparison to the MR as a thick labor market. If L1 is close to one, it means a higher heterogeneity between both groups. The Fig. 11.1 shows a high heterogeneity level for the whole set of regions, the most of L1 statistics are around 0.9. This result supports our hypothesis about the bias when the workers in a region are compared with workers in the MR without a proper matching procedure. Additionally, the statistics is reduced for those regions close to the MR which indicates that a higher bias is carried out for the extremes of the country. In the same Fig. 11.1, the gray line represents the L1 statistics, but when the CEM is applied. Clearly, the new subsets of workers are more comparable and the bias is considerably reduced. These

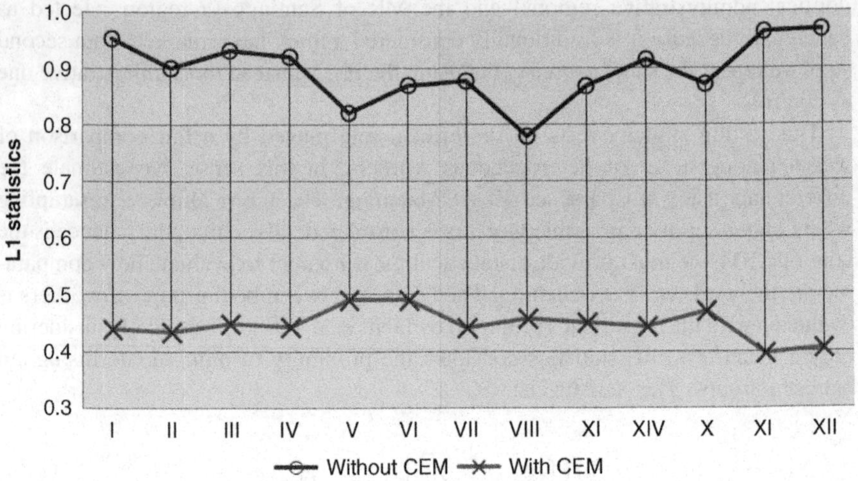

Fig. 11.1 L1 statistics for a particular region and MR with and without CEM

results show the advantages of the matching procedure and how when it is applied the worker's heterogeneity is minimized. After to get an homogenous groups of workers, the next step consists of estimating the wage differential between regions, using both definition of thick labor markets, the administrative one (MR) versus the functional one (FMA). As it was discussed in the introductory section, we expect that using MR is going to bias the UWP estimation because the presence of spatial mismatch affecting the selected control group.

After the matching procedure, each region owns a new subset of workers and the MR also presents a set of comparable workers. This group is equivalent according to the variables used to carry out the CEM. With this new set of comparable workers, we can estimate the wage differential as a simply means test between each thin and thick labor market. However, this approach would assume that there is not any kind of differential in human capital among workers. This strong assumption implies an exact and perfect matching, which is not truth by a statistical definition. For this reason, we estimate the mean test, but controlling again for the same variables utilized in the CEM procedure. This exercise is equivalent to Mincer equation with a dummy indicating if the worker is located in a thin or thick labor market, but assuring set comparable workers. We show the R-square for each Mincer regression using the MR as a thick labor market.

The Fig. 11.2 shows the R squares of the regression after CEM. The fit level of the complete set of regressions is around of 0.5. This implies that almost the 50 % of the wage variance is correlated with the variance provided by the set of independent variables plus the dummy indicator. This level of fit is increased for those regions located on the extremes, Norther and Souther part of Chile, and it shows the importance of controlling by economic sector such as mining or fishing, that are highly located in those extreme regions. Our final results are displayed in the Figs. 11.3 and 11.4.

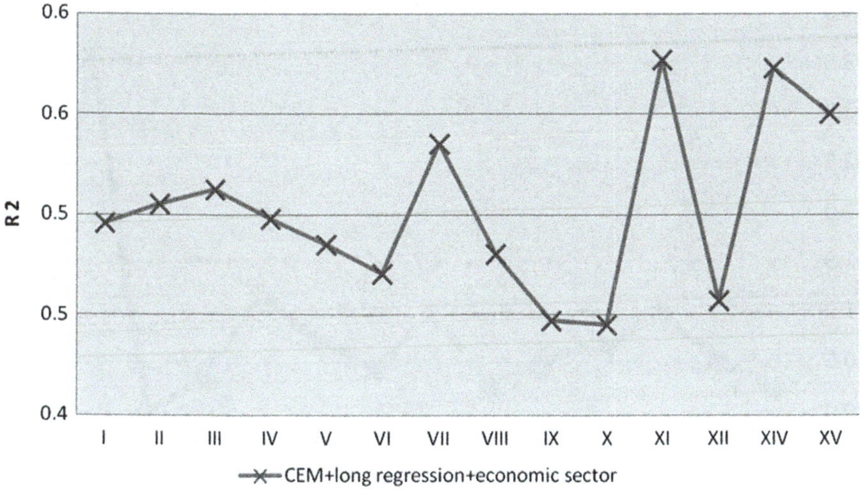

Fig. 11.2 R square for the regression with and without economic sectors

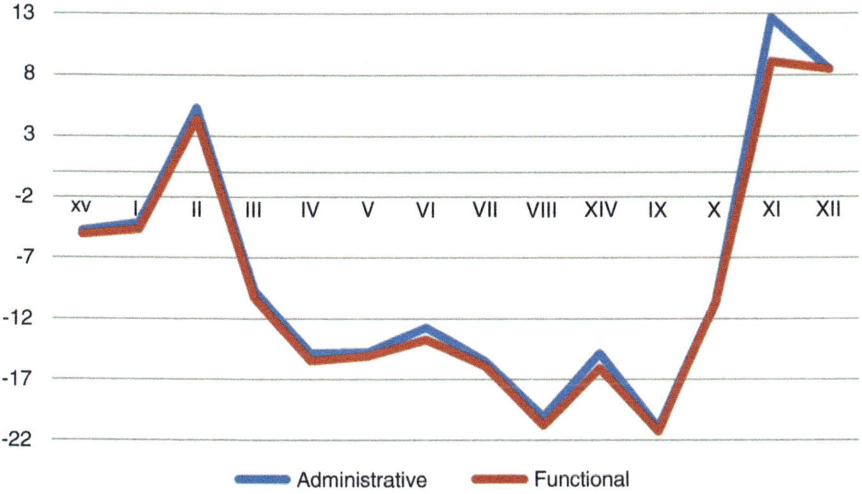

Fig. 11.3 Spatial wage differential using administrative and functional division

The Fig. 11.3 shows the estimation of wage differential using both identified thick labor market areas (RM and FMA). The first result suggests the existence of a wage premium for almost the complete set of regions. In both cases, thick labor markets show a strong wage premium against the regions III, IV, V, VI, VII, VIII, IX and X. It means that any set of comparable workers earns a higher wage in the thick market than in those thin labor markets. This differential is between 10 % and 20 % for that set of regions. In contrast, for those two regions (XV and I) located in the extreme north also present a negative wage effect in comparison with the thick

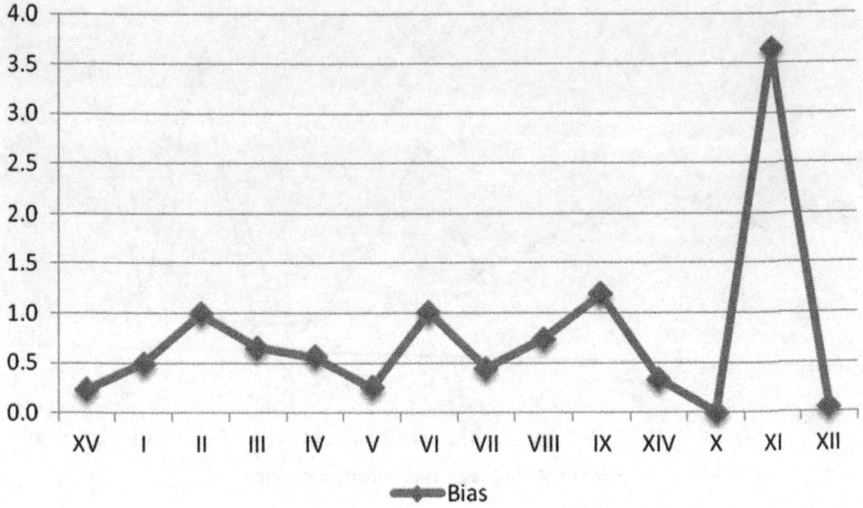

Fig. 11.4 Bias of wage differentials using administrative versus functional division

labor market, but it is small around 4.5 %. Only three regions do not show wage premium, namely the II, XI and XII, but where all of them are characterized by particular industrial conditions. The existence of specialization in mining, fishing and fuel industry generate higher wages that thick markets.

The Fig. 11.4 provides the difference of UWP for both types of regions. The administrative system is always under-estimating the UWP. For the most of the regions, the bias is around 0 % and 1 %. However, the bias increases in the south extreme regions such as X, XI and XII. These results suggest that the spatial sorting of workers implies a high heterogeneity between these workers and those located in the Metropolitan Region of Santiago

11.5 Conclusions

This chapter contributes in several dimensions. First, a new functional approach is built for the Chilean case. The Metropolitan Region of Santiago is a dense labor market and its geographical scope is broader than administrative boundaries. Moreover, a portion of the Metropolitan Region, particularly the rural areas, does not belong to the new thick labor market. These results suggest evidence how the geographical extent of labor markets is not defined by administrative reasons, but rather by its functionality.

The UWP is estimated using both spatial divisions. The results show how, after a matching procedure, a similar set of workers earn higher wages in the thick labor markets comparison for almost any region. We estimate this premium around 4 % and 21 %. Only three regions show a higher wage than thick market, but its difference is reduced with our functional approach. This exercise provides a first look for the Chilean case and the estimations suggest how the persistent concentration of economic activity increases the wage through the existence of a urban premium.

References

Anselin L (1995) Local indicators of spatial association – LISA. Geogr Anal 27:93–115

Barnow BS, Cain GG, Goldberger AS (1980) Issues in the analysis of selectivity bias. In: Stromsdorfer EW, Farkas G (eds) Evaluation studies: review annual, vol 5. Sage Publication, Beverly Hills

Berry C, Glaeser E (2005) The divergence of human capital levels across cities. NBER working papers 11617. National Bureau of Economic Research

Bonacich P (1972) Factoring and weighing approaches to clique identification. J Math Sociol 2:113–120

Boots B, Kanaroglou P (1988) Incorporating the effects of spatial structure in discrete choice models of migration. J Reg Sci 28:495–507

Boots B, Tiefelsdorf M (2000) Global and local spatial autocorrelation in bounded regular tessellations. J Geogr Syst 2(4):319–348

Boots BN (1984) Evaluating principal eigenvalues as measures of network structure. Geogr Anal 16:270–275

Borgatti S (2005) Centrality and network flow. Soc Networks 27(1):55–71

Castells M (2006) The theory of the network society. MPG Books, UK

CASEN (2009) Encuesta de Caracterizacion Socio Economica Nacional

Combes P, Duranton G, Gobillon L (2008) Spatial wage disparities: sorting matters! J Urban Econ 63:732–742

Elvery J (2010) City size and skill intensity. Reg Sci Urban Econ 40(6):367–379

Fruchterman T, Reingold E (1991) Graph drawing by force-directed placement. Softw Pract Exp 21(11):1129–1164

Gastner M, Newman M (2004) Diffusion-based method for producing density-equalizing. PNAS 101(20):7499–7504

Getis A, Griffith D (2002) Comparative spatial filtering in regression analysis. Geogr Anal 34 (2):130–140

Glaeser E, Mare D (2001) Cities and skills. J Labor Econ 19(2):316–342

Glaeser E, Resseger M (2009) The complementarity between cities and skills. NBER working papers 15103, National Bureau of Economic Research

Green N (2007) Functional polycentricity: a formal definition in terms of social network analysis. Urban Stud 44:2077–2103. doi:10.1080/00420980701518941

Heckman J (2005) Lessons from the technology of skill formation. NBER working papers 11142, National Bureau of Economic Research

Holland P (1986) Statistics and causal inference. J Am Stat Assoc 81:945–970

Iacus S, King G, Porro G (2008) Matching for causal inference without balance checking. UNIMI - research papers in economics, business, and statistics unimi-1073, UniversitádegliStudi di Milano

Iacus SM, King G, Porro G (2011) Multivariate matching methods that are monotonic imbalance bounding. J Am Stat Assoc 106(493):345–361. doi:10.1198/jasa.2011.tm09599

Imbens G (2004) Nonparametric estimation of average treatment effects under exogeneity: a review. Rev Econ Stat 86(1):4–29

Kanbur R, Venables A (2005) Spatial inequality and development overview of unu-wider project. http://www.arts.cornell.edu/poverty/kanbur/WIDERProjectOverview.pdf

Kanbur R, López-Calva LF, Venables AJ (2005) Symposium on spatial inequality in Latin America. Cuadernos de economía 42(125):133–136
Lechner M (1999) Nonparametric bounds on employment and income effects of continuous vocational training in East Germany. Econ J 2(1):1–28
Marshall A (1920) Principles of economics. MacMillan, London
Mion G, Naticchioni P (2009) The spatial sorting and matching of skills and firms. Can J Econ 42 (1):28–55
Puga D (1998) Urbanisation patterns: European vs. less developed countries. J Reg Sci 38:231–252
Puga D (1999) The rise and fall of regional inequalities. Eur Econ Rev 43(2):303–334
Rosenbaum PR, Rubin DB (1983) The central role of the propensity score in observational studies for causal effects. Biometrika 70(1):41–55
Rosenthal S, Strange W (2004) Evidence on the nature and sources of agglomeration economies. In: Henderson J, Thisse J (eds) Handbook of regional and urban economics. Elsevier, Amsterdam, pp 2120–2171
Rubin DB (1986) Which ifs have causal answers. J Am Stat Assoc 81:961–962
Venables A (2005) Spatial disparities in developing countries: cities, regions and international trade. Int Econ Rev 5:3–21

Chapter 12
A Stepwise Procedure to Determinate a Suitable Scale for the Spatial Delimitation of Urban Slums

Juan C. Duque, Vicente Royuela, and Miguel Noreña

12.1 Introduction

The globalisation era in which we live has made the world an interconnected space with several global trends. We find developing countries with very high growth rates, what helps to find world economic convergence. As a complement to this trend, within those countries there is a dramatic growth pattern of cities into megacities, as economic activity concentrates in space to exploit agglomeration economies. According to UN-Habitat, in the next two decades the global population living in urban areas will move from 50 % to 70 %.

According to John Weeks, these trends imply that the classical analysis of the differences between urban and rural regions needs to be completed with the study of intra-urban variability. Inequalities are becoming more important within the cities and metropolitan areas than between the regions (Weeks et al. 2006). This analysis

The authors thank the comments of Jean Paelinck and all other assistants at the Forth Workshop in Spatial Econometrics in Oviedo (Spain) in October 2010. V. Royuela acknowledges support of ECO2010-16006.

J.C. Duque (✉)
Research in Spatial Economics (RISE-group), School of Economics and Finances, EAFIT University, Carrera 49 7sur-50, Medellín, Colombia
e-mail: jduquec1@eafit.edu.co

V. Royuela
AQR Research Group-IREA, Universidad de Barcelona, Avda Diagonal, 690, Barcelona 08034, Spain
e-mail: vroyuela@ub.edu

M. Noreña
Research in Spatial Economics (RISE-group), School of Economics and Finances, EAFIT University, Carrera 49 7sur-50, Medellín, Colombia
e-mail: mnorena1@eafit.edu.co

E. Fernández Vázquez and F. Rubiera Morollón (eds.), *Defining the Spatial Scale in Modern Regional Analysis*, Advances in Spatial Science,
DOI 10.1007/978-3-642-31994-5_12, © Springer-Verlag Berlin Heidelberg 2012

is even more important when urban growth occurs in a disordered way due to rural urban migrations to unplanned urban spaces.

Agglomeration economies derived from city size and density imply absolute advantages compared to rural space. Nevertheless, these advantages can turn into disadvantages or agglomeration diseconomies: increasing transport costs, congestion, higher rent prices and salaries faced by firms, or simply more pollution and crime rates, aspects directly connected to quality of life levels in cities (Royuela et al. 2008). In this line, the cities and their connected regions need to control for the negative aspects related to urban growth in order to succeed and achieve a higher well being of their inhabitants.

The city of Medellín (the second largest city in Colombia) is a good example of the process described above. During the early 1950s, the political violence took place in the rural areas forcing the farmers to move to the cities. Between 1951 and 1973 the population of Medellín grew from 358,189 to 1,077,252 inhabitants. Between the middle 1980s and early 1990s, the toughest period in the urban war against the drug cartels, the city continued growing but at a slower rate (1.38 %). After that period the homicide rates gradually declined from 150 to 33.2 homicides per 100,000 inhabitants in 2005, and the annual population growth went up to 2.6 %. According to the last census (2005), Medellín has a population of 2,223,078 people. This continued and quick demographic growth has been highly concentrated in the slopes of the northwest periphery of the city, exceeded the capacity of the local government to deliver services and infrastructure and caused the appearance of urban slums.

In Medellín the delineation of slummy areas is usually based on subjective aspects. The exact delimitation of the borders of these areas has not been matter of study for local authorities. However, a proper identification and delineation of these areas is a central issue, as it can help to prevent poverty traps and crime nests. Urban regeneration policies usually involve a wide list of policy tools in a single area rather than expanding vague tools to the overall territory. Consequently, identifying spatial priorities arise as a central issue in urban regeneration policies.

Urban policy makers face several options when trying to identify spatial objectives. The first option, and more usual one, is to use administrative, or normatively defined, areas. Data is usually collected at this territorial level and consequently it is easier to determine objectively where the problems are. Nevertheless this procedure can be severely inefficient. In many cases statistical inference based on normative regions may be strongly affected by aggregation problems, such as the ecological fallacy (Robinson 1950), the modifiable areal unit problem (Openshaw 1977a, b, 1984; Openshaw and Taylor 1981; Arbia 1989), aggregation bias (Fotheringham and Wong 1991; Amrhein and Flowerdew 1992; Paelinck and Klaassen 1979; Paelinck 2000), or the small numbers problem, when working with rates (Diehr 1984). The second option consists of using analytical homogeneous regions. Opposite to the normative regions, the analytical homogeneous regions seek to divide the area of study into regions that are homogeneous in terms of a set of socioeconomic variables that are considered relevant to the phenomenon under study (Fischer 1980; Duque et al. 2006).[1] Thus, analytical regions not only provide an alternative to deal with several statistical challenges, but also ensure that specific public policy actions implemented in a region will have an homogeneous impact throughout the region (Fischer 1980).

The focus of this chapter is to present, through a case study, a stepwise procedure that efficiently delimits the intra-urban slums. This procedure will offer policy makers the opportunity to identify slums in a more rigorous way. To identify slum we apply different clustering techniques (local Moran's I and the local G statistic) and we consider their significance through the False Discovery Rate technique.

The chapter is structured as follows. In the next section we describe the case of study, the city of Medellín together with the database and the definition of our variables under study. After this, in Sect. 12.3 we describe the procedure we will be using. Section 12.4 presents the obtained results and Sect. 12.5 concludes.

12.2 The Case of Study: Medellín (Colombia). Database Description

Figure 12.1 shows the location of the studied area. To characterise the city, we use data from the 2007 Quality of Life Survey of Medellín. This survey is developed annually by the municipality of Medellín and is formed by 184 questions on nine dimensions: housing, households, demography, education, social security, income and employment, participation, gender and family violence, and nutrition. The survey includes 21,861 households, representing 79,912 persons.

The survey considers a wide list of questions related with poverty. From that list, we selected a set of questions with which we seek to cover different dimensions of poverty. Among these questions we highlighted the aspects considered by UN-Habitat to construct the slum index. For each question we constructed a binary variable indicating whether or not the housing unit has poverty conditions. Table 12.1 describes the variables and the proportion of housing units in Medellín with poverty conditions.

Figure 12.2 presents the two most important administrative scales in Medellín, Neighbourhoods and Communes.[2] The survey is designed to be representative at

[1] The Statistical Office of the European Communities (Eurostat 2006) provides a clear differentiation between these two types of regions. "Normative regions are the expression of a political will; their limits are fixed according to the tasks allocated to the territorial communities, to the sizes of population necessary to carry out these tasks efficiently and economically, or according to historical, cultural and other factors. Whereas analytical (or functional) regions are defined according to analytical requirements: functional regions are formed by zones grouped together using geographical criteria (e.g., altitude or type of soil) or/and using socio-economic criteria (e.g., homogeneity, complementarity or polarity of regional economies)." (para. 4–5).

[2] The areas represented in white correspond to institutional areas such as parks, Medellín's airport, sporting units, and University campuses, which were not included in this research.

Fig. 12.1 Location of Medellín, Colombia

the Commune level. The Communes are administrative borders that divide the city into 16 spatial units with an average size of 6.30 ± 2.86 km^2. Because of the high level of socioeconomic heterogeneity within each Commune, the required sample size increases considerably. Thus, the average number of surveyed households per Commune is 1,279 ± 312 households. However, the size of these administrative units results extremely large for carrying out studies that intend to delineate intra-urban slums. Moreover, because of the socioeconomic heterogeneity, using the Communes as study unit will imply that each spatial unit will hold a high degree of aggregation bias (Levin 1992).

To gain more spatial detail, the next, and smallest, administrative unit available in Medellín is the neighbourhood. The city is divided in 243 neighbourhoods with an average size of 0.38 ± 0.21 km^2. The average number of surveyed households per neighbourhood is 84 ± 57 households, with values ranging from 3 to 296

Table 12.1 Socioeconomic variables and proportion of people running into poverty

Variable	Description	Dimension	Housing units (%)
Sewer	Absence of sewer system	Housing	1.90
Risk	Housing unit located in area at high risk of natural disaster	Housing	3.70
Bedrooms	Bedrooms for other uses	Housing	22.19
Fridge	Absence of refrigerator	Housing	5.83
Kitchen	No access of kitchen	Housing	3.87
Material	Walls are not of durable material[a]	Housing	0.14
Water	Absence of piped water[a]	Housing	0.01
Toilet	Toilet not connected to a sewer[a]	Housing	3.77
Ownership	Residents are not the owners[a]	Households	35.46
Overcrowding	Three or more persons per room[a]	Households	16.77
Violence	Household victim of violence in conflict and displacement situations	Social Security	1.81
Sisben	The household is beneficiary of SISBEN program[b]	Social Security	30.87
Illiteracy	Illiterate householder	Education	4.38
Income	Household total income below the legal minimum wage	Income	84.03
Nutrition	Household with nutritional problems	Nutrition	13.34
Birth	Householder knows birth control methods	Sex education	10.37

[a]Variables used by the UN-Habitat to define a slum
[b]The System for the Selection of Beneficiaries of Social Programs (SISBEN). For more information about this governmental program see Velez et al. (1998)

surveyed households. Although this scale offers a good level of spatial detail, the fact of having neighbourhoods with such a variability in the number of surveyed households, where 15.64 % of them contain less that 30 surveyed households, would drive to misleading results related to the widely known small numbers problem.[3]

Being the communes too big and internally heterogeneous, and the neighbours to small to guarantee robustness in statistics; it is necessary to design geographical study units that overcome the problems derived from working with the available administrative units. This type of units are known as “analytical regions.” In the literature, there is wide range of algorithms for spatial aggregation (Duque et al. 2007). Each algorithm is designed for serving specific purposes, such as electoral districting (Yamada 2009), school districting (Caro et al. 2004), turfing (Segal and Weinberger 1977), zone design (Openshaw 1995), etc. Amongst them, the Max-p-regions algorithm, devised by Duque et al. (2012) has the characteristics that better fit to our needs.

[3] For a recent debate on the topic see Goovaerts (2009).

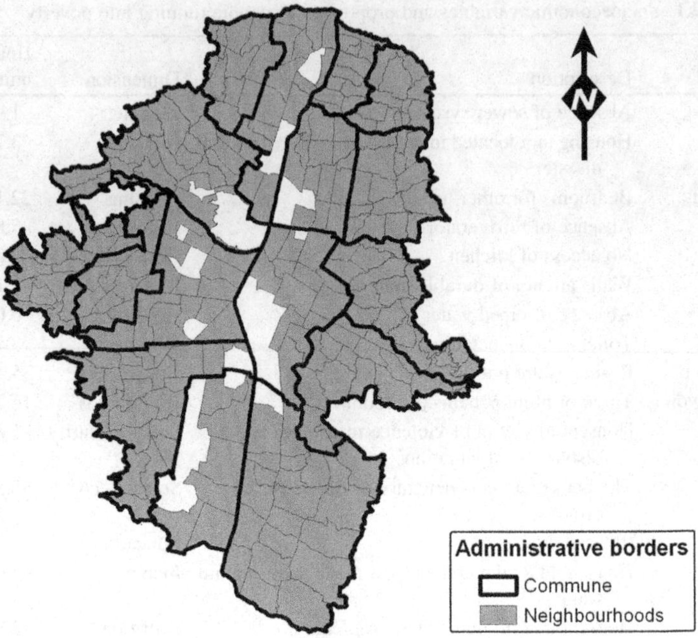

Fig. 12.2 Medellín's administrative borders

12.3 Procedure for Designing Analytical Regions for Studying Poverty

As we stated before, analytical regions are those that aggregate small spatial units into larger spatial units according to predefined criteria. In our context, the study of poverty, we require that (a) each analytical region is homogeneous in terms of a set of socioeconomic characteristics that measure different aspects of poverty, and (b) each analytical region must contain at least 100 surveyed households in order to ensure statistical validity. To satisfy these two requirements we applied the Max-p-regions model. Next, we describe the main features of the algorithm.[4]

The Max-p-regions seeks to aggregate n areas into the maximum amount of spatially contiguous regions, such that each region satisfies a predefined minimum threshold value for some spatially extensive regional attribute (e.g. number of housing units per region). The Max-p-regions also seeks to minimize intra-regional heterogeneity measured as:

[4] The max-p-regions algorithm is available in the software ClusterPy (Duque et al. 2011). ClusterPy is an open source cross-platform library tool for spatial clustering written in Python. Binary installers and source distributions are available for download at http://code.google.com/p/clusterpy.

$$H = \sum_{k=1}^{p} \sum_{(ij \in R_k | i<j)} d_{ij}. \tag{12.1}$$

Where d_{ij} denotes the pairwise dissimilarities between areas i and j assigned to the same region k, R_k. Both objectives, the number of regions and the intra-regional heterogeneity, are merged into a single objective function as follows:

$$Minimize\ Z = \left(-\sum_{k=1}^{n} \sum_{i=1}^{n} x_i^{k_0} \right) \times 10^h + \sum_{i} \sum_{(j|j>i)} d_{ij} t_{ij}. \tag{12.2}$$

With two binary decision variables $x_i^{k0} = 1$ if area i is assigned to region k in order 0 (each region has one and only one area assigned at this order)[5]; and $t_{ij} = 1$ if areas i and j are assigned to the same region. The first term represents the number of regions and the second term represents the intra-regional heterogeneity. Both terms are merged into a single value in such a way that (a) an increment in the number of regions is preferred over any other solution involving a lower number of regions, and (b) for a given number of regions, the solution with lower intra-regional heterogeneity is preferred. This hierarchy in the elements of the objective function are ensured by multiplying the first term by a scaling factor 10^h, with.

The Max-p-regions offers additional characteristics that make it the most suitable algorithm for our purpose:

- The Max-p is capable to create regions of any shape, contrary to the majority of the models for spatial aggregation that guarantee spatial contiguity by maximizing regional compactness. Thus, the shape of the regions depends on the spatial pattern of the aggregation variables.
- From the best of our knowledge, the Max-p-regions is the only partitioning algorithm that allows endogenizing the number of regions. The researcher only has to specify the minimum number of observations per region (i.e., minimum number of surveyed households).
- It minimizes the lost of information by performing the minimum number of aggregations required to create feasible regions. It also minimizes the aggregation bias by minimizing intra-regional heterogeneity.

Although the Max-p-regions is capable to deal with multiple aggregation variables, previous contributions in the area of constrained clustering suggest the use of factor analysis as a way to reduce a large number of variables into a small number of uncorrelated factors to facilitate capturing the most important patterns in the data (Berry 1960; Johnston 1968; Spence 1968; Calciu 1996). The literature

[5] Each one of the remaining non-core areas is assigned to one core area in an order that corresponds to the contiguity order that separates the area to the core. For this reason, the Max-p requires information about the neighborhood structure (also known as the W matrix).

proposes to hold the number of the factors able to capture at least between 60 % and 70 % of total variance.

One additional condition, important for spatial clustering, is that each selected factor should have some spatial pattern, which is verified with a significant level of spatial autocorrelation. Without this last requirement, the spatial contiguity constraint may result conflicting. Lastly, because the variables being reduced with the factor analysis are rates measured at neighbourhood scale, we are facing the small numbers problem. Therefore, before performing the factor analysis each variable should be spatially smoothed. For this purpose we used a technique known as the Global Empirical Bayes.[6] This technique adjusts the raw rate in each area, r_i, by taking into account information of the rest of the sample. The adjustment procedure operates in such a way that the raw rates that come from small populations are moved towards the overall mean, while the raw rates that come from large populations remain unchanged. For each area i, the new estimated rate, , is calculated as follows:

$$\hat{\theta}_i = w_i r_i + (1 - w_i)\mu \tag{12.3}$$

Where:

$$w_i = \frac{\sigma}{\sigma + \frac{\mu}{n_i}} \tag{12.4}$$

$$\mu = \frac{\sum_{i=1}^{n} r_i n_i}{\sum_{i=1}^{n} n_i} \tag{12.5}$$

$$\sigma = \frac{\sum_{i=1}^{n} n_i (r_i - \mu)^2}{\sum_{i=1}^{n} n_i} - \frac{\mu}{\sum_{i=1}^{n} \frac{n_i}{N}} \tag{12.6}$$

Although there exists a wide variety of smoothing methods, we preferred the Bayesian approach because (1) it is widely accepted and applied; and (2) correcting rates towards the overall mean, instead to a local mean, does not generates an artificial increase in the level of spatial autocorrelation.

Figure 12.3 shows a summary of the steps for designing analytical regions for studying poverty. These new regions have the following properties:

- They are homogeneous in terms of the relevant variables, which minimizes the aggregation bias (Fotheringham and Wong 1991).
- Each region has at least 100 surveyed households, which solves problems such as the small numbers problem (Diehr 1984), and ensures that the statistical secrecy will not be violated. Moreover, it reduces the impact of geocoding inaccuracies.

---

[6] See Anselin et al. (2006) for a summary on rate transformation techniques.

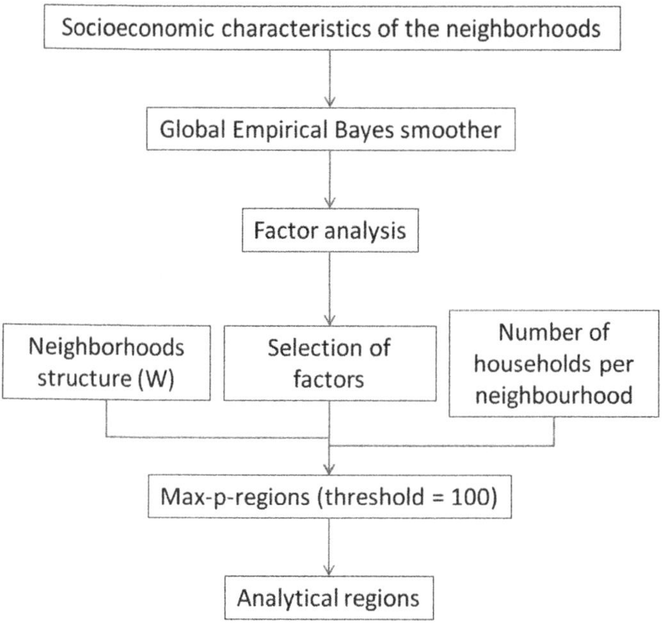

Fig. 12.3 Steps for designing analytical regions

- As pointed out by Weeks et al. (2006), the use of these spatial aggregation techniques is helpful for controlling problems of spurious spatial autocorrelation.
- The number and shape of the regions is endogenously determined, which eliminates conflicts of subjectivity associated with the selection of the scale and aggregation of the data being analysed.

Going back to the case study, we performed a factor analysis on the spatially smoothed rates of the socioeconomic variables, presented in Table 12.1, measured at the neighbourhood level.[7] Table 12.2 presents the rotated factors loading of the first four extracted factors. These factors have eigen values greater than one (Kaiser criterion), and capture 63.20 % of the total variance (variance explained criteria), and they show a significant level of spatial autocorrelation (see last row in the table). The loadings also indicate the correlations between the variables and each factor. They are useful for understanding what aspect of poverty each factor is capturing. Thus, factors 1 and 2 are related to the characteristics of the houses, and factors 3 and 4 are related to characteristics of the members of the households

The spatial distribution of these factors is presented in Fig. 12.4. For each factor the higher the value, the more critical the conditions of the neighbourhood. The

[7] The variable "water" was excluded from the analysis because only two households reported absence of piped water.

Table 12.2 Rotated factor loadings of the principal components results

Variable	Factor 1	Factor 2	Factor 3	Factor 4
Sewer	0.0681	0.8599	0.0844	0.0442
Risk	0.3977	0.6642	0.1528	−0.087
Bedrooms	−0.6561	−0.0142	−0.4025	−0.0516
Fridge	0.7561	0.3573	0.162	−0.0462
Kitchen	0.7531	0.1433	0.1	0.0104
Violence	0.3809	0.0706	0.629	0.0085
Sisben	0.9034	0.1442	0.1947	0.088
Birth	0.0053	0.055	0.1048	0.776
Nutrition	0.234	0.0981	0.8117	0.0882
Illiteracy	0.3248	0.2069	0.0015	0.5397
Income	0.7036	−0.0171	−0.2626	0.3649
Material	0.1903	0.7288	0.0199	−0.0419
Toilet	0.1379	0.6922	−0.0654	0.2441
Ownership	0.0812	−0.2548	0.3424	0.4071
Overcrowding	0.8379	0.132	0.241	−0.0054
z-normalized Moran's I	15.87	3.12	3.88	5.92

distributions are according to our expectations. On the one hand, the factors 1 and 2, related to the characteristics of the houses, show their higher values in the peripheral areas of the city, which are characterized by steep hills and chaotic urban structures. On the other hand, factors 3 and 4, related to characteristics of the members of the households, show high values distributed as small pockets around the city.

All four factors, together with (a) the number of surveyed household per neighbourhood; (b) the neighbouring structure, represented as a queen-based spatial contiguity matrix; and (c) a requirement of at least 100 surveyed households per regions (minimum population threshold); are introduced to the Max-p-regions algorithm. The model aggregates 243 neighbourhoods into 139 analytical regions, both scales are presented in Fig. 12.5, and some basic statistics on the number of surveyed households are provided in Table 12.3. Because of the aggregation criterion of minimizing intra-regional heterogeneity, each region is a conciliatory approach towards the capture of the four different spatial distribution patterns.

12.4 Delineation of Marginal Areas

Once we have more suitable spatial units, the final step seeks to delimitate the intra-urban slum areas. Our analysis is based on Slum Index applied by the UN-Habitat. We also compare our results with those obtained when performing the analysis with two additional strategies: working with raw rates at neighbourhood level; and working with spatially smoothed rates at neighbourhood level.

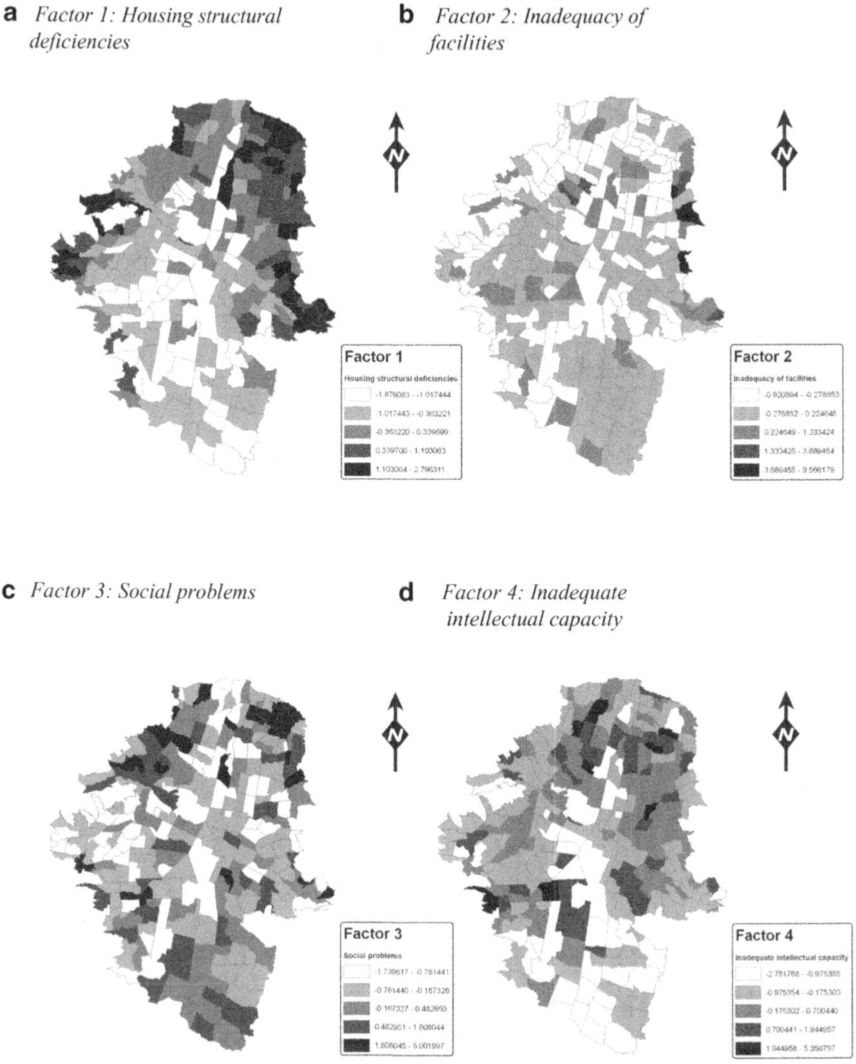

Fig. 12.4 Spatial distribution of the extracted factors

For a housing unit, the Slum Index can be computed as the sum of five dummy variables: (1) walls are not of durable material; (2) absence of piped water; (3) toilet not connected to a sewer; (4) residents are not the owners and (5) three or more persons per room. Each dummy variable takes value '1' if the characteristic is present in the housing unit. Thus, the Slum Index ranges from 0 to 5, being 5 the maximum level of slumness. In our database, no housing unit has a Slum Index of 5; only 0.24 % had an index with a figure of 3 or more; 7.38 % had an index equal to 2;

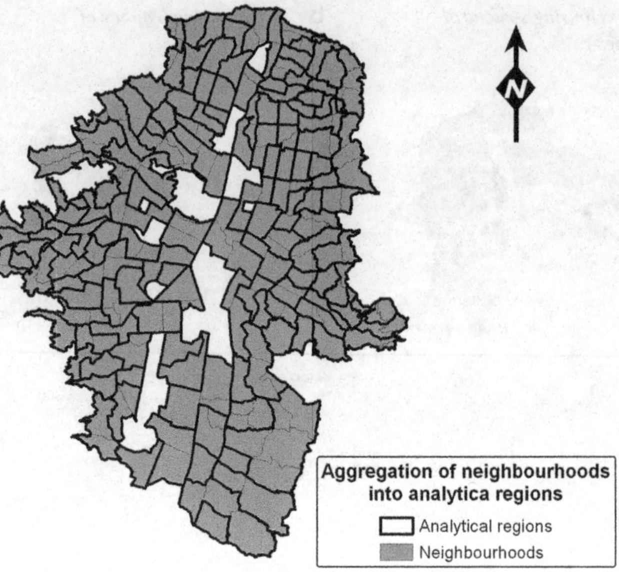

Fig. 12.5 Aggregation of neighborhoods into analytical regions obtained from the Max-p algorithm

Table 12.3 Statistic of number of surveyed households

Statistic	Neighbourhoods	Analytical regions
Minimum	3	101
Maximum	296	308
Mean	84.26	147.30
Std. Deviation	56.66	42.57

40.67 % equal to 1; and 50.71 % had an index equal to 0. The next step consisted in computing the Slum Index for each neighbourhood and for each analytical region. At each scale the Slum Index is calculated as the mean value for all the housing units in the same geographical unit (neighbourhood and analytical region). The choropleth map of the Slum Index is presented in Fig. 12.6a–c.

Finally, to find which areas have a significant level of poverty, defined in terms of the Slum Index, we use exploratory spatial data analysis to identify spatial clusters. Different methods are available for that purpose: Moran's I (Anselin 1994), G statistic (Getis and Ord 1992), AMOEBA algorithm (Aldstadt and Getis 2006; Duque et al. 2010), SaTScan (Kulldorff 1997). Following Burra et al. (2002), researchers should use at least two of the available methods in order to avoid type I errors. We have chosen to use two of the most widely applied statistic procedures of spatial clustering identification: local Moran's I and the local G statistic (Gi*).

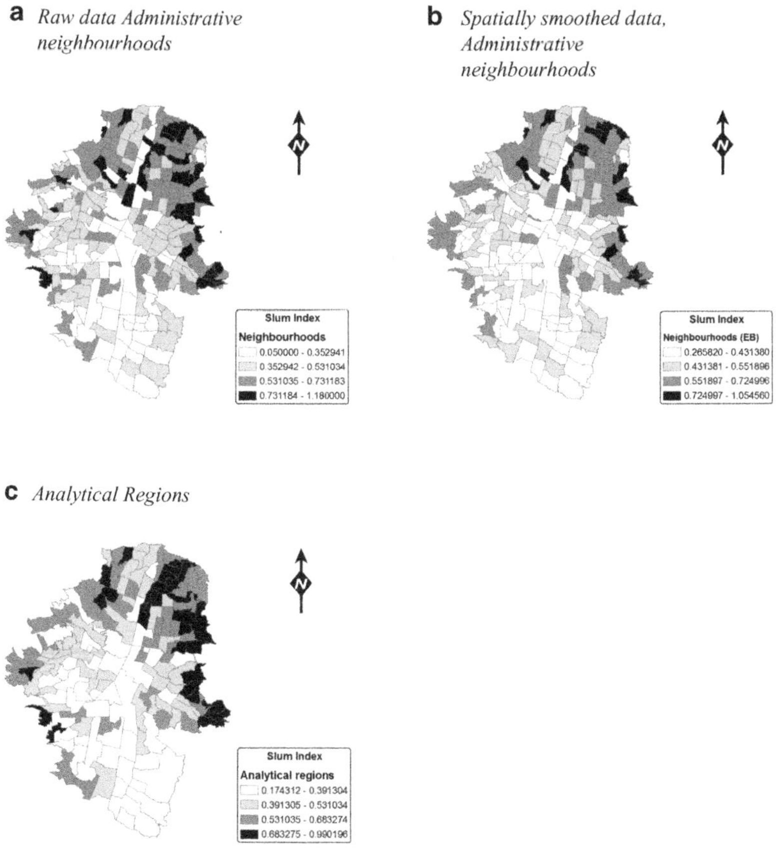

Fig. 12.6 Slum index in Medellin, 2007

Local Moran's I is computed using the following expression:

$$I_i = \frac{z_i}{\sum_i \frac{z_i^2}{N}} \sum_{j \in J_i} w_{ij} z_j \tag{12.7}$$

And the G statistic is computed as follows:

$$G_i^* = \frac{\sum_{j=1}^{N} w_{ij}(d) x_j}{\sum_{j=1}^{N} x_j} \tag{12.8}$$

Where xi and zi are, respectively, the raw and normalized values of the considered variable;wij are the elements of a W contact matrix (using a given distance, d, in the Gi* statistic), and N, the total amount of observations. The local Moran's I

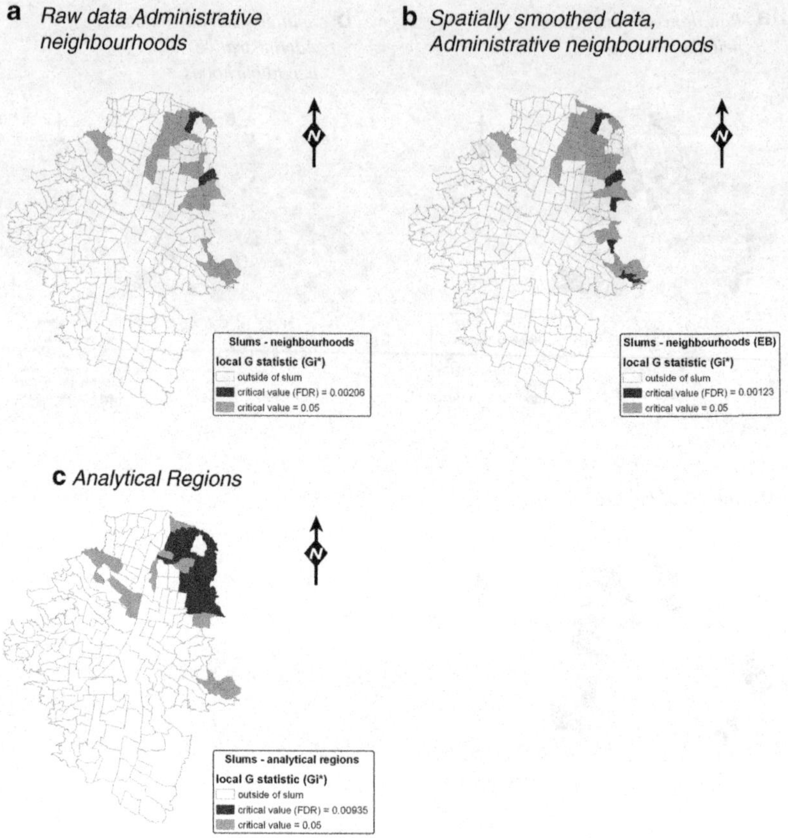

Fig. 12.7 Spatial clusters of the slum index in Medellin, 2007. Getis Gi* Clusters

will inform about the existence of a cluster of similar values, while the Gi* statistic informs about the spatial association. Negative values in the Moran's I statistic indicates if there is a negative association (high values surrounded by low values), while a negative Gi* statistic will tell us if a cluster is formed by similarly low values in the considered variable.

Figure 12.7a–c show the Gi* clusters and Fig. 12.8a–c show the local Moran's I clusters. We apply two strategies to test for the significance of the clusters: (1) conditional permutation based on 10,000 permutations to identify significant clusters at 5 % of pseudo-significance; and (2) the False Discovery Rate (FDR) technique, a conservative approach that was applied in the spatial context by Caldas de Castro and Singer (2006). The use of one criteria or another is a key issue, particularly from a policy perspective. When no control is adopted, false clusters are likely to be identified. If a conservative approach, like Bonferroni or Sidak, is adopted, clusters are only partially identified and true clusters may be missed. In contrast, when the FDR approach is used, clusters are fully identified. Consequently, if we apply no controls on the selection of true clusters, policy makers may be wasting human,

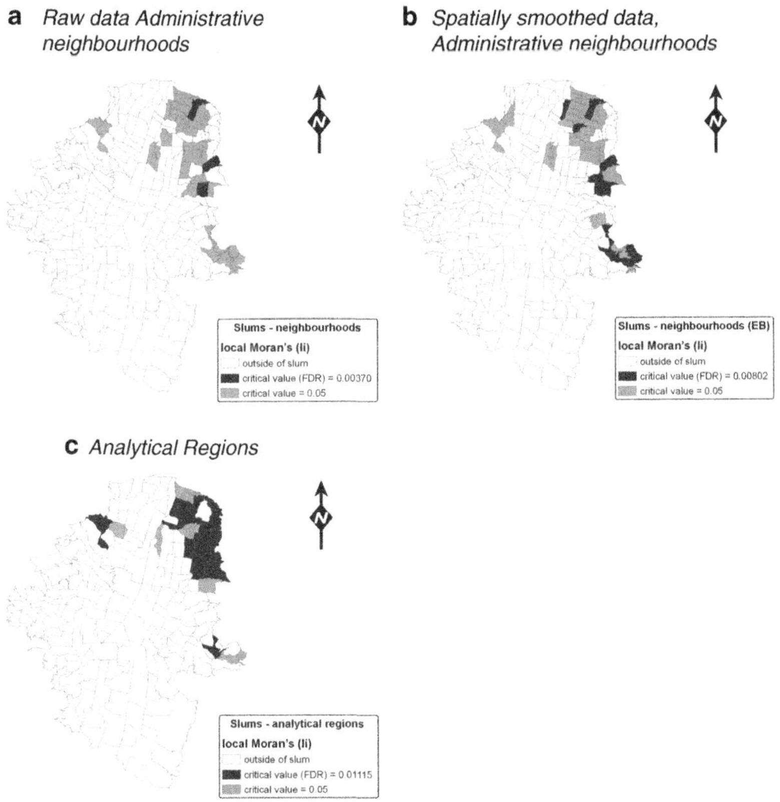

Fig. 12.8 Spatial clusters of the slum index in Medellin, 2007. Local Moran's I Clusters

physical and monetary resources in areas that are not part of the core of the problem. On the contrary, in a conservative approach is followed, it may even happen that no cluster were identified, casting doubts of the existence of a real problem.

Getis' statistic is about spatial concentration, while the Moran's I statistic is about spatial dependence. Both measurements are complementary, but are likely to result into different pictures. Spatial concentration tells us that data is significantly placed in particular areas, while spatial dependence informs about spatial relationships between regions. When a region is significant in any procedure, it will have to be considered by policy makers, either because it is an important part of the problem or because it is affecting or being affected by close regions.

Inspecting Figs. 12.7a–c and 12.8a–b, we see that using analytical regions we find significant clusters using both the local G statistic and the local Moran's I. On the contrary, if we use raw data of administrative neighbourhoods, it is hard to find significant clusters. When data is spatially smoothed, to avoid problems such as the small numbers problem, errors of geocoding, etc., we find small isolated clusters using both statistics.

Once we have found poverty-slum clusters, we believe that the spatial identification matters. We see in all maps that three spatial clusters can be identified. The more important one is located in the north-east. Interestingly, it is almost 100 % coincident in Gi* and local Moran's I results when working with analytical regions. It is worthy to note that only when using analytical regions the local Moran's I is capable to detect two critical slummy areas, one in the west and one in the east.

Having found several spatial clusters should lead policy makers to think on the different nature of poverty in these areas. Probably the causes of poverty may differ in space, and consequently different policies should be applied. Analysis and inspection of these causes would be the next step of any political action and, of course, it overpasses the scope of this chapter.

12.5 Conclusions

The analysis of intra urban socioeconomic differences is placed as a hot topic in economic geography. In this chapter we present a simple step-by-step procedure for identifying slum areas. We prove the goodness of using analytical aggregation methods as the efficient way to create spatial for studying socioeconomic phenomena. We use max-p-regions algorithm to build analytical regions that supersede the use of administrative neighbourhoods. These regions reach a proper balance between the robustness obtained by working with large regions, and maximum spatial detail, which avoids aggregation bias.

By using the false discovery rate procedure we improve the initial exploratory analysis of local statistics and identify the core areas in which the political actions should be applied in the first place. Following the identification of poverty further research should be focused on finding the sources of poverty, which can be different from cluster to cluster.

References

Aldstadt J, Getis A (2006) Using AMOEBA to create a spatial weights matrix and identify spatial clusters. Geog Anal 38(4):327–343

Amrhein CG, Flowerdew R (1992) The effect of data aggregation on a Poisson regression model of Canadian migration. Environ Plann A 24:1381–1391

Anselin L (1994) Local indicators of spatial association—LISA. Geogr Anal 27:93–115

Anselin L, Lozano N, Koschinsky J (2006). Rate transformations and smoothing. GeoDaCenter Research Report

Arbia G (1989) Spatial data configuration in statistical analysis of regional economic and related problems, vol 14. Kluwer Academic, The Netherlands

Berry BJL (1960) An inductive approach to the regionalization of economic development. In: Ginsburg N (ed) Essays on geography and economic development. University of Chicago Press, Chicago

Burra T, Jerrett M, Burnett RT, Anderson M (2002) Conceptual and practical issues in the detection of local disease clusters: a study of mortality in Hamilton, Ontario. Can Geogr/Le Géographe canadien 46(2):160–171
Calciu M (1996) Une méthode de classification sous contrainte de contiguïté en géo-marketing. Institut d'administration des entreprises, Université des sciences et technologies de Lille
Caldas de Castro M, Singer BH (2006) Controlling the false discovery rate: a new application to account for multiple and dependent tests in local statistics of spatial association. Geogr Anal 38:180–208
Caro F, Shirabe T, Guignard M, Weintraub A (2004) School redistricting: eembedding GIS tools with integer programming. J Oper Res Soc 55:836–849
Diehr P (1984) Small areas statistics – large statistical problems. Am J Public Health 74 (4):313–314
Duque JC, Artís M, Ramos R (2006) The ecological fallacy in a time series context: evidence from Spanish regional unemployment rates. J Geogr Syst 8:391–410
Duque JC, Ramos R, Surinach J (2007) Supervised regionalization methods: a survey. Int Reg Sci Rev 3:195–220
Duque JC, Aldstadt J, Velasquez E, Franco JL, Betancourt A (2010) A computationally efficient method for delineating irregularly shaped spatial clusters. J Geogr Syst. DOI: 10.1007/s10109-010-0137-1 13(4):355–372
Duque JC, Dev B, Betancourt A, Franco JL (2011) ClusterPy: library of spatially constrained clustering algorithms, Version 0.9.9. RiSE-group (Research in Spatial Economics), EAFIT University
Duque JC, Anselin L, Rey SJ (2012) The max-p-regions problem. J Reg Sci 52(3):397–419
Eurostat (2006) Nomenclature of territorial units for statistics—NUTS: statistical regions of Europe. http://ec.europa.eu/comm/eurostat/ramon/nuts/basicnuts_regions_en.html. Accessed 19 June 2006
Fischer M (1980) Regional taxonomy. A comparison of some hierarchic and non-hierarchic strategies. Reg Sci Urban Econ 10:503–537
Fotheringham AS, Wong DWS (1991) The modifiable areal unit problem in multivariate statistical-analysis. Environ Plann A 23:1025–1044
Getis A, Ord JK (1992) The analysis of spatial association by distance statistics. Geogr Anal 24:189–206
Goovaerts P (2009) Medical geography: a promising field of application for geostatistics. Math Geosci 41(3):243–264
Johnston RJ (1968) Choice in classification: the subjectivity of objective methods. Ann AAG 58:575–589
Kulldorff M (1997) A spatial scan statistic. Commun Stat Theory Method 26:1487–1496
Levin SA (1992) The problem of pattern and scale in ecology. Ecology 73(6):1943–1967
Openshaw S (1977a) A geographical solution to scale and aggregation problems in region-building, partitioning and spatial modelling. Trans Inst Br Geogr 2:459–472
Openshaw S (1977b) Optimal zoning systems for spatial interaction models. Environ Plann A 9:169–184
Openshaw S (1984) The modifiable areal unit problem, vol 38, Concepts and techniques in modern geography. GeoBooks, Norwich
Openshaw S (1995) Census users handbook. GeoInformation International, Cambridge
Openshaw S, Taylor PJ (1981) The modifiable areal unit problem. In: Wrigley N, Bennett RJ (eds) Quantitative geography. Routledge, London, pp 60–70
Paelinck JHP (2000) On aggregation in spatial econometric modeling. J Geogr Syst 2:157–165
Paelinck JHP, Klaassen H (1979) Spatial econometric. Saxon House, Farnborough
Robinson WS (1950) Ecological correlations and the behaviour of individuals. Am Sociol Rev 15:351–357

Royuela V, Suriñach J, Artís M (2008) La influencia de la calidad de vida en el crecimiento urbano. El caso de la provincia de Barcelona, Investigaciones Regionales 13:57–84
Segal M, Weinberger DB (1977) Turfing. Oper Res 25:367–386
Spence NA (1968) A multifactor uniform regionalization of British counties on basis of employment data for 1961. Reg Stud 2:87–104
Vélez CE, Castaño E, Deutsch R (1998) An economic interpretation of colombia's SISBEN: a composite welfare index derived from the optimal scaling algorithm. Mimeo, Poverty and Inequality Advisory Unit. Inter American Development Bank, Washington
Weeks JR, Hill A, Stow D, Getis A, Fugate D (2006) The impact of neighborhood structure on health inequalities in Accra, Ghana. In: Population association of America 2006 annual meeting program, Los Angeles, USA, pp 1–34
Yamada T (2009) A mini-max spanning forest approach to the political districting problem. Int J Syst Sci 40(5):471–477

Chapter 13
Capturing Spatial Clusters of Activity in the Spanish Mediterranean Axis

Fernando A. López, Ana Angulo, and Andres Artal

13.1 Introduction

Concentration of economic activity constitutes a stylized fact of social sciences, giving birth to a very fertile branch of research since the writings of Marshall (1890) until today. Many researchers have been working on how to measure and explain concentration patterns of activity, trying to face that challenge and transpose it to a tractable argument in terms of modeling. Cluster analysis is one of the most salient efforts in this direction, with different contributions defining measures of concentration that range from the simplest indexes of inequality of Theil and Gini (Krugman 1992), until the more elaborated measures due to the recent work of Henderson (Henderson 1974, 1988; Henderson and Venables 2009), and Ellison and Glaeser (1997). In general, advances in this literature have focused on refining the construction of concentration indexes for identifying clusters of employment or firms in a certain territory. For example, the pioneer work of Ellison and Glaeser (1997) developed an agglomeration index (EG), together with a co-location one, that has been generalized in the literature as a reference. Its main contribution is that

Fernando A. López and Ana Angulo wish to thank financial support from project ECO2009-10534/ECON of the Ministerio de Ciencia y Tecnología of the Reino de España; Andrés Artal-Tur wish to thank financial support from the Spanish Ministry of Science and Innovation (Project ECO2011-27169).

F.A. López (✉)
Dpto de Métodos Cuantitativos e Informáticos, Universidad Politécnica de Cartagena, C\Real, 3, Cartagena 30201, España
e-mail: fernando.lopez@upct.es

A. Angulo
Dpto. de Análisis Económico, Universidad de Zaragoza, Gran Vía, 2-4, Zaragoza 50005, España

A. Artal
Dpto. Economía, Universidad Politécnica de Cartagena, C\Real, 3, Cartagena 30201, España

E. Fernández Vázquez and F. Rubiera Morollón (eds.), *Defining the Spatial Scale in Modern Regional Analysis*, Advances in Spatial Science,
DOI 10.1007/978-3-642-31994-5_13, © Springer-Verlag Berlin Heidelberg 2012

it derives from an explicit theory of firm location behaviour (the random-dartboard approach), controls for differences in the size distribution of establishments among industries, and appears to be robust to the level of spatial aggregation at which industry data are available. Other novel studies in this direction as Feser and Bergman (2000) test if the EG index is sensitive to the scale of data employed (at the level of counties, commuting sheds, and zip codes), showing that changes in the spatial scale of data can introduce non-trivial ambiguities in the usual application of the EG index. Because of that, they recommend considerable caution when employing the index in comparative space-time studies about the concentration of industries. Braunerhjelmy and Johansson (2003) employ the EG and Gini locational indexes to evaluate the degree of concentration in 143 industries (at a four-digit level) for Sweden between 1975 and 1993, while Midelfart-Knarvik et al. (2004) use Gini locational index to analyze 36 industrial activities and 5 of services, with both works showing a more disperse pattern for services in comparison with industries. In addition, other locational studies also try to disentangle the forces driving important international flows such as FDI, population, or migrants (see, i.e., Blonigen et al. 2008; Baltagi et al. 2005; Kaushal 2005).

Alternatively to those initial studies employing basic or more elaborated statistics for concentration, new methodologies haven been appearing building on geo-referenced data and introducing new issues and methods of analysis taken from the spatial econometrics literature.[1] Usual contributions in this new strand of the literature take now explicit account of the impact of the geographical dimension of data in the decisions of agents, reintroducing the concept of distance while measuring activity clusters. Those contributions also coincide in that they borrow some techniques of analysis traditionally employed in other disciplines, as ecology, biology or epidemiology, in order to quantify the degree of dispersion shown by spatial data in comparison with purely random distributions. And equally they build on the concept of spatial dependence that used to characterize spatial processes. In general, this approach employs geo-referenced data in the analysis of concentration, and relies in a definition of space as a non-homogeneous variable, in contrast to pioneer studies, now dealing with the overall spatial distribution of the variable of interest, thus automatically taking into account the many inhomogeneities that geography introduce in the location decisions observed in reality (Jensen and Michel 2011).

Duranton and Overman (2005) and Marcon and Puech (2010) contributions become a good and up-to-date example of this new geographical statistics approach. They develop a methodology for measuring spatial concentration or dispersion sharing a number of advantages, two of the most important being that allow for treating space as a inhomogeneous underlying space, and employ precise location geo-data (latitude and longitude coordinates). Their main difference is the integrative or differential approach they use given that while first authors focus on scrutinizing the distribution between two distances r and r + δr,

[1] See, for example, the critique made by Arbia (2001) to previous contributions on cluster analysis.

second group integrate the distribution from 0 to r. In that way meanwhile the differential coefficient allows for zooming on precise distances, and measure differences from randomness in more detail, Marcon and Puech's proposal is simpler to interpret because the coefficient converges to 1 as r approaches the system size, thus allowing to readily quantify deviations from randomness. Other contributions include that of De Dominicis et al. (2006), whom study the degree of concentration characterising 24 industries and 17 services activities in Italy for three levels of spatial aggregation (Local Labour Areas, NUTS-3 and NUTS-2 classifications). After employing EG and Moran's I indexes for capturing cluster formation, results of the investigation show again more concentration for industries versus services activities, concluding that new methods of local and global dependence analysis are necessary to complement previous measures in cluster analysis. Also Brülhart and Traeger (2005) introduce an entropy measure for evaluating spatial concentration of employment in eight sectors for 236 European regions between 1975 and 2000. The authors find no relevant changes in concentration of employment, neither in industries or services along the period of analysis, with industry appearing to be more concentrated in space than services. Desmet and Fafchamps (2006) by their part apply sigma and beta convergence indexes to compare 13 sectors in US counties along 1970–2000 years. Kang (2010) introduce space-time methods in order to measure the degree of spatial concentration of industries and their evolution in time by applying usual methods taken from epidemiology, i.e. the K-functions, as an alternative to classical indexes for spatial data analysis.

As a summary, it has been obvious that important efforts have been devoted to obtain good measures of concentration patterns in regional and urban studies, given that achieving a clear identification of clusters still remains an issue. But one important problem still present in this empirical literature is the one referring to limitations in official data building, particularly those due to administrative classifications characterising official registers of activity. Traditional studies on cluster analysis used to employ official data on employment levels, number of enterprises, or business volume (sales, production levels), which ultimately is compiled for geographical units such as the county, municipality, region or country, all of them being administratively-defined classifications of space. In this way, exercises directed to identify the existence of concentration levels are, consequently, confined to the limitations arising from data compilation. Clusters of activity used not to be confined to administrative barriers, spilling over spatial boundaries and extending in many (if not all) cases by the neighboring areas to that of reference. In traditional studies, agglomeration economies and other externalities has been usually approached as local in nature, and then confined to local boundaries such as counties and municipalities, but new contributions of the spatial econometrics literature are increasingly detecting that this seems not to be the case. Moreover, agglomeration effects arising from the activity in nearby localities seem to clearly enter the decision set of companies and people in their location choices.

Alamá et al. (2011) provide an interesting approach to such an issue, concluding that inter-municipal spillovers are a major force driving firms' choices for the case of the Region of Murcia in Spain. The study shows that agglomeration economies arising from nearby territories are nearly (70 %) as important for companies' location choices as those arising in the own locality finally chosen by the firm. Artal-Tur et al. (2012), by refining the previous methodology of analysis, and controlling for other sources of firm heterogeneity, wonder if the socio-economic characteristics of the neighbouring municipalities are important variables for a company when deciding to locate a plant. Employing a sample of more than one million of new created enterprises between 1998 and 2008 for Spain, they find that companies are so aware of characteristics of nearby municipalities when choosing their single location, with all of these spatial effects entering as first-order factors in their decision sets. All of these results show that a good clustering measure should be capable of dealing with spatial units that overlap one another. In this way, accuracy of cluster identification requires the employed method of analysis to be able of detecting which subset of spatial units in the sample, and what is more important, which part of everyone of them, would be included in the cluster, one feature of location analysis that spatial econometric techniques are increasingly allowing for.

As its main objective, the present study continues abounding in the identification of clusters and concentration patterns of activity along the proposals of the spatial econometrics literature. In doing so, we apply a methodology borrowed from epidemiology studies, namely the Kulldorff Scantest, to improve the detection and identification of clusters. The method affords the use of Scan windows analysis, where the researcher can confront space as a continuous dimension, just looking around through that lens until finding unusual concentration of activity in a particular place of the territory, what we will call a cluster. In this manner, the dataset do not confront the problem of administrative barriers faced by traditional studies, given that it employs firm-level geo-referenced data. Another advantage versus traditional concentration indexes is that the researcher can also define the shape and size of the scanwindow to be employed, depending on the variable that one wants to analyze and the spatial dimension endorsed (regions, countries, localities). In this way the Ku-Scantest, allows the researcher to define cluster areas closer to what we observe in reality, avoiding administrative barriers or constrains in its definition, with externalities spilling over from dense urban centers through metropolitan areas and nearby localities. Finally, the Ku-Scantest also allows for testing the significance of the defined cluster area, what constitutes an important novelty in cluster analysis, traditionally based in non-testable statistical concentration measures.

After this introduction, the rest of the chapter is organized as follows: In Sect. 13.2 we concentrate on defining the Scantest of Kulldorff. In Sect. 13.3 we apply this new tool to cluster analysis for the Mediterranean axis of Spain. In Sect. 13.4 we determine the main factors explaining the probability of a cluster to be defined as of high or low incidence. Finally, in Sect. 13.5 we conclude.

13.2 Applying the Scantest of Kulldorff for Cluster Analysis

As mentioned in previous section, several indicators have been defined in the literature to evaluate the degree of concentration of the economic activity. Most of them are indices that do not consider the space in its definition. Therefore, to cope with our objective, in this chapter we use the Scantest of Kulldorff to deal with such an issue. In this section we describe the main aspects of this methodology.

13.2.1 The Scantest of Kulldorff: Methodological Issues

In the spirit of the Ripley's K function (1976, 1977) retrieved by Duranton and Overman (2005), that defines a spatial concentration measure based on point distribution analysis taken from ecological and epidemiological studies, the Kulldorff's Scan test (Kulldorff and Nagarwalla 1995; Kulldorff 1997) has been designed to identify clusters in space where data shows different behaviour than the average observation in the sample. To test for the existence of such concentration pattern, the Ku-test proceeds by scanning the spatial distribution of the analysed variable (number of industries, employees, people, etc.) in a map imposing "windows" of different size and shape in order to compare how these sets of observations behave within and outside the defined window. Once has fixed the window in a position that renders the maximum difference between values inside and outside it for the variable of reference, then identifying a cluster in the map, the next stage is to evaluate whether that difference appears to be or not significant by applying Monte Carlo simulation methods. Further, the process is repeated in search for secondary clusters, always taking care of excluding those previously identified. Finally, the statistic allows for identifying high and low incidence clusters, defined as those exhibiting concentration values well above or below the average sample value along the area of study.

13.2.2 Notation and Building of the Test

Being N the total population observed in the region under analysis G. The total population as the sum of populations (N_i, $i = 1,\ldots,R$) in each of the R geographical units of analysis (postal codes, municipalities, . . .). Similarly, we will use n and ni ($i = 1,\ldots,R$) to denote total number of cases (employments) in the same branch of activity and in each region, respectively.

To elaborate the testing hypothesis, we assume that the distribution of the number of employments for the location i in sector J (denoted by X_i) follows a binomial distribution (N_i,p_i), with p_i being the probability of finding an employment of sector J in location i. Note that the expected number of employments of sector J from location i is $N_i p_i$.

The joint distribution ($X_1,\ldots,X_R$) follows a multinomial (N,p) with $p = (p_1,\ldots,p_R)$.

We suppose that, under the null hypothesis, H_0, $p_i = \frac{n}{N}$ ($\forall i$) while under the alternative H_1 we suppose that there is a subgroup of contiguous locations Z included in G where the probability of finding an industry of sector J is higher than (resp. lower than) the probability outside Z.

Formally, we can write,

H_0: $(X_1,\ldots,X_R) \sim$ Multinomial (N,p) with $p_i = p$
H_1: $(X_1,\ldots,X_R) \sim$ Multinomial (N,p) with $p_i = p$ $(i \in Z)$ and $p_i = q$ $(i \notin Z)$ being $p > q$ (resp $p < q$)

Under H_0 the maximum likelihood estimator of p is n/N. With n being the number of employees of sector J in area Z and N being the total number of employees in Z space. We can estimate the probabilities p and q under H_1 as:

$$p = \frac{n_Z}{N_Z}; \quad q = \frac{n - n_Z}{N - N_Z} \tag{13.1}$$

The likelihood function of the distribution under H_0 is:

$$L_{H_0} = \frac{R!}{n_1!\ldots n_R!} p_1^{n_1} \ldots p_R^{n_R} = \frac{R!}{n_1!\ldots n_R!} \left(\frac{n}{N}\right)^n \tag{13.2}$$

while under the alternative H_1,

$$L_{H_1} = \frac{R!}{n_1!\ldots n_R!} p_1^{n_1} \ldots p_R^{n_R} = \frac{R!}{n_1!\ldots n_R!} \left(\frac{n_z}{N_z}\right)^{n_z} \left(\frac{n - n_z}{N - N_z}\right)^{n-n_z} \tag{13.3}$$

The Likelihood ratio under the null and alternative hypothesis can be expressed as:

$$\lambda_Z = \frac{L_{H_1}}{L_{H_0}} = \frac{\left(\frac{n_Z}{N_Z}\right)^{n_Z} \left(\frac{n-n_Z}{N-N_Z}\right)^{n-n_Z}}{\left(\frac{n}{N}\right)^n} = \left(\frac{n_Z}{E_Z}\right)^{n_Z} \left(\frac{n - n_Z}{N - E_Z}\right)^{n-n_Z} \tag{13.4}$$

where E_Z denotes the number of expected observations (firms, employees, people, etc.) in the area Z under H_0 ($E_{H_0}[X_z] = \frac{N_z n}{N}$). The Kulldorff statistic is then scanning through the whole area under study in search of the window Z that maximizes the likelihood ratio for the point spatial distribution in the sample. Therefore, the final expression of the statistic is as follows:

$$Ku_s = \sup_{Z \in \Theta} \lambda_Z = \sup_{Z \in \Theta} \left(\frac{n_Z}{E_Z}\right)^{n_Z} \left(\frac{n - n_Z}{n - E_Z}\right)^{n-n_Z} I\left(\frac{n_Z}{E_Z} > \frac{n - n_Z}{n - E_Z}\right) \tag{13.5}$$

being I an indicator function. If the objective is to look for a low-incidence cluster it is necessary to exchange the symbol (>) by (<) in the argument of the function l, while if we wish to identify low and high incidence clusters, the indicator I can be eliminated from the likelihood ratio.

Note that the hypotheses tested by this likelihood ratio can also be expressed as:

$$H_0 : E[X_Z] = \frac{N_Z n}{N} \quad \text{versus} \quad H_1 : E[X_Z] > \frac{N_Z n}{N} \tag{13.6}$$

The way we define the windows denoted by Θ determines the value of the Ku-statistic obtained. It is advisable that the number of cases in the windows is to be limited until it accounts for a maximum of 50 % of total employments. The area Z* that maximizes the likelihood ratio is the Most Likely Cluster (MLC). If the MLC is significant, the process can be repeated looking for secondary clusters that do not overlap the MLC, covering by this way all the space available in the sample. The significance of the most likely cluster is obtained by Monte Carlo Techniques (Dwass 1957). The procedure is implemented in the free software SatScan (http://www.satscan.org).

13.3 Identifying Clusters of Employment in the Spanish Mediterranean Axis

In this section we present the results of applying Scan test to our data set comprised by all municipalities located in the Spanish Mediterranean Axis (SMA).

13.3.1 Spanish Mediterranean Axis: Environment and Data Set

The chosen geographical space comprises the majority of industrial and services establishments of Spain, accounting for more than 40.9 % of total population of Spain (approximately 19 million inhabitants) (3.8 % of EU-27), 18.9 % of the surface area of Spain (2.2 % of EU-27), and with a relative GDP of 40.6 % of that of Spain (3.7 % of EU-27). Figure 13.1 show the SMA area with the subdivisions in eight administrative provinces (Nuts III in Eurostat terminology). The SMA as a geographical unit concentrates more than 40 % of Spanish population and economic activity in less than 20 % of the country territory, so this has been a good reason for deciding to observe the existence of significant patterns of concentration in this area of the country. Through the last decades the area has registered an important demographic growth (boosted by migratory flows), resulting in high population densities, particularly on the seaboard. Other distinctive characteristics of the SMA include a strong specialization in tourism and leisure related activities which exploit environmental advantages (climate, landscape, etc.). Moreover, the manufacturing sector rests on SMEs mainly concentrated on traditional activities. A detailed analysis of the spatial configuration of the SMA reveals the existence of two territorial imbalances, which are in turn reinforced by a still deficient articulation of transport infrastructures. First, there is a remarkable contrast between the active and densely populated seaboard and the rather inhabited inlands. Second,

Fig. 13.1 Spanish mediterranean axis

there exists a discontinuity in the urban network. Southwards, it appears the urban agglomeration of Barcelona, next to the metropolitan area of Valencia, and a set of coastal cities from Benidorm (Alicante) to Cartagena (Murcia). At this point takes place a marked decline of urbanized areas in the extension of the SMA to Andalusia. Summarizing, we can state that in the whole area under study two cities, Barcelona and Valencia, make the difference in terms of global connectivity, both of them becoming well consolidated urban structures connected to the rest of Europe, one location and specialization pattern that we expect to be captured by the following cluster analysis.

Our data set include total employment for two-digit NACE data, taken from SABI database ('Sistema de Análisis de Balances Ibéricos'), that allow collecting geo-referenced data for industrial and services establishments at the level of municipalities in the SMA. Our sample amounts for 3,358,902 employees distributed along 1,446 municipalities. In running such exercise, we have defined several groups of control according to the classification made by Eurostat[2] by considering the different level of technology and knowledge insensitiveness of the industrial and service activities, assuming the following six categories: low-technology level, low&medium-tech level, medium&high-tech level and high-tech level for industries, together with knowledge-intensive and non-knowledge intensive levels for services activities.

13.3.2 Cluster Analysis when Applying the Scantest of Kulldorff

Now we apply the Kulldorff Scantest (Ku-Scantest) to identify emerging clusters in our geo-data set, constructing separate maps for every group of industries

[2] http://epp.eurostat.ec.europa.eu/cache/ITY_SDDS/Annexes/htec_esms_an2.pdf

Table 13.1 Spatial clusters in industry sector by level of technology

Level tech		Id	# Clust	Ku	p-value	Obser.	Expec. H_0	Obs/Exp
Low	HI	1	138	27,906.6	0.001	77,016	29,546.0	2.61
		2	149	8,424.1	0.001	26,624	10,873.2	2.45
		5	81	4,076.7	0.001	11,889	4,543.9	2.62
		8	3	1,974.1	0.001	4,617	1,555.3	2.97
		9	24	1,140.3	0.001	20,428	13,629.8	1.50
	LI	3	4	4,876.1	0.001	6,958	18,439.1	0.38
		4	2	4,293.2	0.001	2,461	10,167.6	0.24
		6	29	2,625.5	0.001	16,865	28,069.1	0.60
		7	178	2,563.8	0.001	24,343	38,887.8	0.63
Low-medium	HI	1	77	21,477.3	0.001	35,479	9,303.8	3.81
		2	38	8,325.5	0.001	26,986	11,185.1	2.41
		3	236	4,331.4	0.001	37,252	22,632.0	1.65
		4	142	4,145.5	0.001	31,898	17,985.3	1.77
	LI	5	5	4,094.0	0.001	3,725	12,049.0	0.31
		6	42	2,917.5	0.001	12,932	24,600.2	0.53
		7	5	2,034.5	0.001	5,960	13,405.8	0.44
Medium-high	HI	1	101	19,808.8	0.001	54,581	22,018.1	2.48
	LI	2	50	7,809.4	0.001	6,556	21,637.1	0.30
		3	143	7,785.2	0.001	5,170	19,862.1	0.26
High	HI	1	51	3,935.6	0.001	6,629	1,973.7	3.36
		2	3	3,281.6	0.001	1,652	66.3	24.91
	LI	3	166	1,324.3	0.001	266	2,050.0	0.13
		4	318	730.5	0.001	591	1,986.7	0.30
		5	57	262.5	0.001	1,141	2,050.1	0.56

Id cluster identification, *#Clust* number of municipality inside of cluster, *Ku* value of Kulldorff statistic, *Obser* number of employment inside of cluster, *Expec H_0* expected number of employment inside of cluster under null, *Obs/Exp* ratio observed/expected. If >1 →High incidence; If <1 →Low incidence cluster

previously defined. We also run Monte Carlo simulations to assess the significance of every emerging cluster in the map, with a final measure that allows us to label the cluster as of high or low incidence. Our results show that Low and Low&Medium-tech industries show important concentration patterns in the Mediterranean Axis of Spain (see Table 13.1). Clusters appear to be all significant, although combining high and low incidence behaviour.

High-incidence clusters for low-tech industries emerge in this area around the north of Barcelona province and the south of Girona, including 149 municipalities (Fig. 13.2a). Its main specialisation relies on food and beverages products and metallic industries. Southwards, we find another high-incidence cluster covering the interior of Valencia and Alicante provinces (138 municipalities), with important specialisation in traditional light industries such as textile and home apparels. Both clusters show high significance levels from simulated distributions of the test and the number of observed employment concentration levels over expected in the null hypothesis H_0 (named as Obs/Exp) show clear concentration patterns above the expected level of employees in the identified areas. Low incidence clusters appear

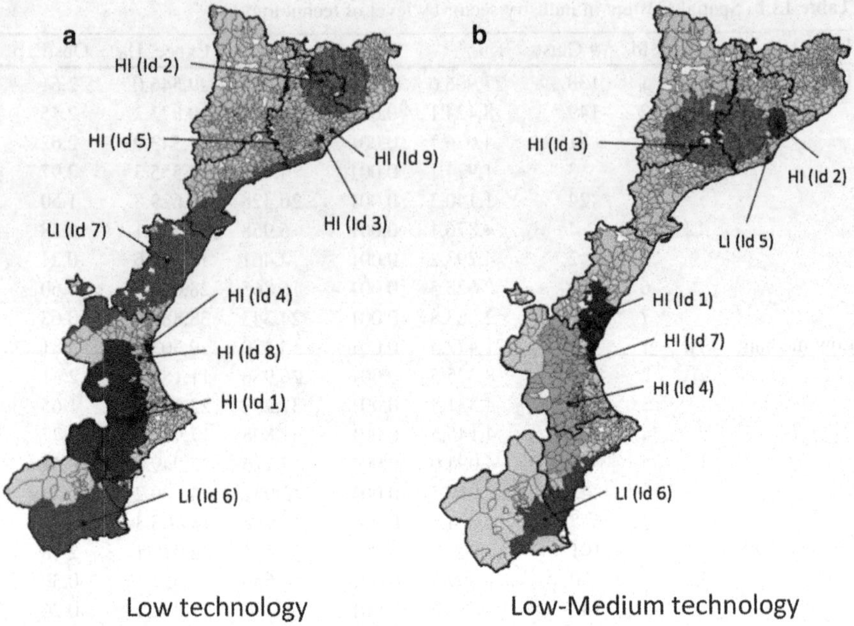

Fig. 13.2 Spatial cluster maps in low and low-medium technology

for low-tech industries around the south of Castellón province and the north of Valencia, with low-tech employment showing levels by far below the expected values. The cluster covering the south of Alicante and the whole coast of Murcia also is revealed through the Ku-test as a low-incidence one in terms of the concentration of industrial employment we observe in data.

According to the Low&Medium-tech industries, the results of the test shows that they concentrate significantly around the boundaries of Barcelona with Tarragona and Lleida provinces all in the centre of Catalonia region, forming a great cluster specialised in oil refining industries, chemical, textile and automobile activities (cluster Id2 and Id3 in Fig. 13.2b). It is interesting to note how the Ku-Scantest allows for mapping clusters in space that do not cover a whole administrative unit of analysis, a municipality in our exercise, this being one of its main advantages for cluster identification. In general, it allows overcoming administrative boundaries in data that constrain identification of clusters in previous studies employing traditional measures of concentration. Low&Medium-tech clusters are also found in the south of Castellón (tile industry) and the interior localities of Valencia province (traditional light industries as textile, wood, leather, etc.).

Regarding Medium&High (M&H) and High-tech industries, Ku-test results show important cluster formation close to the city of Barcelona and its metropolitan area and next to the city of Valencia (Fig. 13.3 and Table 13.2). It includes important concentration of chemical-related companies for Barcelona, together with computer, optical and medical-instrumental industries, and

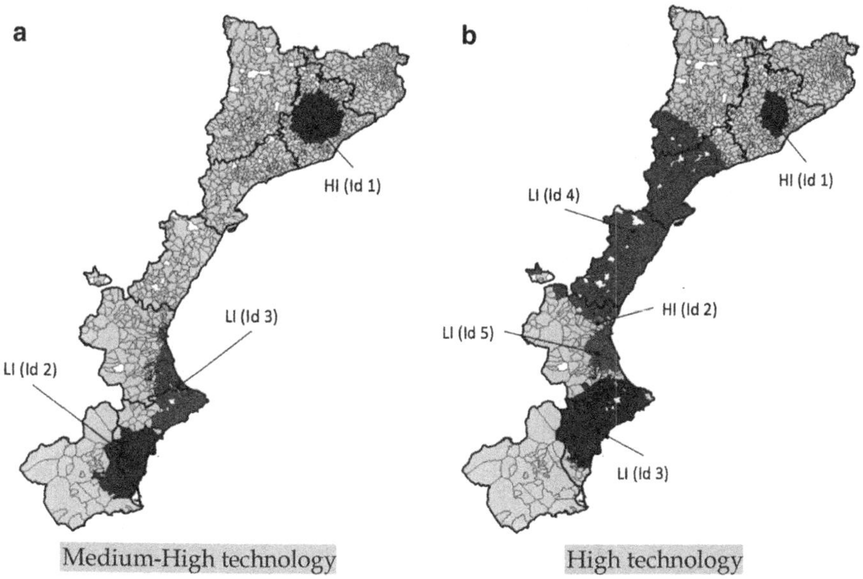

Fig. 13.3 Spatial cluster maps in medium-high and high technology

machinery-manufacturing companies in the urban areas of Valencia and Barcelona. Significance of high-incidence clusters are remarkable in the case of HI(Id1) in Fig. 13.3b, for high-tech industries close to the city of Barcelona, and particularly for the HI (Id2) cluster in Fig. 13.3b next to Valencia city, both including establishments of industries in sectors like chemical, computing, optical, machinery, and medical instruments.

Low-incidence clusters for M&H and High-tech industrial employment appear to be centered in all the territory ranging from Tarragona and south of Lleida in Catalonia, until the south of Alicante, with the mere interruption of Valencia city (cluster Id2 in Fig. 13.3b). In all of this area, observed values of High-tech industrial employment are significantly well below those expected by H_0, what again is in line with low presence of high-tech industries in this Mediterranean axis, mainly specialised in light traditional manufactures and services with low-and-medium knowledge content as we will see in the following paragraph.

In what refers to identification of clusters for services activities, our Cluster Maps (see Fig. 13.4a, b) show that non-knowledge intensive services mainly concentrate around the cities and regions of Barcelona HI (Id6), Tarragona HI (Id12), Castellón HI (Id12), south of Alicante HI (Id14) and Murcia HI (Id14 and Id8) in Fig. 13.4a. These clusters services include services for primary activities, services for the construction and real estate activities, and other manual non-knowledge intensive services. Regarding the main areas where knowledge-intensive services seem to cluster in the SMA we observe, strictly speaking, the relevance of activities established in the city of Barcelona (a red spot located between LI (Id4) and LI (Id2) clusters) and the city of Valencia (see areas determined by LI (Id1) and (Id6)).

Table 13.2 Spatial cluster maps in service sector by level of technology

Level tech		Id	# Clust	Ku	p-value	Obser.	Expec. H_0	Obs/Exp
Non-knowledge intensive	HI	2	111	3,488.9	0.001	65,197	46,060.4	1.42
		3	1	2,482.2	0.001	14,013	7,261.9	1.93
		5	2	1,843.0	0.001	5,224	1,971.6	2.65
		6	10	1,830.4	0.001	37,529	27,132.3	1.38
		8	10	1,326.1	0.001	14,674	9,263.7	1.58
		11	8	425.5	0.001	17,378	13,834.2	1.26
		12	109	355.7	0.001	31,677	26,725.8	1.19
		14	45	332.1	0.001	53,495	47,724.4	1.12
	LI	1	106	3,840.6	0.001	77,561	104,066.2	0.75
		4	59	1,996.4	0.001	25,300	36,523.2	0.69
		7	114	1,348.2	0.001	17,782	25,796.5	0.69
		9	1	805.3	0.001	2,246	4,713.2	0.48
		10	10	623.2	0.001	1,500	3,583.2	0.42
		13	17	353.4	0.001	7,203	9,717.1	0.74
Knowledge intensive	HI	1	1	23271.2	0.001	45,490	13,852.5	3.28
		6	1	3090.7	0.001	42,595	28,686.2	1.48
	LI	2	191	15567.3	0.001	24,443	61,817.6	0.40
		3	203	8374.7	0.001	28,264	57,103.8	0.49
		4	307	6777.9	0.001	21,635	43,060.8	0.50
		5	84	4074.1	0.001	11,028	23,619.9	0.47

Id cluster identification, *#Clust* number of municipality inside of cluster. *Ku* value of Kulldorff statistic, *Obser* number of employment inside of cluster, *Expec* H_0 expected number of employment inside of cluster under null, *Obs/Exp* ratio observed/expected. If >1 →High incidence; If <1 →Low incidence cluster

Low-incidence (LI) clusters in knowledge-intensive services seem to emerge in inland localities of the regions of Catalonia (LI (Id2) and (Id4)), Valencia (LI (Id5) and (Id3)), Alicante (LI (Id3)) and Murcia (LI (Id3)), all of them in Fig. 13.4b in blue colours. It also appears, observing the Ku-Scantest results, that all of these municipalities in the SMA are still in an initial phase of development of knowledge-intensive services, with those advanced services to enterprises only appearing to emerge in the metropolitan areas of the two most developed cities in the SMA: Valencia and Barcelona, as we have observed previously. Knowledge intensive services include postal, communication, education, technology-related and health services, usually agglomerated around big cities (Valencia and Barcelona) in the SMA, as the Ku-Scan test indicates. This graphical mapping of clusters for knowledge-intensive services (Fig. 13.4b) seems a very plausible empirical result, given the economic specialisation characterising this geographical area.

In general terms, we have observed some of the main characteristics and advantages of Ku-Scantest in comparison with previous concentration measures in literature: First of all, it allows overcoming administrative barriers in data, given that it accounts for parts of localities as belonging to the cluster area identified by the software, not necessarily having to take in account all of the municipality surface. In second place, the software is able to define and mark the exact cluster area in the Cluster Map for the variable of interest. Moreover, it computes a statistical measure in

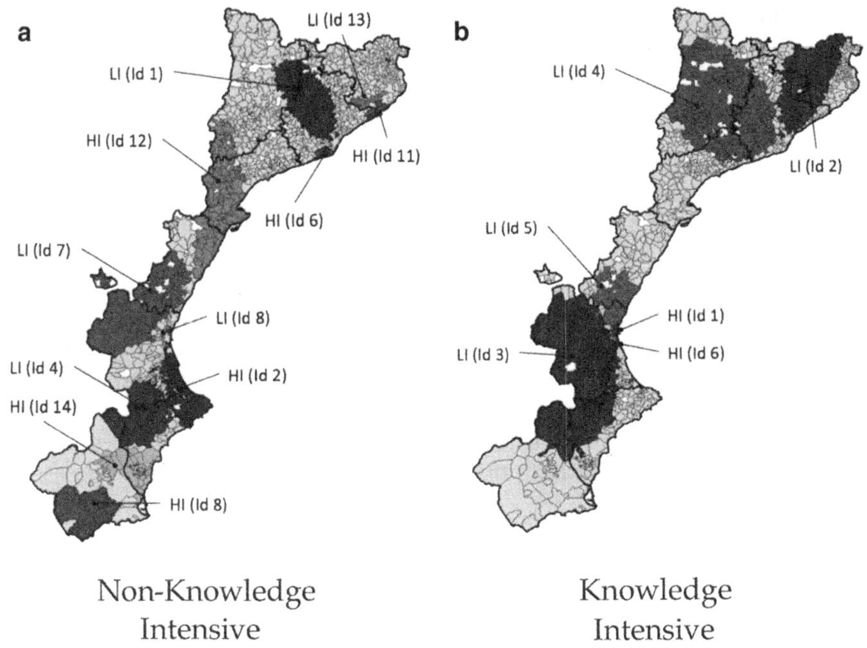

Fig. 13.4 Spatial cluster maps in NKI and KI

order to check out if the identified cluster area includes values of the variable of interest (number of employees in our case) significantly different from those outside the cluster. And third, it also allows labelling the character of every identified cluster area, as high or low-incidence clusters, in terms of the null hypothesis that the software is checking along the process. The software also includes information on the number of observations (municipalities) making the full cluster, the number assigned to the identified cluster, and takes care that one municipality could not belong for the same variable of analysis to two different clusters in the whole data sample.

13.4 Probability Analysis of Clustering Maps: An Ordered Probit Exercise

Next, we go further on the clusters obtained and we will try to explain which variables determine the constitution of the different types of clusters, namely high and low incidence clusters. A high-incidence cluster is the one that significantly shows the presence of a relatively high number of observed employees over expected number in terms of the Ku-Scantest; low-incidence cluster is that showing a low number of observed employees in the territory over the expected ones. In order to identify the main determinants of high and low incidence cluster we will run an ordered discrete choice model, defining our endogenous variable (y) as a categorical one taking three possible outcomes, that is:

$$y_{(Rx1)} = (y_1, \ldots, y_i, \ldots y_R)'$$
$$\text{with } y_i = 0 \text{ if municipality i belongs to a low incidence cluster}$$
$$= 1 \text{ if municipality i does not belong to any significant cluster}$$
$$= 2 \text{ if municipality i belongs to a high incidence cluster}$$
$$(i = 1, \ldots, R) \quad (13.7)$$

In order to explain such variable, as it is commonly treated in the literature, we defined a latent variable, denoted as y^*, which refers to the unobserved tendency to be involved in a high incidence cluster. That is, if the latent variable for one municipality increases, such municipality will be more likely to belong to a high incidence cluster. In other words, the latent variable relates to the original one by the following pattern:

$$y^*_{(Rx1)} = (y^*_1, \ldots, y^*_i, \ldots, y^*_R)'$$
$$with\ y^*_i = \begin{cases} 0 \ if & y^*_i \leq \mu_1 \\ 1 \ if & \mu_1 < y^*_i \leq \mu_2 \\ 2 \ if & y^*_i > \mu_2 \end{cases} \quad (13.8)$$

Where μ_1 and μ_2 are unobserved thresholds. To preserve the ordering, such thresholds must satisfy $\mu_1 < \mu_2$.

Next, we choose several variables that could explain the variation of such latent variable; or, in other words, the variables that determine the probability for each municipality of belonging to each of the three groups (low incidence cluster, no cluster or high incidence one). To this respect, we assume that characteristics of any municipality not only affect its own probabilities but also probabilities of its neighbouring municipalities. For this reason, the spatial lag of previous variables has also been considered in the model, defining the following model, in matrix form:

$$y* = X\beta + \varepsilon$$
$$\text{with } X_{(Rx(2k+1))} = [\iota \quad Z \quad WZ]; \ \beta_{((2k+1)x1)} = [\beta_0 \ \beta_Z \ \beta_{WZ}]' \quad (13.9)$$

Where ι represents a vector of ones; Z is an (R × k) matrix of explanatory variables; and WZ denotes the (R × k) matrix of spatial lag of Z. The explanatory variables conforming Z are the following socio-economic variables: the percentage of specialised employment, measured by the percentage of employment, over the total one, that it is prepared for developing the required work in the respective sector (denoted by PERESPEMPL); total population (POB); the population density of the municipality (POPDEN); nationality, measured as the percentage of Spaniards in the total population (PERSPANISH); unemployment rates (UNEMP); public expenditure per inhabitant (PUBEXP); the price of land (LANDPRICE); and the number of firms per squared kilometre, (FIRMDEN). Furthermore, we complete the specification with: (1) some geographical variables such as the spatial tendency through the coordinates of each municipality (XC and YC), as well as with the distance of municipality to administrative head (DISTANCE); and with (2) two regional dummies to control for potential differences in the institutional environment which are determined by the Autonomous Community to which the municipality belong to. Spanish Autonomous Communities corresponds to NUTS II administrative spatial unit in terms of Eurostat. They are the first-level administrative division and are responsible for relevant aspects of industrial policy, such as innovation policy, taxes, subsidies, etc., which ultimately determine the cost of establishment for new companies. Hence, we consider important to include as control variables two regional dummies, one for Valencia Community (DVALENCIA, including the provinces of Alicante, Valencia and Castellón) and, the other, for Catalonia, (DCATALONIA, including the provinces of Tarragona, Barcelona, Lleida and Girona), leaving the Region of Murcia as the reference category. Table 13.3 collects detail information on all previous variables as well as information on the unit of measure and the data source of all of them.

The weighted matrix (W) necessary to define the spatial lag of socio-economic variables included in the model has been built assuming that two municipalities are neighbours if they are less than 50 km apart. The resulting binary matrix is also far from being a densely connected weight matrix: the average number of connections is 164 and the percentage of non-zeros values is 11.3. Afterwards, the binary matrix has been row-standardised.

The estimation of the proposed model has been carried out by Maximum Likelihood procedure (ML). From estimated parameters, it is possible to evaluate the probability of belonging to any of the groups by:

$$\begin{aligned}
\Pr\{y=0|X\} &= F(\mu_1 - X\beta) \\
\Pr\{y=1|X\} &= F(\mu_2 - X\beta) - F(\mu_1 - X\beta) \\
\Pr\{y=2|X\} &= 1 - F(\mu_2 - X\beta)
\end{aligned} \tag{13.10}$$

Where F denotes the cumulative distribution function of ε.

Furthermore, the effect on probabilities of a unitary change in one explicative variable (marginal effect) can be calculated as follows:

$$\begin{aligned}
\frac{\partial \Pr\{y=0|X\}}{\partial x_h} &= -f(\mu_1 - X\beta)\beta_h \\
\frac{\partial \Pr\{y=1|X\}}{\partial x_h} &= [f(\mu_1 - X_\beta) - f(\mu_2 - X\beta)]\beta_h \\
\frac{\partial \Pr\{y=2|X\}}{\partial x_h} &= f(\mu_2 - X\beta)\beta_h
\end{aligned} \tag{13.11}$$

denoting by $f(\cdot)$ the density function.

Table 13.3 Definition of variables in the empirical model

Variable	Definition	Unit of measure	Source
PERESPEMPL	Percentage of specialised employment over total employment	%	SABI
POB	Number of inhabitants in the municipality	Thousands of inhabitants	CajaEspaña
POBDEN	Number of inhabitants per squared kilometre	Thousands of inhabitants per km^2	CajaEspaña
PERSPANISH	Percentage of Spaniards in the total population	%	CajaEspaña
UNEMP	Unemployment rate	%	CajaEspaña
PUBEXP	Public expenditure per inhabitant	Thousands of Euros	CajaEspaña
LANDPRICE	Price per unit of urban land	Thousands of Euros	CajaEspaña
FIRMDEN	Number of firms per squared kilometre	Firms per km^2	CajaEspaña
XC	X- coordinate		SABI
YC	Y- coordinate		SABI
DISTANCE	Distance to administrative head	Kms	Own elaboration
DVALENCIA	Dummy for Region of Valencia. The reference category is the Region of Murcia		Own elaboration
DCATALONIA	Dummy for Region of Catalonia. The reference category is the Region of Murcia		Own elaboration

From Eq. 13.11, the effect on the respective probability of a proportional change in one explicative variable (semielasticities) can be calculated by multiplying the corresponding marginal effect by the (mean) value of the x_h variable.

In the present chapter, we assume a normal distribution for the error term ε and, therefore, we estimate the corresponding ordered probit. Results for the estimated parameters together with the semielasticities for low incidence and high incidence cluster are shown, by sectors, in Tables 13.4, 13.5, 13.6, 13.7, 13.8, and 13.9. As regards semielasticities, it is important to note that calculations have been carried out at the mean of all covariates included in X. Furthermore, at the bottom of all tables, we offer some diagnostics referred to the estimation results, including log-likelihood values, Pseudo R^2, and joint significance measure of the whole model. As shown in the tables, the joint model appears to be highly significant in all cases and, consequently, the goodness of fit is appreciable, considering the cross-sectional nature of data.

Regarding effects for the percentage of specialised employment, results show that, in general, municipalities are more likely to be involved in a high incidence cluster if they and their surrounding municipalities count with specialised labour force. The only exception is for the knowledge intensive sectors, since neighbouring labour force represents a threat for a creation of employment in a municipality.

Table 13.4 Results for low technology sector[a]

			Low incidence cluster		High incidence cluster	
	Parameter	z-value	Semi-elasticity	z-value	Semi-elasticity	z-value
PERESPEMPL	0.013*	7.210	−0.019*	−8.810	0.052*	7.250
POB	−0.001*	−2.220	0.003*	3.180	−0.002*	−2.330
POBDEN	−0.155*	−4.400	0.012*	3.170	−0.016*	−5.320
PERSPANISH	0.016*	3.930	−0.210*	−3.990	0.323*	3.900
UNEMP	−0.006	−0.570	0.009	0.570	−0.014	−0.570
PUBEXP	0.014	0.160	−0.002	−0.170	0.003	0.160
LANDPRICE	−0.003*	−2.450	0.016*	2.310	−0.027*	−2.540
FIRMDEN	0.030*	3.630	−0.009*	−4.020	0.019*	3.320
DVALENCIA	0.981*	3.560	−0.058*	−3.670	0.065*	3.510
DCATALONIA	1.921*	4.820	−0.160*	−4.880	0.292*	4.690
XC	0.001	0.630	−0.134	−0.630	0.207	0.630
YC	−0.004*	−2.610	2.986*	2.640	−4.380*	−2.580
DISTANCE	−0.006*	−2.820	0.039*	2.730	−0.053*	−2.890
W*PERESPEMPL	0.046*	2.780	−0.102*	−2.740	0.177*	2.810
W*POB	−0.011	−0.990	0.014	0.980	−0.024	−0.990
W* POBDEN	−14.486*	−14.930	0.758*	12.310	−1.757*	−16.250
W*PERSPANISH	−0.054*	−2.410	0.711*	2.400	−1.053*	−2.410
W*UNEMP	−0.071	−0.990	0.108	0.990	−0.162	−0.990
W* PUBEXP	4.727*	6.090	−0.540*	−6.210	0.833*	6.060
W* LANDPRICE	0.008	0.820	−0.043	−0.820	0.079	0.820
W* FIRMDEN	3.065*	14.950	−0.777*	−12.530	1.928*	15.990
μ_1	−17.928					
μ_2	−15.467					
Log likelihood	−953.413					
Pseudo R^2	0.308					
LR χ^2 (21)	848.560*					

[a]An asterisk means that the corresponding null hypothesis is rejected at the 5% level of significance.
*significant at 1%

These results are as expected since the access to a qualified labour force implies that firms can introduce advanced production techniques faster as well as they can absorb easily new innovations and knowledge.

Results for the unemployment rate variable for industries can be explained in similar lines, since industries can take advantage of the increase in the stock of human capital stock. In general terms, unemployment positively affects the probability of being involved in a high incidence cluster; that is, an increase in the unemployment rate of a municipality favours location choices on such municipality. However, the case of services seems to be different, since for both categories, unemployment rate reduces the probability of developing high incidence cluster not only in the own municipality but also in the surroundings.

Regarding variables referred to population, we must distinguish between the effect of total population and population density. On one hand, total population has been introduced in the model as a measure of market size, since total population in a

Table 13.5 Results for low-medium technology sector[a]

			Low incidence cluster		High incidence cluster	
	Parameter	z-value	Semi-elasticity	z-value	Semi-elasticity	z-value
PERESPEMPL	0.021*	6.920	−0.004*	−6.860	0.044*	7.970
POB	−0.002*	−3.050	0.001*	5.890	−0.003*	−3.230
POBDEN	−0.227*	−5.530	0.009*	2.740	−0.021*	−6.070
PERSPANISH	0.014*	2.880	−0.046*	−2.920	0.300*	2.870
UNEMP	0.049*	3.890	−0.020*	−3.890	0.118*	3.920
PUBEXP	−0.008	−0.070	0.000	0.070	−0.001	−0.070
LANDPRICE	−0.001	−0.990	0.002	0.960	−0.012	−0.990
FIRMDEN	0.009	0.950	−0.001	−0.920	0.005	0.940
DVALENCIA	−0.116	−0.360	0.002	0.360	−0.009	−0.360
DCATALONIA	2.558*	5.700	−0.057*	−4.220	0.389*	5.840
XC	−0.007*	−3.420	0.245*	3.170	−1.338*	−3.480
YC	−0.007*	−3.400	1.329*	3.240	−7.510*	−3.390
DISTANCE	0.008*	3.390	−0.014*	−3.340	0.080*	3.350
W*PERESPEMPL	0.357*	10.160	−0.109*	−7.670	0.749*	10.900
W*POB	−0.053*	−4.150	0.024*	4.130	−0.114*	−4.190
W* POBDEN	−1.910*	−2.110	0.024*	2.050	−0.236*	−2.130
W*PERSPANISH	0.181*	6.310	−0.618*	−5.430	3.731*	6.380
W*UNEMP	−0.331*	−3.910	0.144*	3.790	−0.787*	−3.950
W* PUBEXP	−3.812*	−3.650	0.126*	3.450	−0.679*	−3.690
W* LANDPRICE	0.021	1.790	−0.036	−1.800	0.206	1.780
W* FIRMDEN	0.636*	3.200	−0.038*	−3.020	0.405*	3.280
μ_1	−24.317					
μ_2	−20.168					
Log likelihood	−662.685					
Pseudo R^2	0.413					
LR χ^2 (21)	933.880*					

[a]An asterisk means that the corresponding null hypothesis is rejected at the 5% level of significance.
*significant at 1%

municipality and in its surroundings emerges as a reasonable proxy for potential consumers' demand (Krugman 1992). The expected positive result is found for spill over effects for medium-high and high technology sectors, in the case of industries; and also, for the own population variable in the case of services, for knowledge intensity sector (at the 10 % level of significance). For the rest of categories, the effect of population is not significant or even negative. These last results can be indicating that the relevant market for a large proportion of the firms does not necessarily match the municipality area. On the other hand, population density is introduced in the model to capture, together with the land price variable, the effect of the cost of land on firms' location decision. Although the correlation coefficient between of both variables (POBDEN and LANDPRICE) only reaches a value of 0.13, it is a fact that residential and industrial users can compete for land. As expected, population density reduces the probability of developing high incidence

Table 13.6 Results for medium-high technology sector[a]

			Low incidence cluster		High incidence cluster	
	Parameter	z-value	Semi-elasticity	z-value	Semi-elasticity	z-value
PERESPEMPL	0.028*	3.520	−0.002*	−3.530	0.013*	3.280
POB	−0.001	−0.890	0.001	0.830	−0.001	−0.900
POBDEN	−0.082	−1.280	0.001	0.940	−0.003	−1.360
PERSPANISH	0.034*	2.420	−0.081*	−2.410	0.158*	2.440
UNEMP	0.052*	2.090	−0.019*	−2.070	0.027*	2.080
PUBEXP	0.626*	2.150	−0.011*	−2.130	0.027*	2.190
LANDPRICE	0.002	1.110	−0.002	−1.110	0.008	1.120
FIRMDEN	−0.021	−1.440	0.001	1.300	−0.005	−1.550
DVALENCIA	−1.471*	−2.680	0.036*	2.780	0.000	0.000
DCATALONIA	4.559	0.030	0.000	0.000	0.229	0.030
XC	−0.036*	−7.820	0.685*	7.480	−1.653*	−9.260
YC	0.041*	7.320	−4.790*	−8.130	9.649*	7.750
DISTANCE	−0.013	−1.650	0.015	1.680	−0.027	−1.680
W*PERESPEMPL	0.591*	3.340	−0.044*	−3.280	0.234*	3.530
W*POB	0.291*	7.060	−0.093*	−6.370	0.227*	7.350
W* POBDEN	−7.745*	−4.010	0.104*	3.570	−0.355*	−4.300
W*PERSPANISH	0.519*	6.400	−1.177*	−7.060	2.390*	6.330
W*UNEMP	0.351	1.290	−0.123	−1.300	0.179	1.280
W* PUBEXP	−9.847*	−3.770	0.186*	3.780	−0.421*	−3.790
W* LANDPRICE	0.304*	7.910	−0.317*	−7.210	0.934*	9.310
W* FIRMDEN	−0.716	−1.530	0.052	1.530	−0.168	−1.520
μ_1	208.120					
μ_2	227.702					
Log likelihood	−194.999					
Pseudo R^2	0.788					
LR χ^2 (21)	1448.670*					

[a]An asterisk means that the corresponding null hypothesis is rejected at the 5% level of significance.
*significant at 1%

cluster not only in the own municipality but also in the surroundings. The only exception is for knowledge intensity sector (Table 13.9), where a municipality is more likely to get involved in a high incidence cluster as its population density increases.

In general, land price parameters, when significant, are negative. That is, an increase in the land price of a municipality affects negatively to the establishment of new firms, decreasing the probability of constituting a high incidence cluster in such municipality. Regarding spill over effects, effects for industries are positive, indicating that an increase in the land price of neighbouring municipality affects positively to the establishment of new firms in the own municipality. However, spill over effect for services are of negative sign, meaning that increasing in neighbours' land does not neither attract services to our municipality.

Table 13.7 Results for high technology sector[a]

	Parameter	z-value	Low incidence cluster Semi-elasticity	z-value	High incidence cluster Semi-elasticity	z-value
PERESPEMPL	0.130*	6.860	−0.002*	−8.920	0.007*	6.920
POB	0.001	0.740	−0.001	−0.700	0.001	0.580
POBDEN	−0.065	−1.670	0.004	1.760	−0.004	−1.800
PERSPANISH	−0.010	−1.700	0.134	1.710	−0.044	−1.690
UNEMP	0.001	0.050	−0.001	−0.050	0.000	0.050
PUBEXP	0.073	0.620	−0.008	−0.620	0.003	0.620
LANDPRICE	0.003	1.920	−0.014	−1.910	0.010	1.850
FIRMDEN	0.010	1.070	−0.003	−1.040	0.003	1.010
DVALENCIA	−3.028*	−7.500	0.255*	7.810	0.000	−0.840
DCATALONIA	−3.513*	−6.150	0.218*	6.930	−0.157*	−5.630
XC	−0.006*	−2.750	0.730*	2.730	−0.274*	−2.790
YC	0.015*	6.270	−9.791*	−6.540	3.203*	5.810
DISTANCE	0.020*	5.890	−0.112*	−5.770	0.038*	5.200
W*PERESPEMPL	−0.295	−0.650	0.010	0.650	−0.012	−0.650
W*POB	0.133*	7.770	−0.158*	−7.190	0.114*	6.580
W* POBDEN	−10.832*	−8.820	0.685*	9.900	−0.511*	−8.000
W*PERSPANISH	0.105*	3.390	−1.374*	−3.430	0.449*	3.250
W*UNEMP	0.110	1.050	−0.178	−1.050	0.054	1.040
W* PUBEXP	1.836	1.440	−0.198	−1.450	0.074	1.430
W* LANDPRICE	0.140*	8.740	−0.736*	−9.070	0.417*	7.810
W* FIRMDEN	1.788*	6.970	−0.584*	−7.630	0.427*	6.580
μ_1	74.892					
μ_2	81.344					
Log likelihood	−485.969					
Pseudo R^2	0.581					
LR χ^2 (21)	1349.180*					

[a]An asterisk means that the corresponding null hypothesis is rejected at the 5% level of significance.
*significant at 1%

As for nationality, and regarding technology classification (industries), Spaniards habitants on a municipality generally increase probability of high incidence cluster, both, in the proper location and in the surrounding. However, the opposite occurs if we pay attention to results for knowledge classification (services), since services seem to be targeted to foreigners.

Public expenditure in a municipality generally attracts industries and services to settle in such municipality. That is, as expected, public expenditure is, in general, positive to firms' establishment as it contributes to improve infrastructures and accessibility facilities which benefit firms. As regards spillover effects, results are not so conclusive. For low technology and high technology industries sector (Tables 13.4 and 13.7, respectively) as well as knowledge insensitive service sector (Table 13.9) the effect is positive; therefore public investment in neighbouring municipalities favours location in the own municipality. On the contrary, for the

Table 13.8 Results for non-knowledge intensive sectors[a]

			Low incidence cluster		High incidence cluster	
	Parameter	z-value	Semi-elasticity	z-value	Semi-elasticity	z-value
PERESPEMPL	0.008*	5.490	−0.045*	−6.030	0.047*	5.180
POB	−0.002	−1.700	0.003	1.530	−0.003	−1.820
POBDEN	−0.075	−1.380	0.008	1.380	−0.006	−1.390
PERSPANISH	−0.026*	−5.470	0.484*	5.400	−0.385*	−5.660
UNEMP	−0.034*	−3.100	0.073*	3.060	−0.058*	−3.170
PUBEXP	−0.061	−0.680	0.010	0.680	−0.008	−0.690
LANDPRICE	0.001	0.540	−0.005	−0.540	0.004	0.530
FIRMDEN	0.006*	2.060	−0.014*	−2.190	0.011	1.830
DVALENCIA	−0.923*	−3.100	0.070*	3.080	−0.047*	−3.090
DCATALONIA	2.426*	5.410	−0.309*	−5.400	0.278*	5.360
XC	0.005*	3.130	−0.926*	−3.130	0.778*	3.120
YC	−0.009*	−5.600	8.795*	5.610	−7.359*	−5.600
DISTANCE	0.013*	5.950	−0.111*	−6.210	0.089*	5.810
W*PERESPEMPL	0.103*	4.850	−0.612*	−4.990	0.538*	4.760
W*POB	−0.041*	−3.160	0.082*	3.120	−0.060*	−3.140
W* POBDEN	−2.085	−1.910	0.250	1.930	−0.131	−1.880
W*PERSPANISH	−0.145*	−6.510	2.700*	6.480	−2.184*	−6.640
W*UNEMP	−0.008	−0.100	0.017	0.100	−0.014	−0.100
W* PUBEXP	−6.399*	−7.280	1.042*	7.450	−0.855*	−7.280
W* LANDPRICE	−0.055*	−5.670	0.516*	5.710	−0.385*	−5.660
W* FIRMDEN	0.189*	3.470	−0.508*	−3.550	0.286*	3.330
μ_1	−56.394					
μ_2	−53.991					
Log likelihood	−965.132					
Pseudo R^2	0.311					
LR χ^2 (21)	870.030*					

[a]An asterisk means that the corresponding null hypothesis is rejected at the 5% level of significance.
*significant at 1%

rest of the cases the effect is negative, and public investment in neighbouring municipalities attracts firm to establish in such neighbouring municipalities, converting neighbours in our competitors.

Results for firm density variables are of great interest. As expected, firm density positively affects the probability of constituting a high incidence cluster, both, in the own municipality and its surrounding. Furthermore, these results occur for all types of industries and services. The explanation of this positive effect can be found in the fact that firms' decision on their localization is also determined by industry agglomeration effects or externalities, which can be generated by the same type of industries (intra-industry) or by different types (inter-industries). To this respect, a firm located in close proximity to other firms in the same industry can take advantage of a range of intra-industry benefits, including access to specialized know-how, sharing of sector specific skilled

Table 13.9 Results for knowledge intensive sectors[a]

			Low incidence cluster		High incidence cluster	
	Parameter	z-value	Semi-elasticity	z-value	Semi-elasticity	z-value
PERESPEMPL	0.004	1.570	−0.008	−1.580	0.000	1.100
POB	0.002	1.790	−0.004	−1.900	0.001	1.350
POBDEN	0.119*	2.030	−0.014*	−2.210	0.001	1.000
PERSPANISH	−0.015*	−2.920	0.378*	2.940	−0.003	−1.620
UNEMP	−0.024*	−2.000	0.069*	2.010	−0.001	−1.440
PUBEXP	0.369*	3.570	−0.081*	−3.640	0.001	1.150
LANDPRICE	0.001	1.110	−0.016	−1.100	0.000	0.900
FIRMDEN	0.003	0.870	−0.007	−0.880	0.000	0.840
DVALENCIA	−1.301*	−3.710	0.119*	3.630	−0.002	−1.320
DCATALONIA	−2.050*	−4.490	0.379*	4.550	−0.001	−0.590
XC	0.002	0.860	−0.421	−0.860	0.005	0.780
YC	−0.002	−0.900	2.303	0.900	−0.027	−0.800
DISTANCE	−0.003	−1.250	0.031	1.250	0.000	−0.980
W*PERESPEMPL	−0.194*	−4.870	0.443*	5.000	−0.006	−1.870
W*POB	−0.019	−1.110	0.044	1.100	−0.001	−1.090
W* POBDEN	−6.489*	−5.220	0.798*	5.390	−0.010	−1.890
W*PERSPANISH	−0.176*	−7.040	4.372*	7.350	−0.047	−1.840
W*UNEMP	−0.199*	−2.400	0.565*	2.410	−0.008	−1.340
W* PUBEXP	7.123*	7.680	−1.555*	−8.180	0.019	1.800
W* LANDPRICE	−0.063*	−5.420	0.740*	5.490	−0.009	−1.830
W* FIRMDEN	0.379*	5.970	−1.085*	−6.040	0.014	1.890
μ_1	−24.954					
μ_2	−20.510					
Log likelihood	−715.351					
Pseudo R^2	0.292					
LR χ^2 (21)	590.430*					

[a]An asterisk means that the corresponding null hypothesis is rejected at the 5% level of significance.
*significant at 1%

labour, integration in buyer–supplier networks, opportunities for efficient subcontracting, etc. As a result, the co-location of firms in the same industry generates cluster externalities that enhance productivity of all firms in that industry, increasing local attractiveness for the localization of new firms operating in that industry. Among the benefits from being located in close proximity to other types of industries, inter-industries advantages, we can quote the following: (1) easier access to complementary services, availability of a large labour pool with multiple specializations, and the availability of general infrastructure and a vibrant socio-economic ambient (Jacobs 1969).

Finally, parameters for dummies for Regions of Valencia and Catalonia indicate the following. Regarding industries: (1) for low-technology firms, both Regions, Valencia and Catalonia are more likely to be chosen as location than Murcia; (2) for low-medium technology Catalonia is the preferred region, followed by Murcia and Valencia; (3) for medium-high technology, Catalonia and Murcia are statistically similar, being Valencia a less likely location; finally, for high technology sector, Murcia is the most likely location. As for services, dummies variables indicate the following: (1) for non-knowledge intensity sector, Catalonia is the most likely location, followed by Murcia and Valencia; however, (2) for knowledge intensity sector, Murcia is the most likely location, well above Valencia and Catalonia.

Finally, regarding the effect of municipality distance to its province capital, we can conclude the following. As a municipality gets closer to the administrative head, it is more likely to be involved in a high incidence cluster only for low and medium-high technology sectors.

13.5 Conclusions

As highlighted by Paul Krugman (1992) economic activities are definitely not homogeneously distributed in space. For this reason, evaluating their geographic concentration has been one of the topics that attracted most interest among economists and geographers in recent decades. Since the work of Alfred Marshall (1890) many researchers have been dealing with how to measure and explain concentration patterns of activity. During the last decade, some measures have been developed to deal with identification of such spatial patterns, ranging from the simplest indexes of inequality of Theil or Gini, until the most elaborated M index derived from the recent work of Marcon and Puech (2010). As a result, economists have improved the measurement of the geographic concentration of activities, even defining various criteria that a 'good concentration index' should account for (Combes and Overman 2004).

However, as a general shortcoming in this literature, all these measures do not expressly consider the spatial dimension of data when computed. As a response, new indicators of spatial association have started to emerge building on new developments of the spatial statistics and spatial econometrics fields. All these new methods for measuring spatial concentration or dispersion share two main goals: First, they allow for treating space as a inhomogeneous underlying space and, second, they employ precise location geo-data in defining concentration of activities (building on latitude and longitude coordinates of data). Two examples of these new contributions are the pioneer of Arbia (2001) and the most recent of Kang (2010). In this chapter we employ a traditional concentration measure in epidemiology, the Kulldorff scan statistic, transposing it to the field of economics, in order to define a cluster map. Two have been the main advantages of this method in comparison with previous research: First, it has provided a graphical tool that allows identifying the existence of a cluster in a map by statistical methods while employing geo-referenced data. In this way, the Ku-Scantest have ruled out traditional computation problems of cluster analysis such as the existence of administrative boundaries or other geographical inhomogeneties present in spatial data analysis. And second, it also includes a statistic for testing the significance of the cluster area identified, labelling them as of high or low economic incidence depending on their own characteristics, what opens grounds to the researcher for defining her/his own cluster map and further testing for its significance. For both reasons, the Ku-Scantest continues building on the path started by previous spatial econometrics studies dealing with cluster analysis and improving the definition of new measures of concentration.

We then have applied this methodology for building cluster maps for economic activities located in the Spanish Mediterranean Axis at the municipality level. Our interest has been focused on mapping the distribution of employment in industry and service sectors according to the classification made by Eurostat, based on the technological intensity of activities. In building such categories we assumed four groups in the industrial sector and two in the service activities. Our results confirm the heterogeneity of the spatial distribution of industry and service activities along the SMA, highlighting several patterns of location, namely the presence of: (1) Two important cluster formations around the city of Barcelona and its metropolitan area and for Valencia city for high-tech industries. It includes significant accumulation of chemical-related companies for Barcelona, and computer, optical and medical instrument ones, together with machinery manufacturing companies in the urban areas of Valencia and Barcelona. (2) High-Incidence clusters for low-tech industries emerge in the surrounding area of the north of Barcelona and the south province of Girona, with productions as food and beverages industries, or metallic products. (3) Non-knowledge intensive services around the cities, which mainly concentrate on urban centres of Barcelona, Tarragona, Castellón, Alicante and south of Murcia city. This type of services are usually linked to primary activities, such as agriculture and fishing, and related to services for real estate activities, and other non-knowledge intensive manual services. (4) Regarding the main knowledge-intensive areas where services seem to cluster in the SMA, we observe again the relevance of Barcelona and Valencia cities as main attractors of these kinds of knowledge businesses. (5) The concentration pattern that emerge from this study confirms that location patterns and clustering of activities spill over administrative barriers, so new software and statistics are needed to be developed, in the spirit of the Ku-ScanStat for example, that affords for dealing with identification and inspection issues.

Finally, we have wondered about the factors that could be determining the nature of activity clusters, that is, why one cluster belongs to a high or low incidence group. At this stage we have employed an ordered probit model as a tool to explain the causes that determine the probability of a cluster to be named as of high or low concentration, complementing in this way all location maps provided by the Ku-test. In our probit model we introduce as explanatory variables both the characteristics of each municipality and those of the neighbouring municipalities. Factors such as the percentage of specialised employment, the population density, the price of land, the public expenditure and the firm density in local population seem to play an important role in determining the nature of the selected cluster. In the case of industrial activity, a municipality increases the probability of belonging to a high or low cluster if both the municipality and the municipalities within its surroundings show specialised employment. This does not happen in the services sector where this variable is not so important. Population density also appears as an important factor in determining the nature of cluster. Generally, a municipality with high population density reduces the probability of belonging to a cluster of high incidence. So, the companies show preferences for municipalities with small populations where the price of land is lower. The only exception is for areas classified as knowledge-intensive, where a municipality increases the probability of belonging to a cluster of high incidence in cities with high population density, as shown by urban and regional studies (Hall and Ciccone 1996; Puga 2010, 2011). In the same way, operates firm density that positively affects the probability of belonging to a cluster of high incidence. Furthermore, public expenditure in a municipality generally attracts firms to settle in such municipality, since it contributes to improve infrastructures and public incentives to locate.

References

Alamá A, Artal A, Navarro JM (2011) Industrial location, spatial discrete choice models and the need to account for neighbourhood effects. Ann Reg Sci 47(2):393–418

Arbia G (2001) The role of spatial effects in the empirical analysis of regional concentration. J Geogr Syst 3:271–281

Artal-Tur A, Navarro-Azorín JM, Alamá-Sabater L, García-Sánchez A (2012). The role of destination spatial spillovers and technological intensity in the location of manufacturing and services firms, Environment and Planning B: Planning and Design, advance online publication, doi:10.1068/b38024

Baltagi BH, Egger P, Pfaffermayr M (2005) Estimating models of complex FDI: are there third-country effects? J Econ 127(1):260–281

Blonigen BA, Davies RB, Naughton HT, Waddell GR (2008) Spacey parents: spatial autoregressive patterns in inbound FDI. In: Brakman S, Garretsen H (eds) Foreign direct investment and the multinational enterprise. MIT Press, Cambridge, MA

Braunerhjelm P, Johansson D (2003) The determinants of spatial concentration: the manufacturing and service sectors in an international perspective. Ind Innov 10(1):41–63

Brülhart M, Traeger R (2005) An account of geographic concentration patterns in Europe. Reg Sci Urban Econ 35:597–624

Combes P, Overman HG (2004) The spatial distribution of economic activities in the European Union. In: Henderson JV, Thisse JF (eds) Handbook of regional and urban economics, vol 4. Elsevier, Amsterdam, pp 2845–2909

De Dominicis L, Arbia G, De Groot HLF (2006) Spatial distribution of economic activities in local labour market areas: the case of Italy. In: Paper presented at the ERSA conference 2006, Volos, Greece

Desmet K, Fafchamps M (2006) Employment concentration across U.S. counties. Reg Sci Urban Econ 36:482–509

Duranton G, Overman H (2005) Testing for localization using micro-geographic data. Rev Econ Stud 72:1077–1106

Dwass M (1957) Modified randomization tests for non parametric hypotheses. Ann Math Stat 28:181–187

Ellison G, Glaeser EL (1997) Geographic concentration in US manufacturing industries: a dartboard approach. J Pol Econ 105:889–927

Feser EJ, Bergman EM (2000) National industry cluster templates: a framework for applied regional cluster analysis. Reg Stud 34:1–19
Hall R, Ciccone A (1996) Productivity and the density of economic activity. Am Econ Rev 86(1):54–70
Henderson JV (1974) The sizes and types of cities. Am Econ Rev 64:640–660
Henderson JV (1988) Urban development: theory, fact and illusion. Oxford University Press, England
Henderson JV, Venables A (2009) The dynamics of city formation. Rev Econ Dyn 12:233–254
Jacob J (1969) The economy of cities. Random House, New York
Jensen P, Michel J (2011) Measuring spatial dispersion: exact results on the variance of random spatial distributions. Ann Reg Sci 47:81–110
Kang H (2010) Detecting agglomeration processes using space–time clustering analyses. Ann Reg Sci 45:291–311
Kaushal N (2005) New immigrants´ location choices: magnets without welfare. J Lab Econ 23:59–80
Krugman P (1992) Geography and trade. The MIT Press, Cambridge, MA
Kulldorff M (1997) A spatial scan statistic. Commun Stat Theor Method 26:1481–1496
Kulldorff M, Nagarwalla N (1995) Spatial disease clusters: detection and inference. Stat Med 14:799–810
Marcon E, Puech F (2010) Measures of the geographic concentration of industries: improving distance-based methods. J Econ Geogr 10:745–762
Marshall A (1890) Principles of economics. Macmillan, London
Midelfart-Knarvik KH, Overman HG, Redding SG, Venables AG (2004) The location of European industry. In: Dierx A, Ilzkovitz F, Sekkat K (eds) European integration and the functioning of product markets. E. Edgar Publishing, Massachusetts (USA). Northamton, MA, US
Puga D (2010) The magnitude and causes of agglomeration economies. J Reg Sci 50(1):203–219
Puga D (2011) Learning by working in dense cities. Presentation at 51st European congress of the regional science association international (ERSA), Barcelona, 31 Aug–3 Sept 2011
Ripley B (1976) The second-order analysis of stationary point processes. J Appl Probab 13:255–266
Ripley B (1977) Modeling spatial patterns. J Roy Stat Soc B 39:172–212

Chapter 14
Dealing with Data at Various Spatial Scales and Supports: An Application on Traffic Noise and Air Pollution Effects on Housing Prices with Multilevel Models

Julie Le Gallo and Coro Chasco

14.1 Introduction

In empirical studies dealing with spatial data, researchers are frequently confronted with data available at different spatial scales. For instance, hedonic models on housing prices usually combine individual data pertaining to the price and structural characteristics of the dwelling and socio-economic neighbourhood characteristics that are available at some upper administrative levels. Another frequent issue is the change of support problem or misaligned regression problem (Gotway and Young 2002; Banerjee et al. 2004) when there is a spatial mismatch between the spatial supports of the variables. For instance, the measurement of air quality is based on regular sampling at a few stations in an area whereas socio-economic data are available for aggregate administrative.

Appropriate methodologies must be used to consider these issues. In this chapter, we show how multilevel models combined with kriging allow dealing with the heterogeneity of spatial sources, with different spatial supports and with the fact that the different levels might interact with each other. These methods are applied in the context of an hedonic model aimed at evaluating the households' marginal willingness to pay for better air quality and reduced noise in the areas of downtown Madrid that are the most affected by these problems. A dataset of 3,302 houses is

This work has been carried out with the financial support of the Spanish Ministry of Education and Science ECO2009-10534. Coro Chasco also acknowledges financial support from Xunta de Galicia 2009/AX702.

J. Le Gallo (✉)
CRESE, Université de Franche-Comté, 45D, Avenue de l'Observatoire, Besançon Cedex 25030, France
e-mail: jlegallo@univ-fcomte.fr

C. Chasco
Facultad de Economomicas, Departamento de Economía Aplicada (módulo 14), Universidad Autonoma de Madrid, Avda. Francisco Tomás y Valiente 5, Madrid 28049, Spain

E. Fernández Vázquez and F. Rubiera Morollón (eds.), *Defining the Spatial Scale in Modern Regional Analysis*, Advances in Spatial Science,
DOI 10.1007/978-3-642-31994-5_14, © Springer-Verlag Berlin Heidelberg 2012

used for this purpose in which all the aforementioned issues are present: data available at three different spatial scales (urban tracks, census tracts and individual data) and different spatial supports (areal and point data). Specifically, in this chapter, we focus on the city of Madrid where, as in other urban areas in the world, road noise is the dominant source of nuisance in residential areas. Several Action Plan projects have been implemented in the period 2008–2011, aiming at restoring and revitalizing several areas of downtown Madrid. As these projects are costly, it is therefore of interest for the local government to monetize the social value of changes in pollution and noise levels. We therefore apply hedonic regression to examine the effect of air and noise pollution on property prices in downtown Madrid.

Therefore, the main aim of this study is to show how multilevel models and kriging are useful techniques to deal with data at various spatial scales and supports. It must be said that analysing variables from different spatial levels at one single common level creates two different categories of problems (Hox 1995). One category of problems is statistical since if data was aggregated i.e. to urban tracks, some information would be lost and the statistical analysis would lose power. On the other hand, if data were disaggregated i.e. to individual level, some variables available at upper-scaled units (census tracts and urban tracks) would be subdivided into values for a much larger number of micro-units, which will be treated by ordinary statistical tests as independent observations – while they are probably not. The other set of problems is conceptual, since the interpretation of the results may be affected by the fallacy of the wrong spatial level, which consists of analysing the data at one level, in order to draw conclusions at another level. In fact, the 'ecological fallacy' takes place when interpreting aggregated data at the individual level, while the 'atomistic fallacy' consists of drawing inferences at a higher level from analysis performed at a lower level (see Alker 1969 for an extensive typology).

Another interesting issue in multilevel analysis is the analysis of how a number of individual and group variables influence one single individual outcome variable i.e. housing prices. The goal of the analysis is to determine if the explanatory variables at the group level serve as 'moderators' of the individual-level relationships; i.e. if upper-level variables moderate lower level relationships, this would be because of the existence of a statistical interaction between explanatory variables from different spatial scales. Such data are usually analysed using conventional multiple regression analysis with one dependent variable at the lowest (individual) level and a collection of explanatory variables from all available levels, what implies the already mentioned statistical and conceptual problems. Multilevel models deal with the dependence of observations belonging to a same group (intra-class correlation), as well as other kind of dependences or interactions between variables available at different spatial scales.

We illustrate the performance of this model with an application to a hedonic price model in order to measure the impact of air and noise pollution on housing prices. Though multilevel models have also been applied to hedonic housing price models (Jones and Bullen 1994; Gelfand et al. 2007; Djurdjevic et al. 2008;

Bonin 2009; Leishman 2009), only Beron et al. (1999) and Orford (2000)'s papers use them to measure the role of air pollution on property prices. As we show below, multilevel models are very useful when considering the effects of neighbourhood amenities (operating at upper-scaled spatial level), such as environmental quality, on households' preferences.

There are other two contributions in this chapter. Firstly, we use measures of air and noise pollution. Analyzing both the effect of air and noise pollution on housing prices is not so frequent in hedonic specifications that typically only include air pollutants. Since the seminal studies of Nourse (1967) and Ridker and Henning (1967), many authors have tried to estimate the marginal willingness of people to pay for a reduction in the local concentration of diverse air pollutants (see Smith and Huang 1993, 1995 for a review and meta-analysis). On the other hand, studies focusing on noise are less frequent although they can be traced back to the 1970s (Mieszkowski and Saper 1978; Nelson 1979) in order to measure the economic costs of airports, railroads and motorways. Papers analyzing the effects of both air and noise pollution are scarce (see for instance Li and Brown 1980; Wardman and Bristow 2004; Baranzini and Ramírez 2005; Banfi et al. 2007; Hui et al. 2008).

Secondly, in this chapter we use subjective measures of noise and air pollution. All the studies mentioned above use "objective" air quality and noise variables, such as concentrations of various pollutants and decibel levels. Conversely, "subjective" measures of air pollution and noise have been exceptionally considered in hedonic specifications. These measures are based on people's perceptions and are more difficult to obtain (Murti et al. 2003; Hartley et al. 2005; Berezansky et al. 2010). However, there are some advantages in using subjective measures rather than objective measures. On the one hand, Boyle and Kiel (2001) state that the objective measures of air quality may not be measures that are relevant to homeowners. Berezansky et al. (2010) show that housing prices in urbanized areas can be better explained by subjective evaluation factors rather than objective measurements. On the other hand, using subjective measures in our case may limit measurement error problems. Indeed, objective measures are typically recorded at some monitoring stations and then kriged to obtain individual values for each house in the sample. Whenever the number of stations is low, measurement errors in the kriged air pollution or noise surface may be important, hence requiring instruments to alleviate the attenuation bias caused by measurement errors (Anselin and Lozano-Garcia 2008). In our case, subjective measures are provided for a fine spatial level (the census tract) so that the measurement error implied by the kriging procedure will be lower than if we had used objective measures stemming from monitoring stations.

The chapter is structured as follows. First, we describe the database underlying this study. Second, we present the empirical model providing a short description of multilevel modelling applied to hedonic models. Third, we present the econometric results of the study, which allow us to provide some evaluation of the projects plans implemented by the city of Madrid. Finally, the last section concludes.

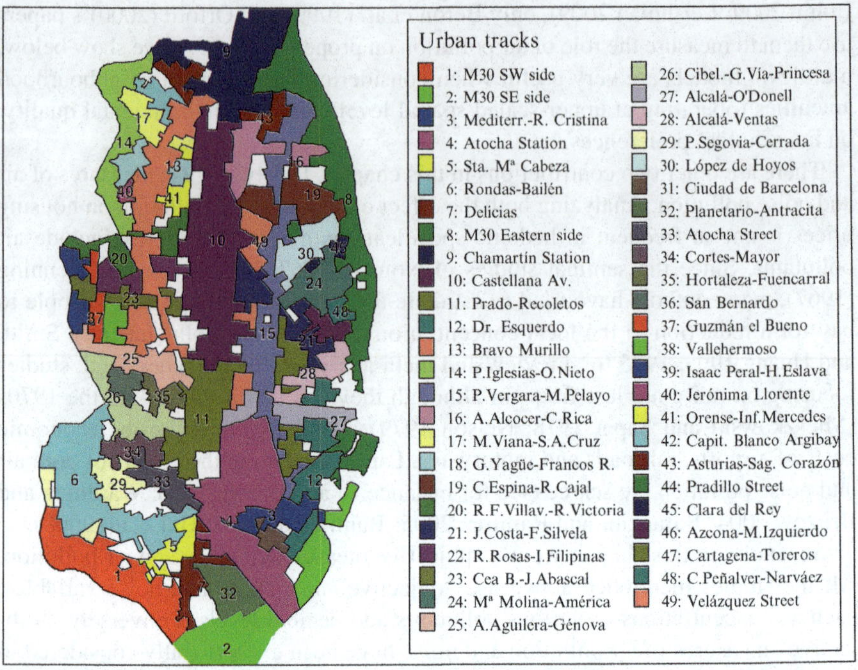

Fig. 14.1 The Central Almond urban tracks

14.2 Data

The city of Madrid is a municipality with a population of roughly 3.3 million inhabitants (as of January 2010). It comprises the city centre or 'Central Almond' and a constellation of 14 surrounding districts. Central Almond is the area formed by seven districts that are encircled by the first metropolitan ring-road (the M30). With more than 30 % of the population and 50 % of GDP of the city, Central Almond is clearly recognized as a unity with its own idiosyncrasy. Indeed, since 2004–2011, the Urbanism and Housing Area of the municipality government has launched two main "action plans" in order to restore and revitalize several areas of Central Almond (Ayuntamiento de Madrid 2009a, 2010; INE 2010). Many of these projects are devoted to construct new green belts with trails and cultural spaces on either decadent residential or industrial enclaves or highly congested tracks, in order to make Central Almond a more friendly and sustainable city.

Our study focuses on 49 main urban tracks present in this specific area (Fig. 14.1). They are the group of avenues and streets with more than 5,000 vehicles per day as reported in the 'Average Daily Intensity of Urban Traffic' report (Ayuntamiento de Madrid 2009b). These different intensities by tracks can be visualized in Fig. 14.2a. Each track is characterized by a similar traffic intensity, spatial continuity and socioeconomic homogeneity of the resident population.

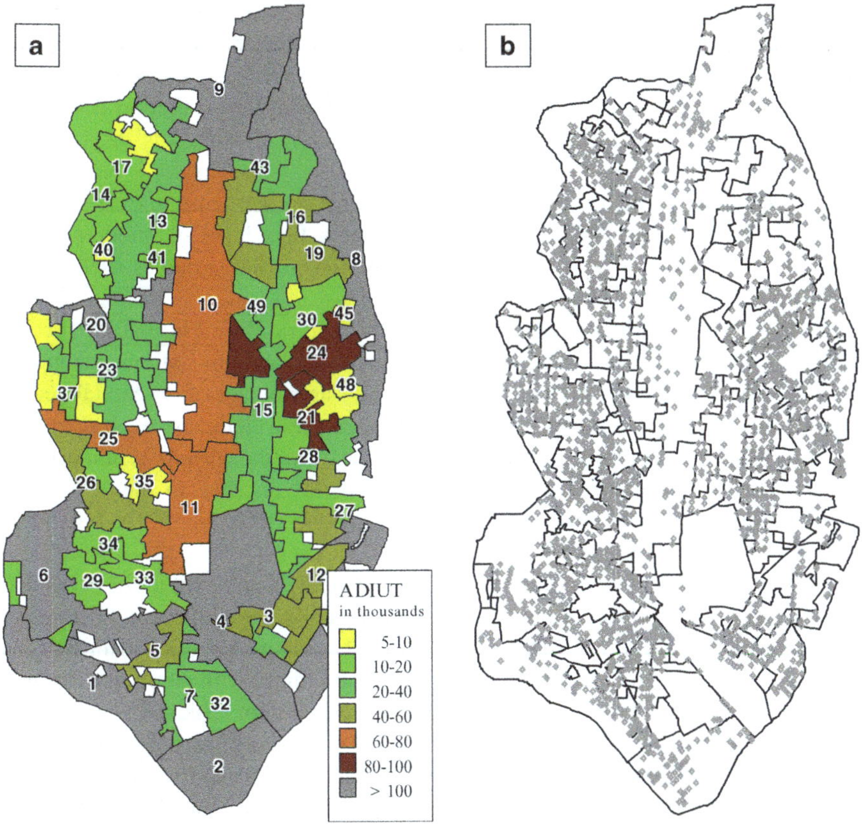

Fig. 14.2 (**a**) Average daily intensity of urban traffic (vehicles/day). (**b**) Sample of houses

We have selected those dwellings located at 150 m (in average, depending on the track range) along either the three tracks of the M30 ring-road adjoined to Central Almond (non-tunnelled Eastern side, semi-tunnelled Southern side and tunnelled South-Western side), as well as 19 main North–south tracks and 27 East–west tracks. As depicted in Figs. 14.1 and 14.2, the consideration of not only the tracks but their corresponding influence area allows us to include in the sample almost the total area of Central Almond. Therefore, our aim is to shed light on an important issue, i.e. the people's marginal willingness to pay for air quality and reduced noise in the most congested avenues of downtown Madrid. Such an evaluation allows a first economic evaluation of certain environmental policies that are being implemented in Central Almond.

Due to confidentiality constraints, it is not easy to obtain housing prices microdata from Spanish official institutions. For this reason, our records were drawn from a well-known on-line real estate database, 'idealista.com'. Since this catalogue immediately publishes the asking price of properties, we extracted the information during January 2008. The asking price has been used as a proxy for the selling price, as it is usual in many other cases (e.g. Cheshire and Sheppard 1998 or Orford 2000). In total, around 3,302 housing prices were finally recorded for the aforementioned 49 main urban tracks after the corresponding consolidation and geocoding processes, which have been performed with the 'Callejero del Censo Electoral' (INE 2008). The geographical distribution of houses is displayed in Fig. 14.2b.

The database 'idealista.com' also provides some property attribute data related to dwelling type, living space, number of bedrooms, floor level and modernization and repair. In Table 14.1, we present the definitions of the variables used in this study.

Proximity of dwellings to enclaves like CBD (Central Business District), accessibility infrastructures (airports, motorways, and metro and rail stations), shopping facilities, parks, etc. is advertised by real estate agents and often capitalized in housing prices. For this reason, in order to capture these elements, we constructed the following accessibility measures: (1) distance to the airport terminals, (2) distance to the nearest metro or railway station, (3) distance to the M30 ring-road, (4) distance to the financial district, (5) distance to the main road-axis and commercial avenues and (6) distance to parks.

From these, only the two first ones were statistically significant in the estimated models, with distance to the financial district the most determinant indicator (see below). In effect, the new CBD, which is located at the geographical centre of the Central Almond, is a huge block of modern office buildings with metro, railway and airport connections beside the government complex of Nuevos Ministerios. Another important variable is nearness to the main road-axis and commercial avenues. We have selected those dwellings located at 250 m (in average) along the main North–south axis (Castellana-Recoletos-Prado) and four East–west axes, i.e. Raimundo Fernández Villaverde-Concha Espina, José Abascal-María de Molina-América, Alberto Aguilera-Bilbao-Colón-Goya and Princesa-Gran Vía-Alcalá. This variable will capture the effects linked to the proximity of accessibility infrastructures.

From an administrative point of view, the Central Almond is divided into 7 districts and 780 census tracts, from which 660 are crossed by the 49 main urban tracks.[1] The 2001 Census provides a series of variables on socioeconomic and demographic characteristics related to home-ownership at the level of census tracts. In Table 14.1, we present the most significant ones: percent of population over 65 years, percent of foreign population, percent of population with secondary and university degrees, unemployment rate and percent of houses built after 1990. Though these variables are all referred to 2001, they are population averages, which are very stable in time. This validates their inclusion in our model.

In order to measure air-quality and noise effects on housing prices, we use two 'subjective' indicators, which are based on the population's perception of pollution

[1] Although road tracks are the sum of census tracts, there are some census tracts, mainly in the track intersections, which are part of more than one track.

Table 14.1 Variable definitions

Variable	Description	Source	Units	Period
Level 1: Houses				
Lprice	Housing price	Idealista	Euros (in logs)	Jan. 2008
(a) Structural variables				
fl_1	First floor and upper	Idealista	0-1	Jan. 2008
Attic	Attic	Idealista	0-1	Jan. 2008
House	House ('chalet')	Idealista	0-1	Jan. 2008
Duplex	Duplex	Idealista	0-1	Jan. 2008
Bedsit	Bedsit	Idealista	0-1	Jan. 2008
State	To be reformed: 0, new houses: 2, others: 1	Idealista	0-1-2	Jan. 2008
Bedr	Bedrooms	Idealista	#	Jan. 2008
lm2	Living space	Idealista	Square meter (in logs)	Jan. 2008
(b) Accessibility variables				
Discen	Distance to the financial district	Self-elab	Km	–
Axis	Distance to the main commercial avenues	Self-elab	Km	–
Disair	Distance to the airport terminals	Self-elab	Km	–
dismetro	Distance to the nearest metro or railway station	Self-elab	Km	–
dism30	Distance to the M30 ring-road	Self-elab	Km	–
dispark	Distance to the nearest park	Self-elab.	Km	–
(c) Air and noise variables				
Pollu	Objective air-pollution indicator	Munimadrid	100 = average	2007
dBA	Objective noise indicator	Munimadrid	dB(A)	2008
Cont	Subjective air-pollution indicator	Census	%	Nov. 2001
Noise	Subjective noise indicator	Census	%	Nov. 2001
Level 2: Census tracts				
p65	Percent of population over 65 years	Padrón, INE	%	Jan. 2008
Forei	Percent of foreign population	Census, INE	%	2001
Educ	Education level (secondary/university)	Census, INE	%	2001
Unem	Unemployment rate	Census, INE	%	2001
ha90	House built after 1990	Census, INE	%	2001

and noise around their residences. They are measured by the 2001 Census for each census tract as the percentage of households that estimate that their homes' surroundings are polluted or noisy.

Therefore, this study must deal with three spatial scales (individual properties, census tracts and urban tracks) and a set of variables, which are available for two of them; i.e. 15 variables for individual properties and 7 variables for census tracts. Nevertheless, the two Census variables for environmental quality – air and noise – variables have been interpolated by ordinary kriging at the level of houses. Kriging is a technique that allows for changing data of one spatial support to another (it is the change of support problem shown in Gotway and Young 2002). In this case, areal-data (census tracts) is "transformed"[2] into point-data (individual properties) in other to assign a value for contamination to each dwelling.

Subjective data of air quality or noise pollution are not always correlated with objective measures of pollutants, which are usually recorded at some fixed monitoring stations. Even though some authors have pointed the limitations of subjective measures based on individuals' perceptions (e.g. Cummins 2000), subjective approaches seem to provide a better perspective for evaluating certain latent variables connected with quality of life (Delfim and Martins 2007). For example, prospective homebuyers most probably evaluate air quality based on whether or not the air 'appears' to be polluted or based on what other people and the media say about local air pollution (Delucchi et al. 2002). The same applies for noise (Miedema and Oudshoorn 2001; Nelson 2004; Palmquist 2005).

Air and noise pollution are similar though not an identifiable phenomenon, even when they are caused by the same font (vehicle flows, industrial activity, etc.). Noise operates at a more local scale depending on the traffic intensity or the time of the day, while airborne pollutants are not so location specific since they are relevant on a global scale (Bickel et al. 1999). For this reason, air-pollution and noise marginal costs in a same place neither coincide nor are equally perceived by population.

In order to analyse these differences for our sample, we represent on a map the values of the four quadrants of a scatterplot of air-pollution versus noise in Central Almond, so that it is possible to identify some peculiar non-coincidences between these variables (Fig. 14.3a). In general, air and noise pollution are reported to be high in the South and South-Western tracks, which are affected by the commercial activity of the historical centre, as well as the accesses to Atocha Station, the M30 and some national radial highways located in this area. However, there are also some interesting disparities in people's perceptions affecting some tracks where noise is considered as excessive, though air-pollution is reported as low level. This

[2] Arbia (1989) uses the term "spatial data transformations" to refer to situations in which the spatial process of interest is inherently of one form but the data observed are of another form, resulting in a "transformation" of the original process of interest. For example, sometimes the data are just not available at the desired scale of interest. Consider meteorologic processes and pollution that occur over a continuum, but only small areal-data aggregates or even point observations along such a surface can be recorded.

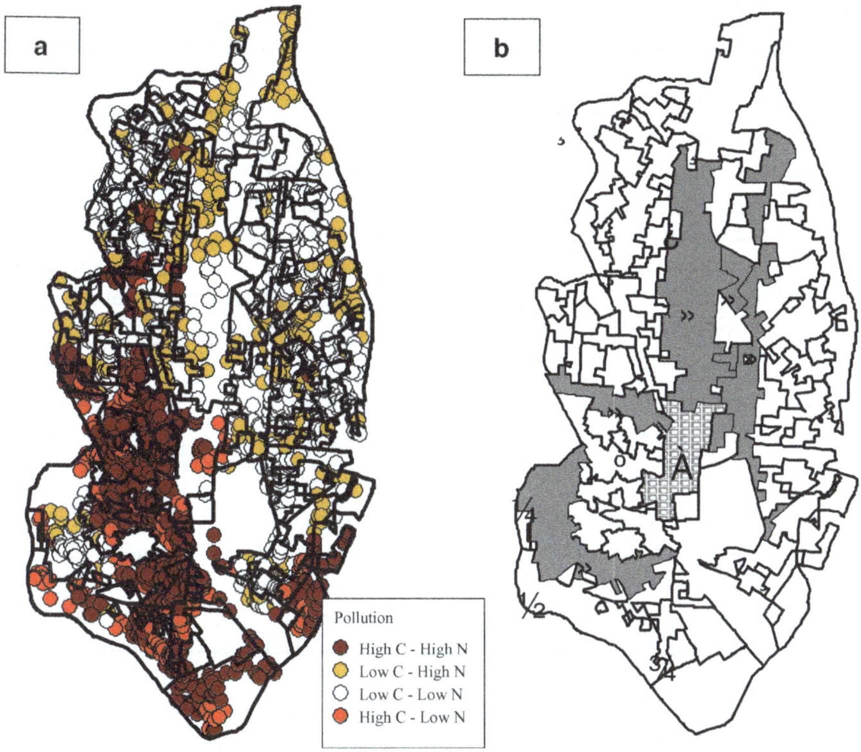

Fig. 14.3 (**a**) Scatterplot map of air-pollution (C) and noise (N). (**b**) Projects for the construction of parks, new green belts with trails and cultural spaces* (*Note*: *1*: Metro depots tunnelling (*Cuatro Caminos*, *Castilla*, *Ventas* and *Cavanilles*). *2*: *Delicias-Méndez Álvaro* especial plan. *3*: *Gran Vía* director plan. *4*: Old boulevards restoration. *5*: *S. Francisco el Grande* revitalization. *6*: *Vicente Calderón-Mahou* restoration. *7*: *Legazpi* subterranean suburban bus station. *8*: *Azca* remodelling. *9*: *Recoletos-Prado* project. *10*: Construction of the *Pº Dirección* green-trail)

is the case of tracks with some relevant value added (such as accessibility to the financial district or to main road-axis), in which their residents -though aware of the drawbacks of noise in their neighbourhoods- do not have the perception of living in a so air polluted area (while, from an objective point of view, they probably are). The M30 Eastern side accesses, as well as certain parts of Castellana, Orense, Príncipe de Vergara, Velázquez, Alcalá or O'Donell tracks, among others (in orange, in Fig. 14.3a), have such location advantages that could be exerting some kind of "halo effect" on the air-pollution people's perception (Brody et al. 2004). In effect, air-pollution is frequently associated with industries, bad odours and somehow depressed neighbourhoods that have nothing in common with the aforementioned residential avenues.

It is interesting to highlight that most projects of the last Urban Action Plan for Central Almond in 2008–2011 (Ayuntamiento de Madrid 2009a) are directly or indirectly oriented to the improvement of air and noise environmental conditions.

As depicted in Fig. 14.3b, these projects are located in highly congested enclaves and tracks where population reports the highest air and/or noise contamination levels. We evaluate the impact that a reduction in the contamination levels has in each place with the help of a multilevel hedonic housing model.

It is interesting to highlight that most projects of the last Urban Action Plan for Central Almond in 2008–2011 (Ayuntamiento de Madrid 2009a) are directly or indirectly oriented to the improvement of air and noise environmental conditions. As depicted in Fig. 14.3b, these projects are located in highly congested enclaves and tracks where population reports the highest air and/or noise contamination levels. We evaluate the impact that a reduction in the contamination levels has in each place with the help of a multilevel hedonic housing model.

14.3 Empirical Multilevel Hedonic Housing Model

In this section, we briefly present multilevel models as a way of combining data available at different spatial scales. These models are adapted whenever the data have a hierarchical structure, where a hierarchy refers to units clustered at different levels. In our case, the houses are nested within census tracts. As detailed above, there is also a higher level, the urban tracks, which are the sum of census tracts.

While many applications of multilevel modelling can be found in education science, biology or geography, economic applications are scarcer, although they have been increasing in the last few years. Among others, multilevel models have been applied to the study of female labour force participation (Ward and Dale 2006), unemployment in Israel (Khattab 2007), wage disparities in Brazil (Fontes et al. 2009), β-convergence in Europe (Chasco and López 2009), internal migration in Estonia (Kulu and Billari 2010) or geography of innovation (Srholec 2010). Some applications of multilevel models to hedonic models can be found in Beron et al. (1999), Orford (2000), Djurdjevic et al. (2008) or Leishman (2009). We follow this latter strand of literature and use multilevel models to evaluate the differential impacts of subjective measures of noise and air quality on housing prices in downtown Madrid.

Indeed, employing multilevel modelling for hierarchical data presents advantages. Firstly, from an economic perspective, taking into account a hierarchical structure makes it possible to analyse more accurately the extent to which differences in housing prices come from differences in housing characteristics and/or from differences in the environment of the transactions, i.e. the characteristics of the census tracts or the urban tracks. In our case, this is an appealing feature, as we integrate in the econometric specification various explanatory factors that operate at the first two spatial levels. It is also possible to capture cross-level effects. Secondly, from an econometric perspective, inference is more reliable than in single-level models that assume independent observations. Actually, if the units belonging to the same group (for instance houses in the same census tract) are associated with correlated residuals, these models are not relevant. More efficient estimates are obtained when this independence assumption is relaxed and when this intra-group correlation is modelled explicitly.

Formally, consider a transaction i, located in census tract j, which is itself located in urban track k. In the most general case, we can specify a three-level model with transactions at level 1 located in census tracts at level 2 and urban tracks at level 3.

At level 1, we specify a linear relationship as follows:

$$y_{ijk} = \beta_{0,jk} + \sum_{s=1}^{S} \beta_{s,jk} x_{s,ijk} + \varepsilon_{ijk} \tag{14.1}$$

where $i = 1, \ldots, N$ refers to the transaction, $j = 1, \ldots, M$ refers to the census tract and $k = 1, \ldots, K$ refers to the urban track. $k = 1, \ldots, K$ is the housing price (or its logarithm) of transaction i in census tract j and urban track k; $x_{s,ijk}$ (with $s = 1, \ldots, S$) are the level 1 predictors; ε_{ijk} is a random term with $\varepsilon_{ijk} : Nid(0, \sigma_\varepsilon^2)$. This is a multilevel model since the intercept $\beta_{0,jk}$ and the slopes $\beta_{s,jk}$ are allowed to vary randomly at the census tract level such as (level 2):

$$\begin{aligned} \beta_{s,jk} &= \gamma_{s0,k} + \sum_{l=1}^{N_s} \gamma_{sl,k} x_{sl,jk} + w_{s,jk} \\ s &= 0, \ldots, S \end{aligned} \tag{14.2}$$

where N_s is the total number of variables operating at the census tract level affecting each transaction-specific parameter $\beta_{s,jk}$; $x_{sl,jk}$ (with $l = 1, \ldots, N_s$) are the level 2 predictors for the parameters $\beta_{s,jk}$; $w_{jk} = (w_{0,jk} \ldots w_{s,jk} \ldots w_{S,jk})'$ is a random term distributed as a multivariate normal with 0 mean and τ_β as a full variance-covariance matrix of dimension $(S+1)$. Finally, the intercept $\gamma_{sl,k}$ and the slopes $\gamma_{sl,k}$ of Eq. 14.2 are themselves allowed to vary randomly at the urban track level such as (level 3):

$$\gamma_{sl,k} = \mu_{sl0} + \sum_{m=1} \mu_{slm} x_{slm,k} + u_{sl,k} \tag{14.3}$$

for $s = 0, \ldots, S$ and $l = 0, \ldots, N_s$ where N_{sl} is the total number of variables operating at the urban track level affecting each census tract-specific parameter $\gamma_{sl,k}$; $x_{slm,k}$ (with $m = 1, \ldots, N_{sl}$) are the level 3 predictors for the parameters $\gamma_{sl,k}$; $u_k = (u_{00,k} \ldots u_{0l} \ldots u_{0N_s} \ldots u_{S0,k} \ldots u_{Sl} \ldots u_{SN_s})'$ is a random term distributed as a multivariate normal with 0 mean and τ_γ as a full variance-covariance matrix of dimension $\sum_{s=0}^{S} (N_s + 1)$. Note that the coefficients in Eq. 14.3 are not random but fixed. Finally, the errors terms (ε_{ijk}, $w_{s,jk}$ and $u_{sn,k}$) are assumed to be independent of each other.

Substituting Eqs. 14.2 and 14.3 in the level 1 model (Eq. 14.1) yields a mixed specification where the dependent variable y_{ijk} is the sum of a fixed part and a random part. The former includes explanatory variables operating at the three different spatial levels ($x_{s,ijk}$, $x_{sl,jk}$, $x_{slm,k}$), together with interactions between these levels, while the latter is a complex combination of the random terms ε_{ijk}, $w_{s,jk}$ and $u_{sl,k}$. This model is usually estimated using restricted maximum likelihood, noted thereafter REML (see for instance Raudenbush and Bryk 2002 or Goldstein 2003 for more details on the estimation method).

The full multilevel model Eqs. 14.1, 14.2 and 14.3 is very general with potentially a high number of unknown parameters to estimate. In practice, simpler models are estimated. In particular, not all parameters at level 1 vary randomly at the census tract level and/or not all parameters at level 2 vary randomly at the urban track level. We specify in the empirical analysis our assumptions concerning the variability of each parameter.

14.4 Results and Discussion

14.4.1 Grand Mean Model

In order to determine how variations in housing prices are allocated across each spatial level, we first specify the grand mean model. This model is fully unconditional in the sense that no predictor variables are specified at any level. Formally, it is represented as the following log-linear model:

$$\begin{cases} lprice_{ijk} = \beta_{0,jk} + \varepsilon_{ijk} \\ \beta_{0,jk} = \gamma_{00,k} + w_{0,jk} \Rightarrow lprice_{ijk} = \mu_{000} + u_{00,k} + w_{0,jk} + \varepsilon_{ijk} \\ \gamma_{00,k} = \mu_{000} + u_{00,k} \end{cases} \tag{14.4}$$

where $lprice_{ijk}$ is the log of price of transaction i in census tract j and urban track k. The coefficients are interpreted as follows: $\beta_{0,jk}$ is the mean log of price of census tract j in urban track k; $\gamma_{00,k}$ is the mean log of price in urban track k; μ_{000} is the grand mean. Finally, the error terms have the following properties: $\varepsilon_{ijk} \rightarrow Nid(0, \sigma_\varepsilon^2)$ is the random term measuring the deviation of transaction ijk's log of price from the mean log of price in census tract j; $w_{0,jk} \rightarrow Nid(0, \sigma_w^2)$ is the random term measuring the deviation of census tract jk's mean log of price from the mean log of price in urban track k; $u_{00,k} \rightarrow Nid(0, \sigma_u^2)$ is the random term measuring the deviation of urban track k's mean log of price from the grand mean.

The REML estimation results are displayed in Table 14.2 (third column). The average house price for the main urban tracks of 'Central Almond' in Madrid amounts to 426,015 € (Table 14.2).[3] The variation around this grand mean can be

[3] Since we use a log-linear model, this figure is the result of computing exp(12.96223).

Table 14.2 The Grand Mean model and Models 1 and 2

Variables			Grand mean model	Benchmark model	Model 2	
					Noise	Air-pollution
Fixed						
		Const.	12.96223***	8.696402***	8.837826***	8.957598***
	Structural	*Floor*	–	0.102336***	0.103926***	0.105367***
		Attic	–	0.036974***	0.032498***	0.037773***
		House	–	0.406189***	0.364783***	0.407730***
		Bedsit	–	0.067244***	0.075373***	0.063030***
		lm2	–	2.100856***	2.123871***	2.098390***
		State	–	0.111625***	0.115538 ***	0.114543***
Accessibility variables		*Discen*	–	–	−0.000092***	−0.000070***
		Axis	–	–	0.057878***	0.043855***
Air and noise variables		*Noise*	–	–	0.000452	–
		Cont	–	–	–	−0.003560***
Random:	Variance (std. error)					
		Tracks	0.067248	0.018476	–	0.008912
			(0.01547)	(0.00405)		(0.00215)
		Census	0. 051929	0.008431	0.013419	0.007277
			(0. 00600)	(0.00090)	(0.00120)	(0.00082)
		Houses	0. 186692	0.026395	0.026362	0.026255
			(0. 00534)	(0.00076)	(0.00075)	(0.00075)
Intra-class (tracks)			22 %	35 %	0 %	21 %
Intra-class (census)			17 %	16 %	33 %	17 %
LIK			−2255.50***	915.51***	869.34***	948.35***
Deviance (H_0: previous -if nested- model)			–	6.342.01***	–	65.69***
LR vs linear model			775.09***	1.141.67***	492.05***	624.71***

*significant at 0.10, **significant at 0.05, ***significant at 0.01

decomposed into variations at the level of the individual transaction, census tract and urban tracks using the variances of the error terms at the different levels.[4] The greatest variation occurs between individual transactions (61 %) although almost one-fourth of the variation takes place between tracks (22 %). Therefore, housing prices vary significantly between tracks. Finally, a multilevel model is indeed relevant since the LR test of absence of random effects strongly rejects the null.

The first column of Table 14.3 describes the price variations around the grand mean (426,015 €) at the urban track level. For instance, transactions in Velázquez, Prado-Recoletos and Castellana Av. are more than 300,000 € more expensive than the average price in Central Almond main tracks, while transactions in P° Delicias, Capitán Blanco Argibay, Pablo Iglesias-Ofelia Nieto and Jerónima Llorente tracks are more than 130,000 € cheaper.

These results are illustrated graphically in the upper left part of Fig. 14.4. The cheapest tracks are concentrated in the southwestern and northwestern parts of the area whereas the tracks with the highest premiums are located along Castellana and Recoletos-Prado tracks as well as some Eastern main streets. The deviations of prices in census tracts compared to the grand mean (right part of Fig. 14.4) follow a similar pattern but displaying some variations in more heterogeneous tracks like those located in the south and western part of Central Almond.

14.4.2 The Benchmark Model

Next, we estimate our benchmark model, labelled as Model 1. It consists in the grand mean model to which only structural attributes of each transaction are included in the level 1 equation:

$$\begin{cases} lprice_{ijk} = \beta_{0,jk} + \sum_{s=1}^{S} \beta_s x_{s,ijk} + \varepsilon_{ijk} \\ \beta_{0,jk} = \gamma_{00,k} + w_{0,jk} \\ \gamma_{00,k} = \mu_{000} + u_{00,k}. \end{cases} \quad (Model\ 1) \tag{14.5}$$

where S is the number of structural attributes. We assume that the associated coefficients are fixed: they do no vary randomly across census tracts and/or tracks.

[4] They are computed respectively as follows:

$$\sigma_\varepsilon^2/(\sigma_\varepsilon^2+\sigma_w^2+\sigma_u^2);\quad \sigma_w^2/(\sigma_\varepsilon^2+\sigma_w^2+\sigma_u^2) \text{ and } \sigma_u^2/(\sigma_\varepsilon^2+\sigma_w^2+\sigma_u^2).$$

The last two equations correspond respectively to the intra-class correlation for tracks and census tracts that are reported in Table 14.2.

Table 14.3 Urban tracks level premiums for the grand mean, benchmark and Model 2

Grand mean model		Benchmark model		Model 2	
Rank order	Price (€)	Rank order	Price (€)	Rank order	Price (€)
Valázquez Street	334,948	Prado-Recoletos	2,586	Prado-Recoletos	2,195
Prado-Recoletos	313,219	Velázquez Street	2,576	Velázquez Street	1,913
Castellana Av.	305,381	Castellana Av.	1,845	Castellana Av.	1,529
Concha Espina-R. Cajal	254,884	Cibeles-Alcalá-O'Donell	1,153	Isaac Peral-H. Eslava	947
Cibeles-Alcalá-O'Donell	229,961	P. Vergara-M. Pelayo	966	A. Alcocer-C. Rica	822
P. Vergara-M. Pelayo	117,759	Concha Espina-R. Cajal	821	P. Vergara-M. Pelayo	786
Isaac Peral-H. Eslava	110,666	Cartagena-Toreros	787	Cortes-Mayor	709
A. Alcocer-C. Rica	82,759	Isaac Peral-H. Eslava	762	Concha Espina-R. Cajal	601
M30 Eastern side	76,409	C. Bermúdez.-J. Abascal	728	Cibeles-Alcalá-O'Donell	575
C. Bermúdez.-J. Abascal	73,149	A. Aguilera-Génova	686	Planetario-Antracita	557
R. Rosas-I. Filipinas	61,507	R. Rosas-I. Filipinas	679	Alcalá-Ventas	429
Asturias-Sag. Corazón	61,097	Alcalá-Ventas	663	Asturias-Sag. Corazón	356
Alcalá-Ventas	55,484	A. Alcocer-C. Rica	546	R. Rosas-I. Filipinas	339
Cartagena-Toreros	46,436	Cortes-Mayor	468	Cartagena-Toreros	330
Orense-Infanta Mercedes	45,096	Guzmán el Bueno	409	Guzmán el Bueno	301
Planetario-Antracita	32,510	Joaquín Costa-F. Silvela	321	M30 SE side	299
Guzmán el Bueno	29,744	Vallehermoso	251	C. Bermúdez.-J. Abascal	260
Chamartín Station	25,973	Orense-Infanta Mercedes	130	Ciudad de Barcelona	188
A. Aguilera-Génova	23,439	Hortaleza-Fuencarral	80	Rondas-Bailén	174
Joaquín Costa-F. Silvela	16,505	Cibeles-G. Vía-Princesa	40	Orense-Infanta Mercedes	163
Mediterráneo-R. Cristina	14,509	Mª de Molina-América	24	A. Aguilera-Génova	163
Vallehermoso	13,857	R.F. Villav.-R. Victoria	−64	Chamartín Station	55
R.F. Villav.-R. Victoria	6,434	Azcona-Mnez. Izquierdo	−125	Dr. Esquerdo	−2
Mª de Molina-América	388	M30 Eastern side	−129	M30 Eastern side	−8
Dr. Esquerdo	−2,112	Dr. Esquerdo	−139	Vallehermoso	−22
Cibeles-G. Vía-Princesa	−6,863	San Bernardo	−152	P. Segovia-Cerr.-Toledo	−67
	−9,911		−180	Atocha Street	−162

(continued)

Table 14.3 (continued)

Grand mean model		Benchmark model		Model 2	
Rank order	Price (€)	Rank order	Price (€)	Rank order	Price (€)
Conde Peñalver-Narváez		Ciudad de Barcelona			
Cortes-Mayor	−11,025	Bravo Murillo	−195	Mediterráneo-R. Cristina	−178
Hortaleza-Fuencarral	−11,958	Asturias-Sag. Corazón	−243	M30 SW side	−192
M30 SE side	−14,743	Rondas-Bailén	−277	Joaquín Costa-F. Silvela	−194
Azcona-Mnez. Izquierdo	−23,636	Atocha Street	−289	Hortaleza-Fuencarral	−267
Pradillo Street	−29,888	Conde Peñalver-Narváez	−332	Atocha Station	−303
Ciudad de Barcelona	−40,572	P. Segovia-Cerr.-Toledo	−373	Sta. Mª Cabeza	−315
Atocha Station	−50,607	Pradillo Street	−385	Cibeles-G. Vía-Princesa	−316
López de Hoyos	−56,702	Mediterráneo-R. Cristina	−388	San Bernardo	−392
P. Segovia-Cerr.-Toledo	−61,895	Planetario-Antracita	−406	R.F. Vill.-R. Victoria	−416
Clara del Rey	−65,817	López de Hoyos	−432	Bravo Murillo	−419
San Bernardo	−79,978	Clara del Rey	−466	Pº Delicias	−503
Rondas-Bailén	−83,247	Chamartín Station	−470	Azcona-Mnez. Izquierdo	−536
G. Yagüe-Francos Rgz.	−88,731	M30 SE side	−596	M. Viana-S.A. Cruz	−557
Bravo Murillo	−89,368	Atocha Station	−611	Capit. Blanco Argibay	−597
Atocha Street	−89,947	G. Yagüe-Francos Rgz.	−720	Pradillo Street	−626
Sta. Mª Cabeza	−92,251	M. Viana-S.A. Cruz	−852	Conde Peñalver-Narváez	−633
M30 SW side	−101,645	Sta. Mª Cabeza	−904	Mª de Molina-América	−696
M. Viana-S.A. Cruz	−110,390	M30 SW side	−905	G. Yagüe-Francos Rgz.	−715
Pº Delicias	−131,708	Pº Delicias	−961	López de Hoyos	−758
Capit. Blanco Argibay	−132,078	P. Iglesias-O. Nieto	−981	Clara del Rey	−805
P. Iglesias-O. Nieto	−136,130	Capit. Blanco Argibay	−1,040	P. Iglesias-O. Nieto	−999
Jerónima Llorente	−163,490	Jerónima Llorente	−1,334	Jerónima Llorente	−1,533

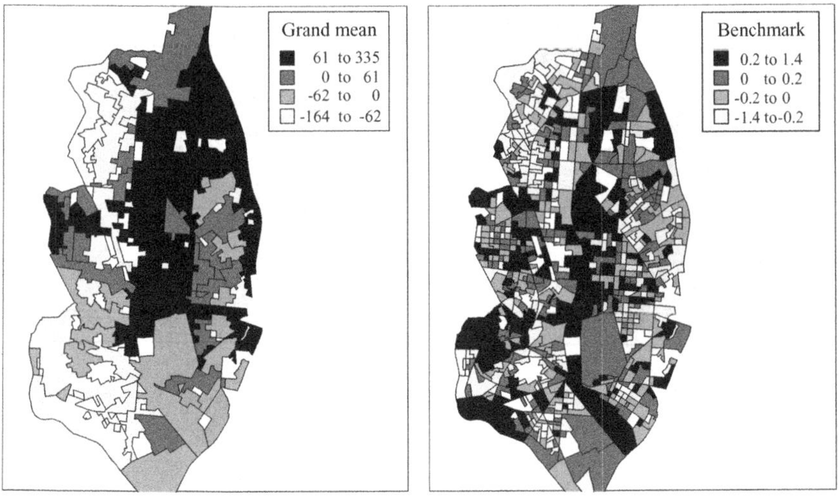

Fig. 14.4 Urban track (*left*) and census tract-level (*right*) premiums (mile €)

Note that this assumption will be relaxed in the following section for some variables. The REML results are reported in Table 14.2. Among all structural variables considered, we only report the coefficients that are significant at 5 %. All the structural attributes coefficients estimates show the expected sign. They are strongly statistically significant at 1 % with the exception of 'duplex' and 'number of bedrooms'. Specifically, the number of bedrooms variable, which is usually relevant in hedonic price models, is not significant even at the 5 % level because it shows a strong correlation with 'floor area'.

The deviance or likelihood test, i.e. the difference in the likelihood ratio statistic of this model and the grand mean model, is equal to 6,342.01. Under the null hypothesis, it follows a chi-squared distribution with degrees of freedom equal to 6, i.e. the number of new parameters (Woodhouse et al. 1996). The p-value is less than 0.001: the structural attributes therefore significantly explain house price variation in the model.

Looking at the intra-class correlations, we see that the inclusion of structural attributes implies a strong decline of the transaction-level variance. Consequently, a large part of price differences between individual transactions is a result of differences in these attributes. In contrast, 35 % of the total variation now occurs between tracks, compared to 22 % in the grand mean model. This result is reflected by the analysis of the track-level differences (Table 14.3) as both the rank of tracks and the size of their contextual effects are modified. For instance, two of the previous most expensive urban tracks, Asturias-Sagrado Corazón and Chamartín Station are now closer to Central Almond average, while a previously below-average track, Cortes-Mayor, is now significantly above average. Much more evident are the modifications in the rank of the census tracts (right part of Fig. 14.5). There still exists some concentration of higher premiums in part of the

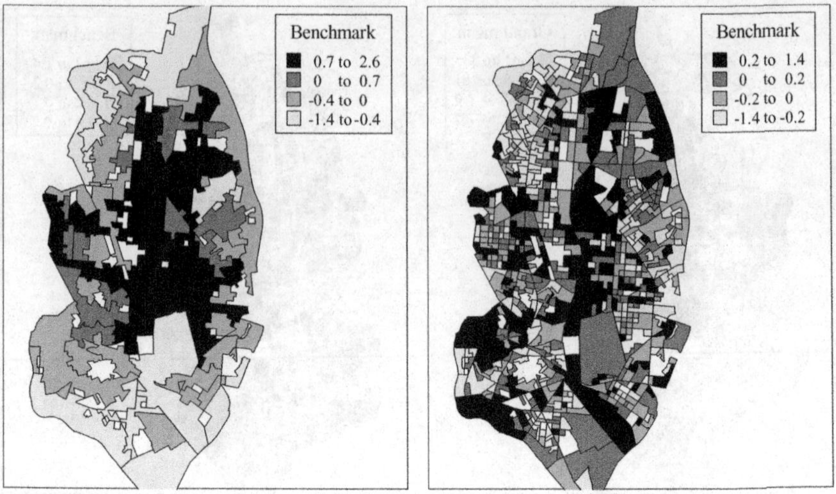

Fig. 14.5 Road track (*left*) and census tract-level (*right*) premiums (mile €) in the benchmark model

census tracts of the central axis (Castellana and Prado-Recoletos), with the rest of the values more or less scattered all over the Central Almond. Also, the size of the neighbourhood and census tract premiums has declined substantially, meaning that they were previously mainly capturing the effects of structural attributes. Furthermore, buyers are getting much less for their money in tracks like Prado-Recoletos and Velázquez than in areas like *Jerónima Llorente* and *Capitán Blanco Argibay*.

14.4.3 Model with Structural and Accessibility Variables

Model 2 includes additional fixed accessibility indicators and pollution variables (noise or air pollution) at the individual level. Among all the accessibility variables that we tried, only two accessibility indicators are significant at 5 %: distance to the CBD (*discen*) and distance to the main city axis (*axis*). Concerning the analysis of the impact of noise and air pollution on housing prices, we have specified two different models depending on the selected pollution variable: (1) model 2N includes the subjective measure of noise (*noise*) and (2) model 2C includes the subjective measure of air pollution (*cont*). Due to the high correlation between air and noise pollution levels (Li and Brown 1980), it is necessary to sort out these separate effects in order to measure their marginal effect on housing prices.

The REML estimation results are displayed in Table 14.2. The inclusion of these accessibility and pollution variables does not alter either the values or the sign of the structural attributes, which are all significant at 5 %. Distance to the CBD (*discen*) and distance to main axis (*axis*), they are significant at 1 %. We note that the former plays a negative effect on housing prices while the latter has a positive effect. Households therefore have a higher willingness to pay whenever they are located closer to the CBD and farther away from the main axis.

The coefficient for air pollution is significant at 5 % while the noise parameter is positive and it does not seem to have any impact on housing prices. The deviance statistic (with Model 1 as the null hypothesis) indicates that the addition of accessibility and air-pollution attributes has a significant effect on housing prices. Interestingly, we find that in the model for noise, the track-level random effect is no longer significant[5] resulting in the census tract level now explaining 33 % of house price variations. This result means that noise seems to be a more "local" phenomenon than air quality so that random variations at the census tract level are enough to capture price variability. Indeed, according to Falzone (1999) and Bickel et al. (1999), noise is often transitory and seldom catastrophic. It is therefore considered as an environmental intrusion with a very local effect, which depends, among other things, on the time of the day or the distribution and distance of exposed persons from the source.

Finally, we analyse how the track premiums have changed following the inclusion of accessibility and air-pollution variables (last column of Table 14.3). First, the effects of area are not significantly different from the benchmark model, meaning that this one was not capturing the effects of accessibility and/or air-pollution variables. However, there are interesting changes in rank, as the promotion of the *M30 East side*, *Planetario-Antracita* and *Rondas-Bailén*, between others. These tracks command a higher premium, given the accessibility and air-pollution attributes of the areas, which may be caused by other features, such as social class. Other tracks have experienced an important decrease in the premium rank with respect to the previous models. This is the case of *María de Molina-América*, *Conde de Peñalver-Narváez* and *Azcona-Martínez Izquierdo*, in which buyers are getting much more for their money.

14.4.4 Model with Structural, Accessibility and Census Tract Variables

We now estimate a model with the same random and transaction level fixed terms as in the previous model, but which further incorporates attributes available at the census tract level (Model 3):

[5] This is why the deviance statistic has not been computed in as Model 1 is not nested in this model.

$$\begin{cases} lprice_{ijk} = \beta_{0,jk} + \sum_{s=1}^{S} \beta_s x_{s,ijk} + \varepsilon_{ijk} \\ \beta_{0,jk} = \gamma_{00,k} + \sum_{l=1}^{N_0} \gamma_{0l} x_{0l,ij} + w_{0,jk} \quad (Model~3) \\ \gamma_{00,k} = \mu_{000} + u_{00,k}.) \end{cases} \tag{14.6}$$

We assume that these N_0 census tract level variables only affect the intercept of the level 1 model ($\beta_{0,jk}$) and that they remain fixed across census tracts, i.e. they do not vary randomly at the track level. Among all the attribute variables that we tried, only two of the census tracts variables shown in Table 14.1 are significant in the following models: *p65* and *educ*.

The REML estimation results are displayed in Table 14.4. Compared to model 2, since the census tract variables do not vary at the level of houses, the fixed and random estimates for the transaction-level attributes remain more or less unchanged, mainly for the structural attributes. However, the census tract-level and urban track-level random effects have decreased, so that the transaction level now explains more than two thirds of house price variations (67 % for noise and 72 % for air-pollution). Again, the track random effect is not significant for the model with noise.

As the census tract variables act as a proxy for social class, they have a significant effect upon house price differentials with the expected sign. This result is confirmed by the computation of the deviance statistic with Model 2 as the null hypothesis. The differential impacts of noise and air pollution on house prices remain unchanged: noise coefficient is positive and statistically non-significant while air-pollution still exerts an inverse significant effect on house prices.

14.4.5 Model with Varying Slopes for lm2 and, in Case, Noise/Cont

All the previous models assumed that the structural attributes and the pollution variables are constant across downtown Madrid. However, in Model 3, approximately one third of house price variation occurs between census tracts and/or urban tracks (28 % for the noise equation and 33 % for the air-pollution one). These unexplained variations might in fact be caused by variation in the implicit prices of structural attributes and/or pollution variables at both spatial levels. Therefore, we finally estimate models in which some level 1 coefficients are allowed to vary randomly at higher spatial levels.

Since floor area (*lm2*) is the main structural attribute, it is allowed to vary randomly at the census tract level. The measures of noise and air-pollution are also allowed to vary randomly at the census tract and urban track levels. After several tries, we found that the measure of noise does not allow random variations at the urban track level, either in its own coefficient or in the constant term. In the

Table 14.4 Models 3 and 4 with structural attributes, accessibility variables and census tract level variables, as well as varying slopes

			Model 3		Model 4	
			Noise	Air-pollution	Noise	Air-pollution
	Constant		8.253940***	8.531377***	8.333635***	8.602855***
Structural	*Floor*		0.103200***	0.105547***	0.113124***	0.109873***
	Attic		0.037431***	0.040765***	0.043705***	0.044288***
	House		0.365515***	0.401797***	0.361952***	0.364110***
	Bedsit		0.071369***	0.060026***	0.066536***	0.057481***
	lm2		2.098307***	2.082467***	2.076103***	2.061383***
	State		0.117685***	0.114747***	0.120419***	0.117618***
Accesibility	*Discen*		−0.000066***	−0.000059***	−0.000056***	−0.000060***
	Axis		0.047643***	0.036963***	0.038183***	0.038063***
Census tracts	*Educ*		0.008429***	0.006765***	0.007602***	0.006181***
	p65		–	−0.003885***	–	−0.002730***
Pollution variables	*Noise*		0.000158	–	−0.000294	–
	Cont		–	−0.002535**	–	−0.003073***
Variance and covariance (standard error)	*Neighb.*	*constant*	–	0.0069519 (0.00171)	–	0.005040 (0.00134)
	Census	*constant*	0.010638 (0.00102)	0.006020 (0.00075)	0.235241 (0.05514)	1.65e-19 (–)
		noise/air-pollution	–	–	0.000018 (8.04e-06)	0.000041 (0.00001)
		lm2	–	–	0.071494 (0.01121)	0.016990 (0.00310)
		noise/airpoll.-lm2	–	–	−0.000033 (0.00024)	−0.000820 (0.00018)
		noise var-constant	–	–	−0.00082 (0.00057)	–

(continued)

Table 14.4 (continued)

	Model 3		Model 4	
	Noise	Air-pollution	Noise	Air-pollution
lm2-constant	–	–	−0.116629	–
			(0.02258)	
Houses	0.026837	0.026241	0.023663	0.024613
	(0.00074)	(0.00075)	(0.00070)	(0.00071)
Intra-class (tracks)	0 %	18 %	–	–
Intra-class (census)	28 %	15 %	–	–
LIK	917.28***	976.25***	1,010.53***	1,025.00***
Deviance (H_0: Model 2)	95.89***	55.81***	–	–
LR vs linear model	374.28***	473.12***	560.78***	570.62***

**5 % and *1 %

case of air-pollution equation, only the constant term significantly varies at the level of urban tracks. These results would be confirming the local nature of noise with respect to air-pollution.

Formally, for the noise measure, our final specification is as follows:

$$\begin{cases} lprice_{ijk} = \beta_{0,j} + \beta_{1,j} lm2_{ijk} + \beta_{2,j} noise_{ijk} + \sum_{s=3}^{S} \beta_s x_{s,ijk} + \varepsilon_{ijk} \\ \beta_{0,j} = \gamma_{00} + \sum_{l=1}^{N_0} \gamma_{0l} x_{0l,jk} + w_{0,jk} \\ \beta_{1,j} = \gamma_{10} + w_{1,jk} \\ \beta_{2,j} = \gamma_{20} + w_{2,jk} \end{cases} \tag{14.7}$$

For air pollution, our final specification is as follows:

$$\begin{cases} lprice_{ijk} = \beta_{0,j} + \beta_{1,j} lm2_{ijk} + \beta_{2,j} cont_{ijk} + \sum_{s=3}^{S} \beta_s x_{s,ijk} + \varepsilon_{ijk} \\ \beta_{0,j} = \gamma_{00,k} + \sum_{l=1}^{N_0} \gamma_{0l} x_{0l,jk} + w_{0,jk} \\ \beta_{1,j} = \gamma_{10} + w_{1,jk} \\ \beta_{2,j} = \gamma_{20} + w_{2,jk} \\ \gamma_{00,k} = \mu_{000} + u_{00,k} \end{cases} \tag{14.8}$$

where *cont* is the variable for air-pollution.

The REML estimation results are displayed in Table 14.4. All the structural, locational and pollution variables are strongly significant, with the exception of noise. Interestingly, this model is the only one -for noise- in which the coefficient associated to this variable is negative, as expected, although it still remains non-significant, once higher-level interactions at the level of census tracts are explicitly considered. In spite of not being statistically significant, we find that this coefficient significantly varies at the level of census tracts, showing positive and negative values depending on the spatial location. Therefore, noise and air pollution have a negative influence on housing prices which is only globally significant for the second one.[6]

Figure 14.6a displays the change implied by the inclusion of the random floor area and air pollution terms on the implicit prices by census tracts in Model 4. Compared to the grand mean model (Fig. 14.4), the ranks greatly change.

[6] As stated in Bickel et al. (1999), noise costs are extremely variable since they depend on several factors and exhibit large non-linearities. This is why it is more difficult to find a generalization for marginal noise costs than for air pollution.

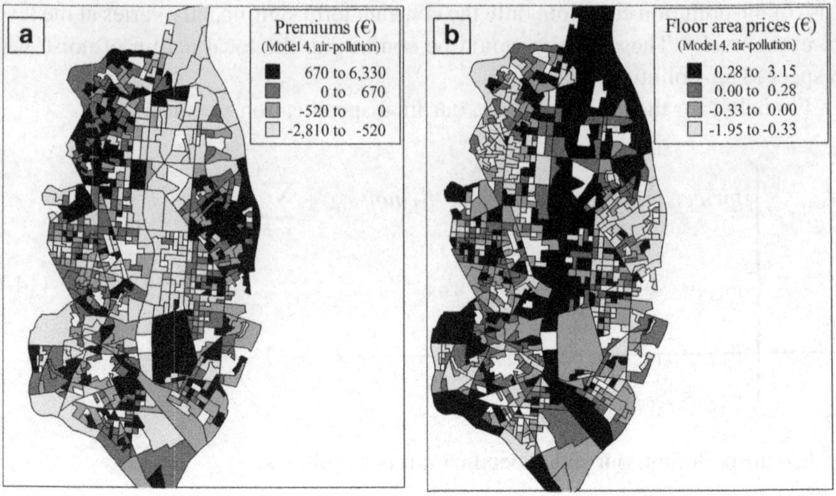

Fig. 14.6 Model 4: (**a**) Census tract-level premiums, (**b**) census tract-level floor area differential prices of floor area

Figure 14.6a shows the existence of a split between the census tracts located along the vertical *Castellana-Recoletos-Prado* axis joint with the northeast edge of Central Almond, and the peripheral areas (northwestern and eastern edges), becoming the first less expensive once the differential prices of floor area and perceived air-pollution have been taken into account, and vice versa. Overall, the map gives an idea of which areas require an additional premium, what could be viewed as a measurement of 'desirability'. This is the case of the urban tracks located in the district of *Tetuán* (northwest) and the districts of *Salamanca* and *Retiro* (east). These two big clusters are characterized for being closed to either the CBD -in the first case- and the most exclusive shopping centres -in the second case-, as well as having an excellent accessibility to both inner and outer city.

Model 4 also allows the implicit price of floor area to vary at the census tract level. The variance reported in Table 14.4 measures the variation in the price of floor area between census tracts and the covariance term measures the relationship between other variables (average census tract-level house price, noise and air-pollution) and the implicit price of floor area. With the exception of the covariance term between air-pollution and floor area price, these terms are all significant. On average across Central Almond, an extra square metre of floor area raises house price by 7.86 € (a general slope of 2.06 in the equation for air-pollution), but this figure varies from place to place in a range from 12.40 € (in *Recoletos-Prado* urban track) to 5.91 € (in *Jerónima Llorente* urban track). These differences reflect the differences in supply and demand that operate in these tracks. In Fig. 14.6b, we have represented the differential implicit prices of floor area by census tracts. This variable shows a more or less opposite spatial distribution than the census tract-level premiums; i.e. the 'discount' of floor area variations from house prices spatial distribution reveals a different pattern for the urban tracks premiums or the

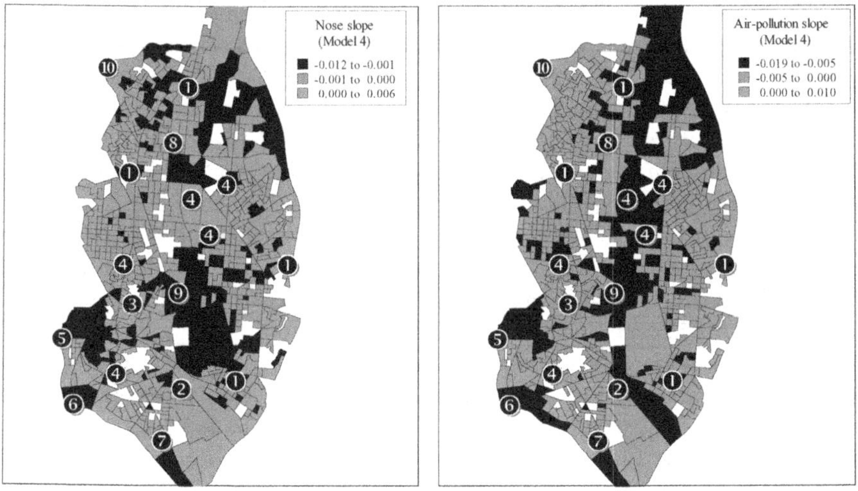

Fig. 14.7 Changes in census-tract-level prices due to noise and air-pollution

desirability that people show for certain sub-markets. This result highlights the importance of considering floor area spatial variations in order to explain house prices distribution in Central Almond.

Finally, in Model 4 the marginal prices of noise and air pollution (parameters) are allowed to vary at the census tract level. We see that the effects of noise and air-pollution per se vary quite significantly between census tracts, though with a different sign (Fig. 14.7). We observe a spatial coincidence between clusters of lower negative coefficients of both noise and air-pollution. They are mainly located along some of the busiest tracks of Central Almond (see Fig. 14.2a), like the *Atocha* and *Chamartín Station* areas, the *M30* Southwestern side, *Rondas-Bailén*, *Recoletos-Prado*, *Castellana*, *Raimundo Fernández Villaverde-Reina Victoria*, *Cibeles-Alcalá-O'Donell* and *Alberto Alcocer-Costa Rica*. There are also other coincident clusters of positive coefficients, which are more or less concentrated in the Northwestern, Eastern and Southern parts of Central Almond. In these places, residents are willing to pay more money for houses with more perceived noise and air-pollution. This counter-intuitive sign could be the result of the influence of other variables acting in people's mind, such as location in wider streets and proximity to principal or historical places, like the Royal Palace. In fact, these areas are characterized by high population density living in old buildings, which are in narrow streets with traffic access and no good visibility. Probably, this is one of the reasons why people are willing to pay more money for houses located in open spaces like avenues or squares, though they are perceived as being noisier.

At this point, it is possible to make a first evaluation of the Action Plan projects for Central Almond in the period 2008–2011 (see Fig. 14.3b). It must be said that not all the projects have been finished (even started in some cases) because of the severe constraints imposed by the economic crisis. For this reason, the computation of the air-pollution costs from the significant coefficient variations across census tracts can help us to establish a first rank of priorities. Building a buffer of 500 m around each project, it is possible to analyze the impact of air-pollution on housing prices in each area. We can detect four projects almost totally located in those census tracts registering the highest cost in air-pollution (the ones in dark blue, in Fig. 14.7): *Recoletos-Prado* project (9), *Delicias-Méndez Álvaro* especial plan (2), *Castellana* and *Velázquez* boulevard restoration (4). These are the areas in which a reduction of air-pollution will have the maximum impact.

Secondly, there are another four projects partially located in census tracts with the highest air-pollution cost: tunnelling of metro depot in *Plaza de Castilla* (1), *Vicente Calderón-Mahou* restoration (6), *San Francisco el Grande* revitalization (5) and *Alberto Alcocer-Génova* boulevard restoration (4). From the environmental point of view, these eight projects should have the highest priority for the local authorities. In contrast, there are two projects mostly located in census tracts for which air-pollution is an amenity (with a positive slope in the model): the metro depot tunnelling in *Cuatro Caminos* and *Ventas*. In terms of air-pollution costs exclusively, these projects should have the least priority in the future action plan.

14.5 Conclusions

The aim of this chapter was to to show how multilevel models and kriging are useful techniques to deal with data at various spatial scales and supports, in order to evaluate the households' marginal willingness to pay for reduced noise and better air quality in the most congested areas of downtown Madrid. For that purpose, we have applied hedonic regression to a sample of 3,302 houses. Contrary to most hedonic studies in the literature, we use subjective rather than objective measures of noise and air pollution. Moreover, as the dataset has a hierarchical structure, i.e. the houses are nested within census tracts and urban tracks, we use multilevel modelling. These models allow using variables operating at different scales and allow the marginal prices of air and noise pollution to vary randomly at higher levels. Moreover, by taking into account the clustered nature of the residuals, more reliable inference can be achieved.

We found that noise does not seem to influence housing prices globally but that it significantly varies randomly at the level of census tracts, showing positive and negative values depending on the spatial location. The impact of air pollution is negative and significant and also randomly varies at the level of tracks. Spatial coincidence of lower negative coefficients of both noise and air pollution are found located along some of the busiest tracks of Central Almond. These results allow making a first evaluation of the Action Plan projects for Central Almond that have been undertaken in the period 2008–2011. Using the estimation results, we detect the projects located in the areas in which a reduction of air pollution will have the maximum impact and that should therefore have the highest priority in the future action plan.

Future research should concentrate in checking the robustness of our results by means of including new observations to the sample. For instance, adding a time dimension to the spatial multilevel model would allow checking the possible evolution of marginal prices for air and noise pollution in Central Almond. We could also consider the peripheral districts information in order to detect possible spatial discontinuities between our model results in Central Almond and the vast area outside the M30 belt, which is known to operate as a strong barrier in the city of Madrid.

References

Alker HR (1969) A typology of fallacies. In: Dogan M, Rokkan S (eds) Quantitative ecological analysis in the social sciences. M.I.T. Press, CambridgeMA

Anselin L, Lozano-Gracia N (2008) Errors in variables and spatial effects in hedonic house price models of ambient air quality. Empirical Economics, Springer 34(1):5–34

Arbia G (1989) Statistical effect of data transformations: a proposed general framework. In: Goodchild M, Gopal S (eds) The accuracy of spatial data bases. Taylor and Francis, London, pp 249–259

Ayuntamiento de Madrid (2009a) Plan de Acción del Área de Gobierno de Urbanismo y Vivienda para la Revitalización del Centro Urbano 2008–2011, Área de Gobierno de Urbanismo y Vivienda, Madrid

Ayuntamiento de Madrid (2009b) Anuario Estadístico Municipal 2009, Madrid

Ayuntamiento de Madrid (2010) Contabilidad Municipal de la Ciudad de Madrid. Base 2002. Serie 2000–2008(1ªe). Dirección General de Estadística, Madrid

Banerjee S, Carlin BP, Gelfand AE (2004) Hierarchical modeling and analysis for spatial data. Chapman and Hall/CRC Press, Boca Raton

Banfi S, Filippini M, Horehájová A (2007) Hedonic price functions for Zurich and Lugano with special focus on electrosmog. CEPE working paper no. 57

Baranzini A, Ramírez JV (2005) Paying for quietness: the impact of noise on Geneva rents. Urban Stud 42:633–646

Berezansky B, Portnov BA, Barzilai B (2010) Objective vs. perceived air pollution as a factor of housing pricing: a case study of the greater haifa metropolitan area. J Real Estate Lit 18:99–122

Beron KJ, Murdoch JC, Thayer MA (1999) Hierarchical linear models with application to air pollution in the South Coast Air Basin. Am J Agric Econ 81:1123–1127

Bickel P, Schmid S, Tervonen J, Hämekoski K, Otterström T, Anton P, Enei R, Leone G. van Donselaar P, Carmigchelt H (1999) Environmental marginal cost case studies. Working funded by 5th framework RTD programme. IER, University of Stuttgart, Stuttgart. Available from www.its.leeds.ac.uk/unite

Bonin O (2009) Urban location and multi-level hedonic models for the Ile-de-France region. In: Paper presented at ISA international housing conference, Glasgow, 1–4 Sept 2009. Available from www.gla.ac.uk/media/media_129762_en.pdf

Boyle MA, Kiel KA (2001) A survey of house price hedonic studies of the impact of environmental externalities. J Real Estate Lit 9:117–144

Brody SD, Peck BM, Highfield WE (2004) Examining localized patterns of air quality perception in Texas: a spatial and statistical analysis. Risk Anal 24:1561–1574

Chasco C, López A (2009) Multilevel models: an application to the beta-convergence model. Région et Développement 30:35–58

Cheshire P, Sheppard S (1998) Estimating the demand for housing, land, and neighbourhood characteristics. Oxford Bull Econ Stat 60:357–382

Cummins R (2000) Personal income and subjective well-being: a review. J Happ Stud 1:133–158
Delfim L, Martins I (2007) Monitoring urban quality of life: the Porto experience. Soc Indic Res 80:411–425
Delucchi MA, Murphy JJ, McCubbin DR (2002) The health and visibility cost of air pollution: a comparison of estimation methods. J Environ Manag 64:139–152
Djurdjevic D, Eugster C, Haase R (2008) Estimation of hedonic models using a multilevel approach: an application to the Swiss rental market. Swiss J Econ Stat 144:679–701
Falzone KL (1999) Airport noise pollution: is there a solution in sight? Boston Coll Environ Aff Law Rev 26:769–807
Fontes GG, Simões RF, Hermeto Camilo De Oliveira AM (2009) Urban attributes and wage disparities in Brazil: a multilevel hierarchical model. Reg Stud 44:595–607
Gelfand AE, Banerjee S, Sirmans CF, Tud Y, Ongd SE (2007) Multilevel modelling using spatial processes: application to the Singapore housing market. Comput Stat Data Anal 51:3567–3579
Goldstein H (2003) Multilevel statistical models. Arnold, London
Gotway CA, Young LJ (2002) Combining incompatible spatial data. J Am Stat Assoc 97–458:632–648
Hartley PR, Hendrix ME, Osherson D (2005) Real estate values and air pollution: measured levels and subjective expectations. Discussion paper, Rice University
Hox JJ (1995) Applied multilevel analysis. TT-Publikaties, Amsterdam
Hui ECM, Chau CK, Pun L, Law MY (2008) Measuring the neighbouring and environmental effects on residential property value: using spatial weighting matrix. Build Environ 42:2333–2343
Instituto Nacional de Estadística, INE (2008) Callejero del Censo Electoral 2007. Madrid
Instituto Nacional de Estadística, INE (2010) Padrón de Habitantes. Madrid
Jones K, Bullen N (1994) Contextual models of urban house prices: a comparison of fixed- and random-coefficient models developed by expansion. Econ Geogr 70:252–272
Khattab N (2007) Ethnic and regional determinants of unemployment in the Israeli labour market: a multilevel model. Reg Stud 40:93–105
Kulu H, Billari F (2010) Multilevel analysis of internal migration in a transitional country: the case of Estonia. Reg Stud 26:243–255
Leishman C (2009) Spatial change and the structure of urban housing sub-markets. Housing Stud 24:563–585
Li MM, Brown HJ (1980) Micro-neighbourhood externalities and hedonic housing prices. Land Econ 56:125–141
Miedema HME, Oudshoorn CGM (2001) Annoyance from transportation noise: relationships with exposure metrics DNL and DENL and their confidence interval. Environ Health Perspect 109:409–416
Mieszkowski P, Saper AM (1978) An estimate of the effects of airport noise on property values. J Urban Econ 5:425–440
Murti MN, Gulati SC, Banerjee A (2003) Hedonic property prices and valuation of benefits from reducing urban air pollution in India. Institute of Economic Growth, Delhi Discussion Papers 61
Nelson JP (1979) Airport noise, location rent, and the market for residential amenities. J Environ Econ Manag 6:320–331
Nelson JP (2004) Meta-analysis of airport noise and hedonic property values. J Trans Econ Pol 38:1–28
Nourse HO (1967) The effect of air pollution on house values. Land Econ 43:181–189
Orford S (2000) Modelling spatial structures in local housing market dynamics: a multilevel perspective. Urban Stud 37:1643–1671
Palmquist RB (2005) Property value models. In: Mäler KG, Vincent J (eds) Handbook of environmental economics, vol 2. North Holland, Amsterdam
Raudenbush SW, Bryk AS (2002) Hierarchical linear models. Applications and data analysis methods, 2nd edn. Sage Publications, Thousand Oaks, US

Ridker RG, Henning JA (1967) The determinants of residential property values with special reference to air-pollution. Rev Econ Stat 49:246–256
Smith VK, Huang J-C (1993) Hedonic models and air pollution: twenty-five years and counting. Environ Resour Econ 36:23–36
Smith VK, Huang JC (1995) Can markets value air quality? A meta-analysis of hedonic property value models. J Polit Econ 103:209–227
Srholec M (2010) A multilevel approach to geography of innovation. Reg Stud 44:1207–1220
Ward C, Dale A (2006) Geographical variation in female labour force participation: an application of multilevel modelling. Reg Stud 26:243–255
Wardman M, Bristow AL (2004) Traffic related noise and air quality valuations: evidence from stated preference residential choice models. Transp Res Part D 9:1–27
Woodhouse G, Rasbash J, Goldstein H, Yang M (1996) Introduction to multi-level modelling. In: Woodhouse G (ed) Multi-level modelling applications: a guide for users of Mln. Institute of Education, University of London, US, pp 9–57

Chapter 15
Concentration Analysis Using Microgreographic Data

Federico Pablo-Martí and Josep-Maria Arauzo-Carod

15.1 Introduction

Analysis of spatial distribution of economic activity has plenty of implications in several areas like urban planning, infrastructures, firm supporting policies and land use, among others, and is receiving an increasing attention by researchers. Most of analyses of spatial distribution of economic activity have been carried out using extant administrative units (e.g., counties, regions, etc.), but unfortunately, these analyses suffer from the shortcoming that administrative units vary greatly in size and shape, do not always coincide with real economic areas and are sometimes arbitrary.

To deal with such constraints, recent research has started to use ad hoc units (usually smaller than previous ones), as we do in this chapter. These units are created by equally dividing an area into homogeneous squared cells and, therefore, they do not exactly match any extant administrative unit,[1] which could (potentially) be a problem in terms of data collection from these units but, at the same time, they provide more flexibility to the analysis in view that they are not artificially constrained by any administrative unit. Accordingly, this

[1] There are also other approaches such as those that use the stochastic methodology of Point Pattern or those that use Neuronal Networks for pattern recognition. However, these approaches are not able to do the multisectorial analyses that are the goal of this work. For a discussion about whether administrative units match functional or economic units see Parr (2008).

F. Pablo-Martí (✉)
Universidad de Alcalá, C/Alvaro Cumqueio, 7, Alcala de Henares,
Madrid 28805, Spain
e-mail: federico.pablo@uah.es

J.-M. Arauzo-Carod
Facultat d'Economia i Empresa, Universitat Rovira i Virgili (CREIP-QURE),
Avda. de la Universitat, 1, Reus 43204, Spain

E. Fernández Vázquez and F. Rubiera Morollón (eds.), *Defining the Spatial Scale in Modern Regional Analysis*, Advances in Spatial Science,
DOI 10.1007/978-3-642-31994-5_15, © Springer-Verlag Berlin Heidelberg 2012

methodology allows to better control for spatial inequalities and to better capture real distribution of economic activity.

Mapping the spatial distribution of economic activity is of key importance for policy makers, since they do need precise and accurate data in order to, firstly, identify real spatial distribution of economic activity and, secondly, to design public policies. Unfortunately, nowadays there is no a clear agreement among academics about which methodology is much suitable for identifying this spatial distribution of firms. Currently, there are two main approaches at the empirical literature: Industrial Districts and Clusters. While the former is more popular[2] mainly due to the Sforzi-ISTAT methodology (Sforzi 1990; ISTAT 2006), the latter is potentially easier to use because of its lower data requirements. Nevertheless, our approach differs from previous ones since we are not identifying number, industry specialisation and geographical position of clusters, but whether firms for specific industries follow a concentrated or a dispersed locational pattern.

The methodology proposed here tries to identify manufacturing and service[3] concentration vs. dispersion processes in Spain and aims to overcome previous constraints, to obtain more precise results and, as a consequence, to improve the design of public policies. Concretely, by dividing Spain into homogeneous cells we check whether each industry follows a concentrated or dispersed pattern and, later, whether collocation exists for pairs of industries, so clusters made by different industries can also be identified. Although it is not exactly the aim of this chapter, there are related issues that arise after concentrated and dispersed industries are identified, as grouping previous industries according to the reasons that explain their clusterization processes; that is, whether firms tend to locate together because they look for the same types of site (regardless of the industry to which they belong), or whether firms look to be located close to their suppliers/customers in order to optimise commercial exchanges but, as we have stated previously, we left this topic for further research.

Nevertheless, mapping economic activity is not enough since it is just the first step to get a clear portrait of its spatial distribution. What is really important is to identify whether extant location implies a concentrated/dispersed pattern and, in a second step, to identify reasons explaining this locational pattern. Additionally, it is also important to check whether this is a specific pattern or it is common at national level. In this sense, Arbia et al. (2008) classify two main approaches followed by scholars trying to identify clustering of economic activities: first one is based on a hypothetical representative unit, a site, (Jaffe et al. 1993; Rauch 1993; Ciccone and Hall 1996; Henderson 2003), while second one tries to modelise the whole spatial distribution of economic activities according to a set of hypothesis (Duranton and Overman 2005).

---

[2] See, among others, the applications of Boix and Galletto (2008) for Spain and De Propris (2005) for the UK.

[3] Applications for services are scarce. See, for instance, Sforzi (1999).

In this chapter we follow the latter approach, since it is more realistic by assuming that what is really important is to analyse which the joint behaviour of different units across the space is, instead of just focusing on a single representative unit. Concretely, Duranton and Overman (2005) summarize how localization's tests have evolved and identify three generations of spatial concentration measures. A first one (e.g., Gini Index) in which spatial issues are roughly taken into account (Krugman 1991). A second one departing from Ellison and Glaeser (1997), that approaches space in a discrete way and, consequently, suffers from MAUP problems since calculations are made using administrative units (Devereaux et al. 2004; Maurel and Sédillot 1999).[4] This approach has advantages in terms of computational feasibility but, at the same time, wastes a lot of useful information (data is aggregated) and implies to arbitrary decide the spatial aggregation level to be used. Finally, a third one (Duranton and Overman 2005; Arbia 2001) that considers spatial continuity, and where MAUP problems have been successfully overtaken. At this point is where we intend to place our contribution.

This chapter is organised as follows. In the next section we review the main literature on the spatial distribution of economic activity and the spatial units used in empirical analysis. In the third section we explain the data set and we describe the spatial distribution of firms in Spain. In the fourth section we define the methodology used for identifying concentration/dispersion patterns. In the fifth section we present and discuss our main empirical results. In the final section we present our conclusions.

15.2 Spatial Distribution of Economic Activity: Theories and Empirical Approaches

There is plenty of empirical evidence regarding the uneven spatial distribution of economic activity and the way how firms tend to cluster to each other, sometimes due to urbanization economies and sometimes due to localization economies. Among most relevant contributions that have reported and analysed this phenomena there are those of Polèse and Shearmur (2006), Duranton and Overman (2005), Devereux et al. (2004), Maurel and Sédillot (1999) and Ellison and Glaeser (1997). There is also a large list of contributions for the Spanish case (Boix and Galletto 2008; Paluzie et al. 2004; Viladecans 2004), where clusterization of economic activity is quite important in some regions.[5]

[4] See Guimarães et al. (2011) for an application about how to solve biases caused by previous measures that do not take into account spatial position of units.

[5] Concretely, Boix and Galletto (2008) identify four axes where specialized industrial district are of great importance: the main axis runs across the Mediterranean coast from the north of Catalonia to the south of Murcia; the second one links the south of Catalonia to the Basque Country and North-East of Castile and León; the third one goes South from Madrid to the provinces of Toledo,

Therefore, firms look to be close to other firms. As we have said before, some firms prefer proximity with similar firms (localization economies), while others just want to be close to other firms, no matter their activities (urbanization economies), but in any case clusters favours entrepreneurship (Rocha 2004). Additionally, in order to better optimise external resources, to enhance innovation (Audretsch 1998; Baptista and Swann 1998) and to increase productivity firms need neighbours, and usually these neighbours are firms that share characteristics as size, markets, industry, technological level, supply needs, type of workforce or use of infrastructures. So there are plenty of reasons to cluster with similar firms (even if they belong to different industries) that have similar needs (Steinle and Schiele 2002).

But apart from the fact that firms cluster into geographical areas, there is not still a consensus about where to look for such clusterization processes. Consequently, what we do have is some arbitrariness about geographical units to be used. So, while some papers look for clusters at a regional level (Baptista and Swann 1998) others prefer smaller areas like cities (Rodríguez-Pose 2001). This is an important issue to be taken into account when discussing how economic activity agglomerates, because there is plenty of empirical evidence showing that economic activities do not perfectly fit into extant administrative borders (e.g., regions or cities), rather they tend to spread across neighbour areas. This phenomenon implies that contiguous areas could share a single agglomeration of firms without internal borders, making difficult to precisely identify where to analyse this agglomerative process. At the lowest geographical level where the spillover effects dissolve internal borders or at a higher level (combining smaller units) where the phenomenon exists only for a small part of the analysed area? There is still no clear answer to this question.

Previous shortcomings illustrate that it is important to accurately analyse implications of spatial aggregation issues and which spatial areas are to be used, in view that using non appropriate areas could tend to biased results, as several scholars like Arbia (2001, p. 414) ("(...) any statistical measure based on spatial aggregates is sensitive to the scale and aggregation problems") and Duranton and Overman (2005, p. 1079) ("(...) any good measure of localization must avoid these aggregation problems") point out.

This spatial aggregation problem is known as *Modifiable Area Unit Problem* (MAUP),[6] and it is clearly illustrated by Arbia (2001) when showing that depending on how spatial borders are designed, the same spatial distribution of, for instance, firms, could result in a minimum concentration pattern, in a maximum concentration pattern or an intermediate concentration pattern. Unfortunately, these issues have not been a major concern for scholars,[7] usually due to the lack of sufficiently spatial disaggregated data, but this situation has started to change several years ago with the

Ciudad Real, Jaen and Córdoba; and the last one is scattered across the provinces of Pontevedra and A Coruña (North-West of Spain).

[6] See Openshaw and Taylor (1979) for a detailed analysis and Wrigley (1995) for a further review.

[7] See, nevertheless, papers by Arauzo-Carod and Manjón-Antolín (2004) and Arauzo-Carod (2008) about the implications for industrial location analysis. See also Olsen (2002) for a broad discussion of the units to be used in geographical economics.

spread of spatially disaggregated datasets and the extended use of raster data with GIS packages. Consequently, several scholars have started to demonstrate how inaccuracy on spatial aggregation level's selection strongly bias econometric results (Briant et al. 2010).[8]

According to previous considerations, in this chapter we aim to explicitly address such spatial aggregation issue when analysing concentration vs. dispersed location patterns in Spain. As we will explain in Sect. 15.4, our spatial unit is not an administrative one, but a cell of 10 × 10 km that covers all mainland Spain.[9] This microgeographical approach has been used previously (with some variations) by Duranton and Overman (2008, 2005), who used British postcodes.

15.3 Data

Our data set refers to 2006 and comprises Spanish firms[10] from manufacturing and service industries. The source of this data base is SABI (*Sistema de Análisis de Balances Ibéricos*), which uses data from the Mercantile Register including balance sheets and income and expenditure accounts. For each firm we also know the number of employees, the industry to which it belongs (the four digit NACE code), and its sales and assets, among other variables. Although SABI provides data at three-digit level, we have decided to use data at two-digit level because it is more reliable and because there are certain computational constraints when working with a high number of industries.

We also have detailed information about the firm's geographical location; that is, information which is particularly relevant for the purposes of this chapter. Nevertheless, the SABI dataset also has two important shortcomings. The first concerns the sample. Although the number of firms is very high (e.g., 581,712 service firms for the 2007 edition), microfirms and self-employed individuals are not considered, despite that fact that it is reasonable to assume that the spatial distribution of such activities is similar to that of the firms included in SABI. The second concerns the nature of the units; that is, SABI only covers firms, not establishments,[11] the latter being more appropriate for analyzing the spatial distribution of economic activity. In any case, since SABI covers most of the economic activity carried out in Spain,

[8] See Páez and Scott (2004) for a detailed report of techniques whose results are affected by MAUP problem.

[9] We have omitted non mainland parts of Spain due to lack of continuity of Balearic Islands and Canary Islands with the rest of the country, which could cause important biases in our results.

[10] It is important to notice that SABI data set is about firms, not establishments, so each firm could have more than one establishment, although most of firms have only one establishment.

[11] Other alternative statistical sources such as *Censo de Locales* (INE) are not currently updated. Although having firms as observation units instead of establishments, the *Censo de Locales* also provides useful information for locational analysis.

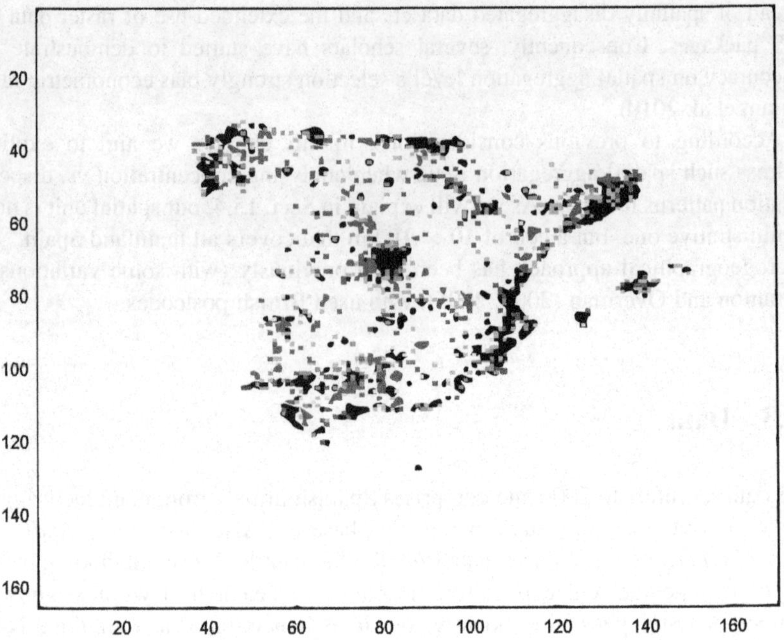

Fig. 15.1 Spatial distribution of manufacturing and service firms (Source: Own elaboration from SABI)

these disadvantages are easily overcomed[12] and, consequently, SABI has extensively been used for analysing spatial distribution of economic activity in Spain (see, for instance, Albert et al. 2012).

Spatial distribution of manufacturing and service firms in Spain is depicted at Fig. 15.1. Concretely, it is shown how firms agglomerate around main urban areas (namely Madrid and Barcelona) and most important economic regions like Mediterranean seaside and Basque Country, among others.

15.4 Methodology for Identifying Patterns of Concentration/Dispersion

Our methodology is a third generation approach (Arbia et al. 2008) that departs partially from previous contributions based on distribution comparisons (Brenner 2006, 2004; Ellison and Glaeser 1997) and on distance distributions (Duranton and Overman 2005) but we introduce some variations that allow us to better portrait single-industry clusters at a very detailed spatial disaggregation level. What do we

[12] There are alternative datasets such as DIRCE (INE) but their data is presented only at two-digit level and geographical location of the firms is also highly spatially aggregated.

do is (1) to use homogeneous cells of 10 × 10 km instead of administrative spatial units, (2) to use firm's georeferenced data to be more precise on firm's location, (3) to take into account total number of firms both at each cell and at a national level, (4) to compare real distribution of firms with a random estimated distribution using Montecarlo procedures and (5) to use multivariate Ñ-functions.

First, as explained above, instead of using administrative units[13] (e.g., municipalities) we use homogeneous cells of 10 × 10 km. [14] This size was decided in terms to avoid computational constraints (smaller sizes implied a huge increase in computational capacity in order to deal with a larger number of spatial units) and to the need to get a cell's size big enough to contain several firms from different industries. Even if alternative sizes were also feasible (e.g., 5 × 5 km, 20 × 20 km) and, consequently, were also tested, we considered that the selected size was appropriate both from a computational and economic point of view. In any case, although we concentrate on 10 × 10 km cells, in order to avoid MAUP it is possible to use any cell size.

By this way, we can overcome several shortcomings like (López-Bazo 2006) the inability to take into account the precise location of firms, the limitations resulting from the special administrative aggregation levels in each country, the difficulties in comparing the results obtained for different levels of administrative aggregation, the non-economic nature of such administrative units, the size differences across administrative units, the modifiable areal unit problem (MAUP) and the existence of neighbour effects across units.

Second, using firms' georeferenced data (point pattern analysis) allows us to precisely identify the exact site where each firm is located, although we only care about whether a cell is occupied or not, not about the exact location of the firm inside the cell, which is not an important issue in view of the small cell's size.

Third, we built up conterfactuals by assuming that total number of firms in each industry remains constant and that total number of firms in each cell also remains constant.

This latter requirement implies that firms localise randomly inside "occupied" cells (i.e., areas where real firms are located) as stated by Duranton and Overman (2008). This approach means that firms are expected to be located only in those places that are available for economic activity (as firms do). Unfortunately, a major shortcoming of this approach is that it assumes that firms could be located elsewhere with other firms, regardless of the industry they are involved in, which is not as realistic (especially at a two/three digit level). An extension of this work (and a possible solution for this shortcoming) would be to regard manufacturing, services

[13] See, among others, Brenner (2006, 2004) and Ellison and Glaeser (1997) for empirical applications with such administrative units.

[14] In view that some of MAUP problems come from size and shape of administrative units we should tackle both issues. While size implications are analysed in detail at beginning of Sect. 15.4, shape problems can be overcomed by using neutral cells, like squares, so we avoid problems linked to ad-hoc designs of geographical units (gerrymandering).

and agricultural firms as being located with other firms from the fields of manufacturing, services and agriculture respectively.

This strategy allows us to compare the same number of firms but with different industry distributions (we expect to find the same industry distribution at each cell that at that of the whole sample). Thus, if the real data shows a cell with only one firm, our simulations will also show this cell with one firm, although the industry will appear as a random variable depending on industry distribution.

Fourth, we compare the actual number of cells with firms (real distribution) with the expected number of cells with firms (random distribution), and obtain a concentration index similar to that of Ellison and Glaeser (1997), except that (1) we focus on industry shares instead of agglomeration and (2) our index is centred at 1 (values below 1 indicate concentration and values over 1 indicate dispersion), while Ellison and Glaeser's (1997) index ranges between zero and infinite (i.e., they arbitrarily define the concentration threshold).

Fifth, even though this approach shares some common features of previous analysis based on K-functions (mainly Arbia et al. 2008), there are several important specificities. In this sense, K-functions compare mean number of points inside a circle with radius r around a "typical" individual, which implies that a K value is obtained for each radius distance selected (Rot 2006). Alternatively, our approach implies to compare (at a cell level) the real distribution with a randomly generated distribution using Montecarlo techniques, as explained above. As well as with K-functions, Ñ-functions depend on distance used (in this case, cell size).

$$\tilde{N}_{ic} = \frac{\#\ of\ cells\ occupied\ by\ firms\ of\ i}{E\{\#\ of\ cells\ occupied\ by\ firms\ of\ i\}}$$

where i is the industry and c is the cell size.[15] According to the stochastic dimension of this index, significance bands are required: values significantly larger than one indicate that the industry has a wider distribution than expected according to a random process and, consequently, has a dispersed pattern; by the contrary, values significantly less than one indicate a concentrated pattern.

It is important to notice that the index reaches values equal to one for extremely big cells (i.e., all the firms get located inside the same cell) or extremely small cells (i.e., cell sizes are small enough to contain just one firm).

Advantages of this approach are twofold: data requirements are lower (as well as intensive computing) and no assumptions about boundaries are required. Concretely, such lower computational constraints allow to increase scale of analysis. As an example, while Arbia et al. (2008) uses level 2, our methodology allows to use level 3 and even higher, although we are not using this level in this contribution.[16] Thereby, bidimensional case can be represented as:

[15] Similarly the index could be calculated for bigger cell sizes.

[16] In any case, upper levels could be needed in order to analyse multisectorial clusters *à la* Porter (1998).

$$\tilde{N}_{ijc} = \frac{\#\ of\ cells\ occupied\ by\ firms\ of\ ij}{E\{\#\ of\ cells\ occupied\ by\ firms\ of\ ij\}}$$

Values larger than one indicate that industries tend to collocate at the same place more often than expected in a random distribution, while values less than one, on the contrary, indicate that firms tend to locate in different places. At this point, cell size is of extremely importance in view of possible spatial heterogeneities inside cells. Let's imagine what happens inside a metropolitan area, where while some firms prefer sites at downtown areas, while others use to locate at the outskirts, but being all of them inside the same metropolitan area.

As we have said before, algorithms based on K-functions are quite sensitive to the size of the considered area (Lotwick and Silverman 1982), mainly if they are n-dimension K-functions. Additionally, there are specific circumstances to be considered, as lack of points at the other side of the border, for instance. Let's consider what happens at the seaside, where there are no firms located into the sea, a spatial distribution that could be incorrectly interpreted as a low concentration pattern. In these circumstances some possible solutions arise, like (1) to use areas small enough to not include such sea areas or, alternatively, (2) to assign at the opposite side of the border a distribution of points (i.e., firms) similar to that in analysed area. In any case, according to previous shortcomings, use of these techniques can produce biased results, while our approach allows us to properly take these border effects into account.

15.5 Main Results

Our main results indicate that there are important industry differences regarding concentration patterns. While some industries (9 out of 28) show an extensive location pattern being spread across space (i.e., they appear in more cells than expectations according to a random process, so they are dispersed), most of industries (17 out of 28) behave in the opposite way, so they are concentrated. Table 15.1 illustrates these results using cells of 100 km^2 (10 $\times$ 10 km).

In particular, Table 15.1 shows how many cells (X) contain firms from industry y (i.e., this is the "real" spatial distribution of firms); the expected number of cells (Mean) where firms from industry y should appear if they were randomly spatially distributed (according to the total number of firms in each industry); and a collocation index (Index) that relates these measurements to each other (i.e., Index $= X/$ Mean). This index can be understood in the following way: an Index < 1 means that the industry y appears in fewer cells than expected (i.e., this industry is spatially concentrated in a smaller number of cells); and an Index > 1 means that the industry y appears in more cells than expected (according to a random distribution),

Table 15.1 Concentration patterns of firms at a single industry level (100 km^2 cells: 100 × 100 km)

Code	Industry	X	Mean	STD	Index	X − 2S	X + 2S	Concentrated	Dispersed
22	*Financial intermediation*	882	1480.11	17.6811804	0.59590166	1444.74764	1515.47236	True	False
6	*Paper and publishing*	947	1494.58	17.9619013	0.63362282	1458.6562	1530.5038	True	False
13	*Office machinery, computers and medical equipment, precision and optical instruments*	324	502.86	13.3553001	0.64431452	476.1494	529.5706	True	False
26	*Education*	790	1209.17	17.5580164	0.65334072	1174.05397	1244.28603	True	False
14	*Electrical machinery and apparatus*	520	782.36	14.6890463	0.66465566	752.981907	811.738093	True	False
24	*Business services*	1,360	1979.03	21.1557261	0.68720535	1936.71855	2021.34145	True	False
23	*Real estate activities*	1,957	2803.29	18.8970069	0.69810829	2765.49599	2841.08401	True	False
28	*Other services*	1,375	1819.52	21.5638493	0.75569381	1776.3923	1862.6477	True	False
12	*Machinery and equipment*	820	1076	17.2533118	0.76208178	1041.49338	1110.50662	True	False
4	*Textiles, leather clothes and shoes*	1,169	1523.26	17.384319	0.76743301	1488.49136	1558.02864	True	False
27	*Health and veterinary activities, social services*	1,122	1458.21	20.5029168	0.7694365	1417.20417	1499.21583	True	False
8	*Rubber and plastic products*	698	903.5	19.1498609	0.77255119	865.200278	941.799722	True	False
25	*Public administration*	141	179.3	7.24812759	0.78639152	164.803745	193.796255	True	False
7	*Chemical products*	734	837.17	14.8691634	0.87676338	807.431673	866.908327	True	False
10	*Basic metals*	567	629.55	16.7460986	0.90064332	596.057803	663.042197	True	False
15	*Transport and communications*	668	726.47	16.8111645	0.91951491	692.847671	760.092329	True	False
19	*Trade and repair*	2,888	3035.78	16.6336521	0.95132058	3002.5127	3069.0473	True	False
16	Recycling	349	359.69	9.90020406	0.97027996	339.889592	379.490408	False	False
11	Fabricated metal products	1,682	1701.7	19.8267751	0.98842334	1662.04645	1741.35355	False	False
21	**Transport and communications**	2,090	2034.14	19.9479221	1.02746124	1994.24416	2074.03584	False	True
17	**Construction**	2,706	2585.57	21.9674944	1.04657774	2541.63501	2629.50499	False	True
20	**Hotels and restaurants**	2238	2136.5	20.4181045	1.04750761	2095.66379	2177.33621	False	True
18	**Electricity and water distribution**	795	739.43	15.2674838	1.07515248	708.895032	769.964968	False	True
5	**Wood, furniture and other manufactures**	1,734	1610.89	20.5956232	1.07642359	1569.69875	1652.08125	False	True
9	**Non-metallic mineral products**	1,297	1125.88	18.1566027	1.15198778	1089.56679	1162.19321	False	True
2	**Extractive activities**	1,152	823.16	15.7015858	1.39948491	791.756828	854.563172	False	True
1	**Agriculture and fishing**	2,409	1691.54	20.5354682	1.42414604	1650.46906	1732.61094	False	True
3	**Food, beverages and tobacco**	2,236	1540.31	20.5001577	1,45165584	1499.30968	1581.31032	False	True

Note: X-2S equals X minus 2 standard deviations and X + 2S equals X plus 2 standard deviations. *Italic* shows concentrated industries and **bold** shows dispersed industries

Source: own calculations

which implies that this industry is spatially dispersed. We have considered that index's value was significantly higher (lower) if it was over expected value plus two standard deviations.

On a technological level, it seems that the lower the technological level of the industry, the higher the spatial dispersion. Thus, high-tech firms tend to be more spatially concentrated than low-tech firms,[17] but concentration patterns also exist for non high-tech firms, as other services or textiles, among others. In any case, concentrated location behaviour of high-tech firms appears to be reasonable since the markets and resources of such firms tend to be concentrated in a few areas, not existing logical reasons for a dispersed pattern.

If our results are approached in terms of the differences between manufacturing and services, they are even clearer than those of previous studies and show that whereas most services activities show high concentration levels (e.g., financial intermediation, education, business services, etc.), manufacturing activities are, generally speaking, more dispersed (agriculture and fishing, food, beverages and tobacco, etc.). These results reflect the spatial distribution of population and economic activity and the production and distribution requirements of manufacturing and services. Specifically, most services need face-to-face interactions and thus their location decisions are strongly motivated by the location of their customers (both firms and individuals). In contrast, manufacturers can easily transport their output, being that such interactions are not essential and that these firms can locate (almost) elsewhere.

So far, we have analysed the spatial distribution of firms at a single industry level and we have shown that looking at certain industry specificities (i.e., manufacturing vs. services and high-tech vs. low-tech) helps us to better understand such location patterns, although some industry specificities already persist after controlling for such issues.

Nevertheless, previous results help only partially to understand the concentration vs. dispersion divide, whilst it is also necessary to check what happens among groups of industries. In this sense, bivariate analysis allows to identify which industries use to locate together (i.e., they appear together more times than expected according to a random distribution) and which ones use to repel to each other (i.e., they appear together fewer times than expected according to a random distribution).

Table 15.2 depicts results of such Bivariate analysis showing mainly that (1) the industries identified as a concentrated under univariate analysis tend to repel the rest of concentrated industries and that (2) dispersed industries have a propensity to locate in the same areas.

Previous results seem to identify specific location patterns at an industry level. Firstly, there are some (usually high-tech) industries that tend to cluster in a small number of sites, surely due to specific characteristics of these areas. Among them, it

[17] As an example, indices of high-tech industries such as office machinery, computers and medical equipment, precision and optical instruments (0.644) and electrical machinery and apparatus (0.664) are clearly lower than those of some low-tech industries such as food, beverages and tobacco (1.452) and agriculture and fishing (1.424).

Table 15.2 Collocation patterns of firms for pairs of industries (100 km^2 cells: 100 × 100 km)

Code industry *j*	Code industry *i*	*X*	Mean	STD	Index	X − 2S	X + 2S	Repulsion	Attraction
Top-10 industries with the lower values of the collocation index									
Textiles, leather clothes and shoes	*Financial intermediation*	639	1092.83	12.749	0.585	1067.333	1118.327	True	False
Financial intermediation	*Real estate activities*	835	1424.44	16.580	0.586	1391.280	1457.600	True	False
Electrical machinery and apparatus	*Financial intermediation*	391	662.3	12.630	0.590	637.039	687.561	True	False
Financial intermediation	*Business services*	748	1259.35	15.338	0.594	1228.675	1290.025	True	False
Electrical machinery and apparatus	*Education*	361	606.94	10.773	0.595	585.394	628.486	True	False
Paper and publishing	*Financial intermediation*	651	1080.17	14.169	0.603	1051.833	1108.507	True	False
Machinery and equipment	*Education*	464	769.81	13.134	0.603	743.542	796.078	True	False
Textiles, leather clothes and shoes	*Education*	574	948.17	14.318	0.605	919.534	976.806	True	False
Trade and repair	*Financial intermediation*	878	1449.2	17.590	0.606	1414.021	1484.379	True	False
Financial intermediation	*Education*	569	934.68	13.485	0.609	907.709	961.651	True	False
Top-10 industries with the higher values of the collocation index									
Agriculture and fishing	**Construction**	2,013	1572.54	18.303	1.280	1535.934	1609.146	False	True
Agriculture and fishing	*Trade and repair*	2,120	1648.27	20.232	1.286	1607.805	1688.735	False	True
Extractive activities	**Non-metallic mineral products**	785	609.29	11.853	1.288	585.584	632.996	False	True
Extractive activities	**Construction**	1,059	799.41	15.180	1.325	769.049	829.771	False	True
Extractive activities	*Trade and repair*	1,100	815.4	15.300	1.349	784.801	845.999	False	True
Food, beverages and tobacco	**Construction**	1,957	1446.37	19.204	1.353	1407.963	1484.777	False	True
Food, beverages and tobacco	*Trade and repair*	2,042	1506.4	19.019	1.356	1468.361	1544.439	False	True
Agriculture and fishing	**Extractive activities**	990	723.86	13.682	1.368	696.496	751.224	False	True
Extractive activities	**Food, beverages and tobacco**	985	700.7	13.457	1.406	673.787	727.613	False	True
Agriculture and fishing	**Food, beverages and tobacco**	1,773	1193.15	15.684	1.486	1161.782	1224.518	False	True

Italic shows concentrated industries and **bold** shows dispersed industries
Source: Own calculations

is interesting to notice that activities like financial intermediation and education are likely to repel other concentred activities. Secondly, there are also some typically concentrated industries that use to locate together. They are often traditional industries heavily dependent on first-nature location determinants (e.g., agriculture and extractive activities) or industries with strong market linkages (e.g., construction, food processing and retail).

15.6 Conclusions

We have suggested a methodology for empirical analysis of spatial concentration of economic activities that overcomes previous drawbacks linked with the use of *K*-functions in terms of, among others, data constraints, computing limitations and border's definitions.

The methodology proposed in this chapter allows to better explaining the main reasons driving cluster formation, but much more work needs to be done in this area, particularly in order to identify cluster size and thus better capture cluster borders. As we have explained previously, our methodology involves dividing spaces into homogeneous cells of equal size, a procedure that must be handled with care because cell size influences results in terms of identification of dispersed vs. concentrated industries. Specifically, bigger cells are more likely to contain groups of firms forming a cluster, whereas smaller cells are more likely to have fewer inter-industrial clusters because the number of firms in each cell will be smaller. Given that we have assumed equal sizes for all cells, it would appear that using flexible sizes could fit better with real distribution of economic activity, being this flexibility a promising line for future research.

Given that this is just a first attempt to better identify the forces driving firm concentration patterns, there is still room for further insights into areas like specific firm location patterns. Although there is a large amount of empirical publications covering such issues, we strongly consider that it is important to accurately connect contributions from industrial location literature with those of geographical concentration of economic activities. Additionally, as we mentioned beforehand, industry aggregation is also important and, despite the computational constraints that make unfeasible to work with so much disaggregated industry-levels, we need to carry out further research to accurately determine whether our results are robust to different industry aggregation levels and also about the identification of the more appropriate levels to be used.

Finally, there are some important policy implications that arise from our results. Most of them are about improvements in policy measures (in terms of their specificities) once policy makers have access to a more detailed and more precise information regarding spatial distribution of economic activity and spatial distribution of clusters. In this sense, policy design improves from a broader knowledge of industry concentration drivers.

Appendix 1 – List of Industries

Code	Industry
1	Agriculture and fishing
2	Extractive activities
3	Food, beverages and tobacco
4	Textiles, leather clothes and shoes
5	Wood, furniture and other manufactures
6	Paper and publishing
7	Chemical products
8	Rubber and plastic products
9	Non-metallic mineral products
10	Basic metals
11	Fabricated metal products
12	Machinery and equipment
13	Office machinery, computers and medical equipment, precision and optical instruments
14	Electrical machinery and apparatus
15	Transport materials
16	Recycling
17	Construction
18	Electricity and water distribution
19	Trade and repair
20	Hotels and restaurants
21	Transport and communications
22	Financial intermediation
23	Real estate activities
24	Business services
25	Public administration
26	Education
27	Health and veterinary activities, social services
28	Other services

Source: SABI

References

Albert JM, Casanova MR, Orts V (2012) Spatial location patterns of Spanish manufacturing firms. Pap Reg Sci 91(1):107–136

Arauzo-Carod JM (2008) Industrial location at a local level: comments on the territorial level of the analysis. J Econ Soc Geogr (Tijdschrift voor Economische en Sociale Geografie) 99:193–208

Arauzo-Carod JM, Manjón-Antolín M (2004) Firm size and geographical aggregation: an empirical appraisal in industrial location. Small Bus Econ 22:299–312

Arbia G (2001) Modeling the geography of economic activities on a continuous space. Pap Reg Sci 80:411–424

Arbia G, Espa G, Quah D (2008) A class of spatial econometric methods in the empirical analysis of clusters of firms in the space. Empirical Econ 34:81–103
Audretsch D (1998) Agglomeration and the location of innovative activity. Oxford Rev Econ Pol 14(2):18–29
Baptista R, Swann P (1998) Do firms in clusters innovate more? Res Pol 27:525–540
Boix R, Galletto V (2008) Marshallian industrial districts in Spain. Ital J Reg Sci (Scienze Regionali) 7(3):29–52
Brenner T (2004) Local industrial clusters: existence, emergence and evolution. Routledge, London
Brenner T (2006) Identification of local industrial clusters in Germany. Reg Stud 40(9):991–1004
Briant A, Combes P-P, Lafourcade M (2010) Dots to boxes: do the size and shape of spatial units jeopardize economic geography estimations? J Urban Econ 67:287–302
Ciccone A, Hall RE (1996) Productivity and the density of economic activity. Am Econ Rev 84(1):54–70
De Propris L (2005) Mapping local production systems in the UK: methodology and application. Reg Stud 39(2):197–211
Devereaux M, Griffith R, Simpson H (2004) The geographic distribution of production activity in the UK. Reg Sci Urban Econ 34(5):533–564
Duranton G, Overman HG (2005) Testing for localization using microgeographic data. Rev Econ Stud 72:1077–1106
Duranton G, Overman HG (2008) Exploring the detailed location patterns of U.K. manufacturing Industries using microgeographic data. J Reg Sci 48(1):213–243
Ellison G, Glaeser EL (1997) Geographic concentration in US manufacturing industries: a dartboard approach. J Polit Econ 195:889–927
Guimarães P, Figueiredo O, Woodward D (2011) Accounting for neighboring effects in measures of spatial concentration. J Reg Sci 51(4):678–693
Henderson JV (2003) Marshall's scale economies. J Urban Econ 53(1):1–28
ISTAT (2006) Distretti industriali e sistemi locali del lavoro 2001. Istat, Collana Censimenti, Roma
Jaffe A, Trajtenberg M, Henderson R (1993) Geographic localization of knowledge spillovers as evidenced by patent citations. Q J Econ 108:577–598
Krugman P (1991) Increasing returns and economic geography. J Polit Econ 99(3):483–499
López-Bazo E (ed) (2006) Definición de la metodología de detección e identificación de clusters industriales en España. Dirección General de la Pequeña y Mediana Empresa (DGPYME), Madrid
Lotwick HW, Silverman BW (1982) Methods for analysing spatial processes of several types of points. J Roy Stat Soc Ser B (Methodol) 44(3):406–413
Maurel F, Sédillot B (1999) A measure of the geographic concentration in French manufacturing industries. Reg Sci Urban Econ 29:575–604
Olsen J (2002) On the units of geographical economics. Geoforum 33:153–164
Openshaw S, Taylor PJ (1979) A million or so correlation coefficients: three experiments on the modifiable areal unit problem. In: Wrigley N (ed) Statistical applications in the spatial sciences. Pion, London, pp 127–144
Páez A, Scott DM (2004) Spatial statistics for urban analysis: a review of techniques with examples. GeoJournal 61:53–67
Paluzie E, Pons J, Tirado D (2004) The geographical concentration of industry across Spanish regions, 1856–1995. Rev Reg Res 24(2):143–160
Parr JB (2008) Administrative spatial structure: a note on an alternative approach. Ann Reg Sci 42:141–151
Polèse M, Shearmur R (2006) Growth and location of economic activity: the spatial dynamics of industries in Canada 1971–2001. Growth Change 37(3):362–395
Porter M (1998) Clusters and the new economics of competition. Harv Bus Rev 76(6):77–90

Rauch JE (1993) Does history matter only when It matters little? The case of city-industry location. Q J Econ 108(3):843–867

Rocha HO (2004) Entrepreneurship and development: the role of clusters. Small Bus Econ 23:363–400

Rodríguez-Pose A (2001) Local production systems and economic performance in France, Germany, Italy, and the United Kingdom. In: Crouch C, Le Gales P, Trigilia C, Voelzkow H (eds) Local production systems in Europe: rise or demise? Oxford University Press, Oxford

Rot MC (2006) Introducción al análisis de datos mapeados o algunas de las (muchas) cosas que puedo hacer si tengo coordenadas. Ecosistemas 15(3):19–39

Sforzi F (1990) The quantitative importance of marshallian industrial districts in the Italian economy. In: Pyke F, Becattini G, Sengenberger W (eds) Industrial districts and inter-firm co-operation in Italy. ILO, Geneva, pp 75–107

Sforzi F (1999) Economic change. In: Bonavero P, Dematteis G, Sforzi F (eds) The Italian urban system. Ashgate, Aldershot

Steinle C, Schiele H (2002) When do industries cluster? A proposal of how to assess an industry's propensity to concentrate at a single region or nation. Res Pol 31(6):849–858

Viladecans E (2004) Agglomeration economies and industrial location: city-level evidence. J Econ Geogr 4(5):565–582

Wrigley N (1995) Revisiting the modifiable areal unit problem and the ecological fallacy. In: Cliff AD, Gould PR, Hoare AG, Thrift NJ (eds) Diffusing geography. Blackwell, Oxford, pp 49–71

Made in United States
North Haven, CT
04 November 2026

21730346R00187